T0328837

MANAGING LEARNING ORGANIZATION IN INDUSTRY 4.0

PROCEEDINGS OF THE INTERNATIONAL SEMINAR AND CONFERENCE ON LEARNING ORGANIZATION (ISCLO 2019), BANDUNG, INDONESIA, OCTOBER 9-10, 2019

Managing Learning Organization in Industry 4.0

Edited by

Indira Rachmawati & Ratih Hendayani

LONDON AND NEW YORK

Routledge is an imprint of the Taylor & Francis Group, an informa business

© 2020 Taylor & Francis Group, London, UK

Typeset by Integra Software Services Pvt. Ltd., Pondicherry, India

Library of Congress Cataloging-in-Publication Data
Applied for

Published by: CRC Press/Balkema
            Schipholweg 107C, 2316XC Leiden, The Netherlands
            e-mail: Pub.NL@taylorandfrancis.com
            www.crcpress.com – www.taylorandfrancis.com

ISBN: 978-0-367-81920-0 (Hbk)
ISBN: 978-0-367-54581-9 (pbk)
ISBN: 978-1-003-01081-4 (eBook)
DOI: 10.1201/9781003010814 https://doi.org/10.1201/9781003010814

Table of contents

Preface

International Seminar and Conference on Learning Organization (ISCLO) 2019 is presented to expand the discussion on how important it is for an organization to compete in technology era 4.0 by increasing knowledge sharing and culture.

As we have already known, industry 4.0 has specific characteristics such as strong robotics, machine autonomy, internet of things, and artificial intelligence (Pagac, 2015). Industry 4.0 increases the degree of digitalization of the entire supply chain cost and generates inter-connection of many people, gadgets and systems through real-time exchanges. Shamim et al (2016) and Slavik (2015) added a necessary element in implementing industry 4.0, which is highly skilled people. While in reality, according to Mohelska and Sokolova (2018), many organizations in Industry 4.0 focus more on the technical aspects, such as hardwares and softwares, than on reliable human resources. After all, those technical aspects mean nothing without reliable and qualified human resources.

The implementation of industry 4.0 increases the need for qualified staffs equipped with broad sets of competencies (Hecklau et al (2016). The dissemination of industry 4.0 knowledge has become increasingly important, employees should have a comprehensive technical skills to shift from operational tasks to more strategic and digitized tasks in the future, which require staffs with coding skills (Tozkwitalska and Slavik, 2018). Therefore, management's role is important to build a collaborative, explorative and entrepreneurial mindset as a success factor. Effective knowledge sharing can enable reuse and regeneration of knowledge at individual and organizational levels (Saufi and Yazmin, 2010). Therefore, with knowledge sharing about the beneficial natures of industry 4.0, employees would learn and increase their problem-solving skills in their daily activities.

Human resource is the main factor for industry 4.0 success. As human-machine interaction will become more common, learning how to interact with machines becomes one of the most important thing to do, because they will have to perform complex and integrated tasks and work in virtual networks with people from different cultures through media in information sharing process. These are the challenges for employees, and they will always need to learn new things, as technology is never constant, it will keep changing.

This proceeding, International Seminar and Conference on Learning Organization (ISCLO), includes papers related to studies that can improve organization's abilities in knowledge sharing and learning culture in Industry 4.0, so organizations can increase their competitive advantages. In this proceeding, both practitioners and academics are involved to discuss and share knowledges related to industry 4.0. Experienced practitioners & academics from Indonesia, Melbourne, and Malaysia are also invited to give speeches : Tommy Wong, Chairman of Indonesia Learning Network, Founder of Billionaire Mindset Indonesia, and Owner of Victorindo Group; Vanessa Ratten, Associate Professor of Entrepreneurship and Innovation in The Department of Management, La Trobe Business School at La Trobe University, Melbourne, Australia; Riyanarto Sarno, Professor of Software Engineering from Institut Teknologi Sepuluh November, Indonesia; and Indrawati, Associate Professor of Marketing from Telkom University, Indonesia.

We believe the papers included in the proceeding of International Seminar and Conference on Learning Organization (ISCLO) will serve as an excellent references to help both academics and practitioners in learning and making new discoveries for scientific studies in the future.

Conference Chair
Puspita Wulansari, Ph.D.

Scientific committee

Yudi Fernando, Ph.D.
Universiti Malaysia Pahang

Dr. Yuvaraj Ganesan
Universiti Sains Malaysia

Prof. Naili Farida
Universitas Diponegoro

Prof. Sam'un Jaja Raharja
Universitas Padjadjaran

Dr. Astrie Krisnawati, S.Sos., M.Si.M.
Telkom University

Dr. Dadan Rahadian, S.T., M.M.
Telkom University

Dr. Gadang Ramantoko
Telkom University

Dr. Majidah, S.E., M.Si.
Telkom University

Dr.Palti Mt. Sitorus, Drs., M.M.
Telkom University

Dr. Riko Hendrawan, S.E., M.M., ACP., CSCP., QIA.
Telkom University

Organizing committee

Conference Chair
Puspita Wulansari, S.P., M.M

Co-Conference Chair
Sri Widiyanesti, Ph. D

Members
Elvira Azis, S.E., M.T
Nike Mandasari, S.Si.
Husna Rahmi, S.Sos., M.Ikom.
Kharisma Ellyana, S.M.B
Muhammad Azhari, S.E., M.B.A
Ardan Gani Asalam, S.E., M.Ak.
Mediany Kriseka Putri, S.K.G, M.B.A.
Khairunnisa, S.E., M.M
Puspita Kencanasari, S.Kom., M.Ti.
Dedik Nur Triyanto, S.E., M.Acc.
Sisca Eka Fitria, S.T., M.M
Dini Wahjoe Hapsari, S.E., M.Si., Ak
Wulandari Ayungningtyas, S.Ikom.,M.M
Grisna Anggadwita, S.T., M.S.M
Indira Rahmawati, Ph. D
Erni Martini, S.Sos., M.M
Ratih Hendayani, Ph. D
Andrieta Shitia Dewi, S.Pd., M.M
Ir. Tri Djatmiko, M.M
Nensi Damayanti, S.S.
Hani Gita Ayuningtias, S.Psi., M.M
Sri Rahayu, S.E., M.Ak., Ak.
Dr. Adhi Prasetio, S.T., M.M
Asep Sudrajat, S.Kom
Riefvan Achmad Masrury, S.Si., M.B.A
Tieka Trikartika Gustyana, S.E., M.M
Indra Gunawan S.Kom
Harrys Sudarmadji, S.M.B
Setiadi, S.Kom

Acknowledgments

Special thanks to Associate Prof. Vanessa Ratten (La Trobe University) as keynote speaker, Prof Riyanarto Sarno (Institute of Technology Surabaya), Associate Prof. Indrawati (Telkom University), and Tommy Wong (Indonesia Learning Group, Billionaire Mindset Indonesia, and Victorindo Group) as invited speakers. Thank you for sharing your knowledges and inspiring us with your speeches.

Our thanks to Bank Mandiri, PT. Telkom Indonesia, Finnet, Mitratel, Telkom Akses, and Telkom Infra for supporting our conference.

Also, thanks to our co-host universities, Nurtanio University and Galuh University.

Addition of lifestyle compatibility and trust in modified UTAUT2 model to analyze continuance intention of customers in using mobile payment

H. Permana & Indrawati
School of Economics and Business, Telkom University, Bandung, Indonesia

ABSTRACT: This study aims to find the factors that influence the intention to continue the use of m-payment XYZ in Indonesia by using the modified UTAUT2 model. This study uses a questionnaire in data collection from 400 respondents in a metropolitan city and megapolitan areas in Indonesia. The results showed that factors that had the highest influence on continuance intention on users of m-payment XYZ were Habit. The added variables in the research model show the results that Trust and Lifestyle Compatibility are substantial and significant factors that influence after Habit sequentially, followed by Performance Expectancy and Hedonic Motivation. Age only moderates the influence of Lifestyle Compatibility on Continuance Intention. The R-Square in this study was 70.2%, so it was in the "Good" category to predict the intention to continue using m-payment XYZ

1 INTRODUCTION

Based on data from the World Bank, the Association for Indonesian Internet Service Providers and We Are Social, internet usage in Indonesia until 2018 was 143.2 million users of the total population of the Indonesian population of 261.12 million. The development of fintech was also shown by the increasing number of companies engaged in the field of fintech. It recorded an increase from 2016 where there were only 50 Fintech companies until 2018 to 167 fintech companies operating in Indonesia (Fintechnews Singapore, 2018). M-payment XYZ is a mobile payment that has only recently been established but has been able to occupy the second position after Go-Pay with a percentage of 58.42%, and followed by the third position, namely T-Cash with a percentage of 55.52%. The adoption of fintech in Indonesia was mentioned by Kuseryansyah as the head of the Indonesian Financial Technology Association's Daily, quoted from republika.id, who said that the adoption rate of fintech for several countries in Asia has a higher value compared to the adoption rate in Indonesia, for example in Singapore of 23%, India of around 70%, and China up to 80%. While the value for the fintech industry in Indonesia is still low, with a figure below 9%.

Data obtained based on preliminary data were collected through in-depth interviews with several m-payment XYZ users. They state that they usually use the services that support payment methods such as paying an online taxi, paying parking, merchants who cooperate with m-payment XYZ such as shopping center or place to eat. Moreover, information obtained from informants shows that m-payment XYZ services are also used to pay for home needs such as payment of electricity bills. M-payment XYZ service changes the user's lifestyle to become cashless. Therefore, the Lifestyle Compatibility variable was added in this study. Based on the existing conditions and phenomena, it is necessary to conduct a study to find out the factors that are considered to influence the adoption of m-payment XYZ that can be used to maintain and enhance future use.

2 LITERATURE REVIEW

2.1 *Literature review*

Deningtyas and Ariyanti (2017) conducted a study on the adoption of an online transportation service application by using a modified UTAUT2 by eliminating experience as a variable that moderates between the independent variable and the dependent variable. The factors that most influence Behavioral Intention are Habit, Price Value, Social Influence, Performance Expectancy, and Hedonic Motivation. While the main factors affecting Use Behavior are Facilitating Conditions, Behavioral Intention, and Habit. Indrawati and Putri (2018) conducted a study to find out the factors that influence Go-Pay adoption using UTAUT2 modification. Trust was added in this study as an independent variable. The results showed that the factors that influence Continuance Intention of Go-Pay from highest to lowest are Habit, Trust, Social Influence, Price Saving Orientation, Hedonic Motivation, and Performance Expectancy. This model can strongly predict Continuance Intention of Go-Pay services in Indonesia because R2 is 72.8%.

Hussain et al. (2018) conducted a study on the adoption of m-payment using UTAUT2 and Lifestyle Compatibility variable was added to the independent variable. Armstrong in Hussain et al. (2018) says that consumers not only buy products but also buy the values and lifestyles that these products represent. The results show that Performance Expectancy, Effort Expectancy, Facilitating Conditions, Habit, and Social Influence significantly influence the Behavioral Intention of the (Bottom of Pyramid) BoP segment.

2.2 *Research framework*

This study used the UTAUT2 approach because according to Venkatesh et al. (2003) in Indrawati, et al. (2017: 32), UTAUT model can explain 70% variation in interest from the use of a technology. This shows a greater percentage compared to other models which have a predictive power of 17-53% in the test. According to Andurill (2018), m-payment XYZ is a type of non-cash (cashless) payment that is used daily by mobile payment users. This type of non-cash payment using m-payment XYZ is targeted for the people with middle to the upper economic class who are considered to have a lifestyle with considerable daily expenses and routine.

Hussain et al. (2018) conducted a study related to the adoption of mobile payment using the UTAUT2 approach by adding Lifestyle Compatibility as the independent variable. Moreover, it was the strongest predictor after Performance Expectancy. Modifications were also made by replacing Price Value with Price Saving Orientation because m-payment XYZ did not incur costs that must be incurred in the service, but allowed lower prices by using it. Previous studies had also adapted a study by Gupta et al. (2018) to change the Price Value with Price Saving Orientation because various cellular applications in various sectors including hospitality and tourism had introduced innovative pricing strategies and provided value by offering money-back offers and price savings. The change in Behavioral Intention variable to Continuance Intention was also carried out by Indrawati and Putri (2018), Savitri and Indrawati (2019). Shao et al. (2019) also conducted a study using Continuance Intention as the dependent variable and mobile payment as the object of research. This study eliminated Use Behavior because it wanted to know the continued use of m-payment XYZ. The reduction of the Experience moderator variable was also carried out in this study, which was in line with research conducted by Indrawati and Putri (2018). Reduction of the Experience moderator variable was made because the data collection process in this study was a cross-sectional study, not a longitudinal study so that the Experience variable could not be applied in this research model.

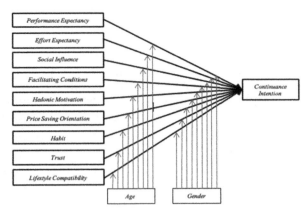

Figure 1.   Research framework.

Source: Modification model of UTAUT2 adapted from Venkatesh et al. (2012); Hussain et al. (2018); Indrawati dan Putri (2018)

2.3 *Hypothesis*

Based on the literature review and research framework, the research hypothesis is shown in Table 1.

Table 1.   Hypothesis.

H1	:	Performance Expectancy has a positive and significant influence on Continuance Intention
H1a	:	The effect of Performance Expectancy on Continuance Intention is moderated by age
H1b	:	The effect of Performance Expectancy on Continuance Intention is moderated by gender
H2	:	Effort Expectancy has a positive and significant influence on Continuance Intention
H2a	:	Effect of Effort Expectancy on Continuance Intention is moderated by age
H2b	:	The effect of Effort Expectancy on Continuance Intention is moderated by gender
H3	:	Social Influence has a positive and significant influence on Continuance Intention
H3a	:	The influence of Social Influence on Continuance Intention is moderated by age
H3b	:	The influence of Social Influence on Continuance Intention is moderated by gender
H4	:	Facilitating Conditions has a positive and significant influence on Continuance Intention
H4b	:	Effect of Facilitating Conditions on Continuance Intention is moderated by age
H4c	:	Effect of Facilitating Conditions on Continuance Intention is moderated by gender
H5	:	Hedonic Motivation has a positive and significant influence on Continuance Intention
H5a	:	The influence of Hedonic Motivation on Continuance Intention is moderated by age
H5b	:	The influence of Hedonic Motivation on Continuance Intention is moderated by gender
H6	:	Price Saving Orientation has a positive and significant influence on Continuance Intention
H6a	:	The effect of Price Saving Orientation on Continuance Intention is moderated by age
H6b	:	The effect of Price Saving Orientation on Continuance Intention is moderated by gender
H7	:	Habit has a positive and significant influence on Continuance Intention
H7a	:	Habit influence on Continuance Intention is moderated by age
H7b	:	Habit influence on Continuance Intention is moderated by gender
H8	:	Trust has a positive and significant influence on Continuance Intention
H8a	:	The influence of Trust on Continuance Intention is moderated by age
H8b	:	The influence of Trust on Continuance Intention is moderated by gender
H9	:	Lifestyle Compatibility has a positive and significant influence on Continuance Intention
H9a	:	The influence of Lifestyle Compatibility on Continuance Intention is moderated by age
H9b	:	The influence of Lifestyle Compatibility on Continuance Intention is moderated by gender

3 METHODOLOGY

3.1 *Measurements*

This study used a questionnaire as the data collection tool. This study conducted a pilot test first to test the questionnaire to meet the validity and reliability test so that the questionnaire could be used in this study. Analysis of data processing in this study used statistical analysis using SmartPLS 3.2.8. All the questionnaires items are shown in Table 2.

Table 2. Questionnaire items.

Item Code	Questionnaire Items
PE1	I feel that m-payment XYZ is useful in daily transaction activities
PE2	I feel that using the m-payment XYZ can help complete the transaction faster
PE3	I can save time when I make payments using m-payment XYZ
PE4	Using m-payment XYZ increases the productivity of my payment transactions
EE1	I find it easy to learn using m-payment XYZ
EE2	I don't need a long time to learn m-payment XYZ
EE3	I find it easy to use m-payment XYZ
EE4	I find it easy to become proficient in using m-payment XYZ
SI1	My family thinks that I should use m-payment XYZ
SI2	My friend thinks that I should use m-payment XYZ
SI3	My family advised me to use m-payment XYZ
SI4	My friend advised me to use m-payment XYZ
SI5	Most people around me use m-payment XYZ
FC1	I have a smartphone that is needed to use m-payment XYZ
FC2	I know how to use m-payment XYZ
FC3	M-payment XYZ is compatible with the smartphone I use
FC4	I get help from others when I find it difficult to use m-payment XYZ
HM1	I find m-payment XYZ service enjoyable
HM2	I feel entertained by m-payment XYZ Points feature
HM3	I feel entertained by m-payment XYZ Deals feature
HM4	I feel entertained by m-payment XYZ Vouchers
HM5	I feel enthusiastic in using m-payment XYZ
PSO1	I can save money by using m-payment XYZ
PSO2	I like looking for cheap offers from m-payment XYZ merchants
PSO3	M-payment XYZ offers more affordable offers than other similar services
PSO4	M-payment XYZ offers an interesting cashback promo for me
H1	Using m-payment XYZ has become a habit for me
H2	Using m-payment XYZ for me is something I do without thinking
H3	Using m-payment XYZ is part of my daily transaction routine
H4	I am addicted to using m-payment XYZ
T1	I am sure that m-payment XYZ is trusted
T2	I am sure there is no fraud in m-payment XYZ service
T3	Even though it isn't supervised, I will believe m-payment XYZ does the right work
LC1	Using m-payment XYZ service fits my lifestyle
LC2	Using the m-payment XYZ service fits the way I like to make non-cash transactions
LC3	Using m-payment XYZ service fits the current payment trend
CI1	I intend to continue using m-payment XYZ
CI2	I will still routinely use m-payment XYZ as I do now
CI3	I intend to continue using m-payment XYZ compared to other alternatives
CI4	I would recommend others to use m-payment XYZ

4 RESULT

4.1 *Profile of respondents*

The respondents' characteristics in this study were dominated by m-payment XYZ users living in Bandung. The characteristics showed that the majority of respondents were female by 69%. The respondents in this study were also dominated by a young age, with a percentage of 72.5%. The characteristics are dominated by the middle class, with an average income of Rp. 2,600,000 - Rp. 6,000,000 and also dominated by respondents with an average use of m-payment XYZ in one month by more than 13 times.

4.2 *Path coefficient and t statistic*

The results of testing the model through bootstrapping show the value of t statistic and path coefficient in Table 5. Based on the value of the path coefficient, it can be concluded that the factors influencing Continuance Intention from the highest value are Habit, Trust, Lifestyle Compatibility, Performance Expectancy, and Hedonic Motivation. Table 3 shows the results of testing the inner model.

4.3 *Moderation test*

Table 4 shows the results of the comparison of the path coefficient for moderation testing.

The results showed that the acceptable moderator variable was age moderator on the influence of Lifestyle Compatibility on Continuance Intention. Based on the comparison of the path coefficient values, adulthood is stronger in moderating the influence of Lifestyle Compatibility, so it can be concluded that adulthood is easier to use m-payment XYZ based on lifestyle suitability among adults. Therefore, the research model in this study is illustrated in Figure 2.

Table 3. Result of inner model by bootstrapping.

	Path Coefficient	T Statistics	Conclusion
EE -> CI	-0,038	0,829	H_0 Accepted
FC -> CI	0,045	1,143	H_0 Accepted
H -> CI	0,414	8,244	H_0 Rejected
HM -> CI	0,105	1,831	H_0 Rejected
LC -> CI	0,139	3,351	H_0 Rejected
PE -> CI	0,137	2,426	H_0 Rejected
PSO -> CI	0,041	0,984	H_0 Accepted
SI -> CI	0,046	1,362	H_0 Accepted
T -> CI	0,163	3,287	H_0 Rejected

Table 4. Result of moderation effect.

	Gender		Age	
Paths	T Statistic	H_0	T Statistic	H_0
PE*Gender → CI	0,8278	Accepted	0,2030	Accepted
HM*Gender → CI	-1,4422	Accepted	0,3845	Accepted
H*Gender → CI	0,7767	Accepted	1,0766	Accepted
T*Gender → CI	0,0273	Accepted	-0,6922	Accepted
LC*Gender → CI	-1,3132	Accepted	1,7272	Rejected

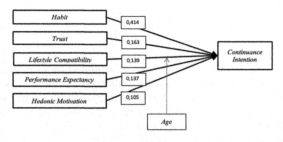

Figure 2. Research model.

5 CONCLUSION AND RECOMMENDATION

Based on the t-statistic value, the results of the study show that there are five factors that are stated to have a positive and significant influence. Hypothesis test shows that there is a positive and significant influence on the variables of Performance Expectancy, Hedonic Motivation, Habit, Trust, and Lifestyle Compatibility on Continuance Intention. Habit is the most powerful factor with a path coefficient of 0.414. The results showed that the age as a moderator variable was found to moderate the effect of Lifestyle Compatibility on Continuance Intention. Based on the value of the path coefficient, it can be concluded that the adult category tends to be stronger in moderating the influence of Lifestyle Compatibility on Continuance Intention. R-Square in this study is equal to 0.702 and is included in the "Good" category. The management of m-payment XYZ should continue to improve one of the services to be launched, Paylater system, by providing different benefits to its users as well as a greater cashback percentage. Thus, it has the potential to make customers feel addicted to using m-payment XYZ for routine payments every month. M-payment XYZ management needs to use the chatbot in the application to be able to help reduce errors that could occur, which can be experienced by users in a fast response. The management can better understand the daily transaction needs, such as monitoring daily transaction activities that people usually do but still use cash and change by providing other electronic transaction services using m-payment XYZ. The more merchants and services that can be used by users through the use of m-payment XYZ, the more user's productivity that can be fulfilled. The content marketing approach can be carried out by the management by adding service content with more interesting features.

REFERENCES

Anduril. 2018. OVO, Dompet Digital yang Praktis dan Banyak Untungnya. [online]. Available: https://jurnalapps.co.id/ovo-dompet-digital-yang-praktis-dan-banyak-untungnya-14835 [9 April 2019].
Daily Social. 2018. Fintech Report 2018. Jakarta: Daily Social.
Deningtyas, F. & Ariyanti, M. 2017. Factors Affecting The Adoption of E-payment on Transportation Service Application Using Modified Unified Technology of Acceptance and Use of Technology 2 Model. *Proceedings of Academic World 64th International Conference.*
Fintechnews Singapore. 2018. Fintech Indonesia Report 2018 – The State of Play for Fintech Indonesia. [online]. Tersedia: http://fintechnews.sg/20712/indonesia/fintech-indonesia-report-2018/[3 Februari 2019].
Gupta, A., Dogra, N. & George, B. 2018. What Determines Tourist Adoption of Smartphone Apps?. *Journal of Hospitality and Tourism Technology* 9(1): 50–64. Retrieved from EmeraldInsight.
Hussain, M., Mollik, A.T., Johns, R., and Rahman, M.S. (2018). M-Payment Adoption for Bottom of Pyramid Segment: An Empirical Investigation. *International Journal of Bank Marketing.* Retrieved from EmeraldInsight.
Indrawati, et al. 2017. *Perilaku Konsumen Individu dalam Mengadopsi Layanan Berbasis Teknologi Informasi dan Komunikasi.* Bandung: PT Refika Aditama.

Indrawati., and Putri, D.A. 2018. Analyzing Factors Influencing Continuance Intention of E-Payment Adoption Using Modified UTAUT 2 Model. *6th International Conference on Information and Communication Technology.*.

Savitri, A.W., and Indrawati. 2019. Measuring Factors Influencing the Adoption of OVO Feature in Grab Application in Indonesia. *The International Journal of Business & Management* 7(1): 70–74.

Shao, Z., Zhang, L., Li, X. & Guo, Y. 2019. Antecedents of Trust and Continuance Intention in Mobile Payment Platforms: The Moderating Effect of Gender. *Electronic Commerce Research and Application.* Retrieved from ScienceDirect.

Venkatesh, V., Morris, M.G., Davis, G.B. & Davis, F.D. 2003. User Acceptance of Information Technology: Toward a Unified View. *MIS Quarterly* 27(3): 425–478.

Venkatesh, V., Thong, J.Y.L., and Xu, X. 2012. Consumer Acceptance and Use of Information Technology: Extending The Unified Theory of Acceptance and Use of Technology. *Forthcoming in MIS Quarterly* 36(1): 157–178.

Tool for analyzing YouTube audience behavior in Indonesia

Indrawati & F. Herbawan
School of Economics and Business, Telkom University, Bandung, Indonesia

ABSTRACT: Using marketing media, the company needs to choose which media are suitable through which to advertise to the target market for the product offered. YouTube is one of the best communication mediums for video advertising, which is currently the most effective medium for communicating a brand. The YouTube application is the most popular medium in Indonesia. More than one-third of Internet users in Indonesia actively use YouTube. In one day, the average Indonesian uses YouTube for 42.4 minutes. To send messages through advertisement effectively, companies must analyze how to package their advertising properly so that potential customers can receive it well. The main purpose of this study was to test new modified research models and to examine whether or not they can be used further. In addition, this study aimed to test the model on the online video ads that are on YouTube using skippable video ads. This model was used to assess the influence of entertainment, informativeness, irritation, and credibility, and added the new variables of novelty, personalization, and perceived social usefulness to study the behavior of the audience through attitude and intention, which is also influenced by flow. The company can determine which important factors can bring added value for its advertisements. The validity results showed three items that are invalid questions: one item on the intention variable and two items on the behavior variable. All variables were declared reliable, but only nine of the eleven variables had good reliability. The remaining three variables were informativeness, personalization, and flow. This study showed that this model can be used in further research; the modifications made were sufficient to meet the need for research. But many new variables can be added with deeper identification. The results of this study indicated that this model can be used to analyze the behavior of ad viewers with factors that have been determined and added. If this study digs deeper, then the results can show the most important factors that can influence attitudes directly through intention then to behavior. With the discovery of the most influential factors, companies need to consider them more closely in order to create effective and efficient advertising.

1   INTRODUCTION

According to research by We Are Social, a British company in collaboration with Hootsuite, in 2019, most Indonesian people spend three hours and twenty-six minutes per day using social media, and they also spend two hours and fifty-two minutes viewing television or other video platforms, including YouTube. YouTube occupies the fourth highest position in monthly traffic accessed by Indonesians, and it ranked first in the spent time per visit at twenty-six minutes seven seconds, and first in the pages viewed per visit. YouTube is in the first position of social media that is always actively accessed with a percentage of 88%; in the second position is WhatsApp with a percentage of 83%, then Facebook with 81% and Instagram trailing with 80%.

Jones (2017) says ads that appear on YouTube are always skipped because content that is displayed does not match what the audience needs or is not in accordance with what they have accessed on YouTube. Research found that as much as 27% of respondents said the audience cannot stand the ads that appear and want to immediately watch the video they want; 22% of respondents said ad content is not according to what they access; 20% of respondents said they don't have time to watch advertisements; 19% of respondents said video advertising is irrelevant; 12% of respondents said the ads that appeared were too focused on the company's needs and does not meet the needs of the audience; 11% of respondents said that advertisements seem to serve the needs of others instead of serving them as viewers; 10% of respondents said they do not believe in the information shown in ads to encourage viewers to buy the product; 10% of respondents said that video content usually conflicts with their personality as an audience; and three respondents gave other reasons.

Seeing a great opportunity in social media and looking for the best formula through which to communicate and advertise a product to consumers easily and on target, many companies are trying to create advertisements that can attract consumers, so that now making interesting advertising content is a race for every company. To optimize online video ads, companies need to understand the desires and expectations of the target audience or prospective consumers. Therefore, a good composition of content is needed to attract the attention of the audience (O'Brien, 2017).

2 LITERATURE REVIEW

According to Ducoffe as cited in Yang, Huang, Yang, and Yang (2017), entertainment, informativeness, and irritation affect consumers' attitudes toward web advertising. Entertainment and informativeness are an important predictor of advertising's value and of the effectiveness of web advertising, while irritation has a negative impact on an audience's attitude (Brackett, Erkan and Evans cited n Yang and colleagues [2017]). *Credibility* refers to whether people trust the contents of advertisements. Researchers have postulated that credibility has a direct relationship with the value of advertising and attitudes toward advertising.

The concept of attitude–intention–behavior states that an individual's motivation to engage in behavior is determined by attitudes that influence behavior. Ajzen and Fishbein cited in Yang et al. [2017]) think that a person's intention is a function of his attitude toward his behavior and his subjective norms. So an action can be predicted from the attitude toward that action. Therefore, there is a high correlation between intention and behavior. Koufaris cited in Yang et al. [2017]) argues that intrinsic enjoyment can have a positive impact on the use of the computer environment. In addition, concentration as a measure of flow has been found to positively influence overall computer user experience (Hoffman and Novak, cited in Yang et al. [2017]) and their intention to use the system repeatedly (Webster at al. quoted in Yang et al. [2017]).

After conducting a more in-depth analysis, we found new variables that we felt needed to be added. Among the variables that we viewed as in accordance with theoretical studies and also in accordance with the object to be examined was novelty. Sheinin and colleagues cited in Feng and Xie (2018), say *novelty* refers to the extent to which ad execution differs from consumer expectations (for example, unusual and surprising designs) and is related to terms like *fresh*, *unique*, and *different*. Lee and colleagues (2016) said in testing the effect of mobile advertising that does not occur in conventional advertisements such as television, a factor related to the context of awareness as long as the user uses his smartphone is personalization. Researchers believe that this makes mobile advertising more able to satisfy users' need for relevant and timely information, as well as entertainment needs by providing an interesting, fun, and emotional experience. Apart from that, mobile users fulfill their needs and develop and enhance their sense of social relations through forwarding and recommending advertising messages that they receive to their social circle (Bauer et al. [2005]; Zhang and Mao cited in Wang and Genc [2019]).

Hypothesis
H1 Entertainment value of an advertisement displayed while viewers watch YouTube affects viewers' attitudes toward the advertisement.

H2 Informativeness value of an advertisement displayed while viewers watch YouTube affects viewers' attitudes toward the advertisement.

H3 Irritation value of an advertisement displayed while viewers watch YouTube affects viewers' attitudes toward the advertisement.

H4 Credibility of an advertisement displayed while viewers watch YouTube affects viewers' attitudes toward the advertisement.

H5 Novelty of an advertisement displayed while viewers watch YouTube affects viewers' attitudes toward the advertisement.

H6 Personalization of an advertisement displayed while viewers watch YouTube affects viewers' attitudes toward the advertisement.

H7 Perceived social usefulness of an advertisement displayed while viewers watch YouTube affects viewers' attitudes toward the advertisement.

H8 The attitudes toward an advertisement affect consumers' intention while watching a YouTube advertisement video.

H9 Consumers' intention affects their behavior while watching a YouTube advertising video.

H10 The perceived level of flow while users watch a YouTube video affects their intention.

H11 The perceived level of flow while users watch a YouTube video affects their behavior.

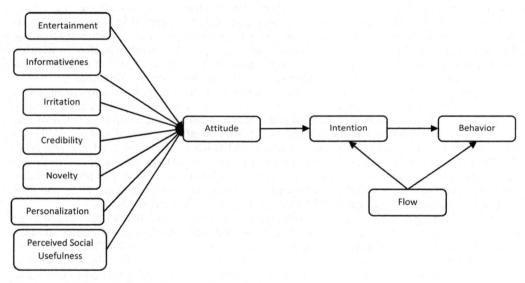

Figure 1. Modified theoretical framework.

3 METHODOLOGY

This study used the stages of validity and reliability, where there are three stages of validity and one stage of reliability. Validity consisted of content validity, readability, and construct validity. Content validity was obtained by looking at the items used in a questionnaire logically in order to determine if they were suitable for measuring the variables to be measured. Consultation with experts in their fields is important; in this research we collaborated with four experts who are lecturers in the field of management. A readability test of the questionnaire was carried out so as to find out whether the questionnaire was easily understood by the target respondents. In this stage the discussion was carried out with three target respondents representing different levels of education. Construct validity tested the validity of items using the Pearson correlation, namely by correlating the score of items with their total score. Then a significance test was carried out with the criteria using r-table and r-statistic comparisons concerning the level of trust, reliability, consistency, or stability of a measurement. In this research we used the Cronbach's alpha value for the reliability test. Respondents for the construct and reliability tests comprised thirty people who came from the three Indonesian provinces that use the Internet the most: West Java, East Java, and Central Java. Respondents who filled in the questionnaire had to know what skippable video ads are and have seen the ads through YouTube.

4 RESULTS AND DISCUSSION

In the content and readability tests several items required a modified sentence structure in the proposed questionnaire; this was because the items used in the research were sourced from research conducted outside Indonesia. A review was conducted in order to adjust the items for an Indonesian audience so that the items used could measure the variables specified.

Table 1. Construct validity and reliability test results.

Variable	Cronbach's Alpha	Reliability	Item	r-stat	r-table	Validity
Entertainment	0.875	Reliable	H1	0.903	0.3061	Valid
			H2	0.915	0.3061	Valid
			H3	0.864	0.3061	Valid
Informativeness	0.571	Reliable	K11	0.820	0.3061	Valid
			K12	0.747	0.3061	Valid
			K13	0.669	0.3061	Valid
Irritation	0.786	Reliable	G1	0.843	0.3061	Valid
			G2	0.770	0.3061	Valid
			G3	0.902	0.3061	Valid
Credibility	0.737	Reliable	KR1	0.791	0.3061	Valid
			KR2	0.815	0.3061	Valid
			KR3	0.829	0.3061	Valid
Novelty	0.907	Reliable	K1	0.845	0.3061	Valid
			K2	0.887	0.3061	Valid
			K3	0.843	0.3061	Valid
			K4	0.869	0.3061	Valid
			K5	0.875	0.3061	Valid
			K6	0.674	0.3061	Valid
Personalization	0.562	Reliable	PS1	0.735	0.3061	Valid
			PS2	0.599	0.3061	Valid
			PS3	0.645	0.3061	Valid
Flow	0.593	Reliable	A1	0.674	0.3061	Valid
			A2	0.677	0.3061	Valid
			A3	0.811	0.3061	Valid
			A4	0.556	0.3061	Valid
Perceived Social Usefulness	0.743	Reliable	SD1	0.850	0.3061	Valid
			SD2	0.817	0.3061	Valid
			SD3	0.796	0.3061	Valid
Attitude	0.814	Reliable	S1	0.850	0.3061	Valid
			S2	0.882	0.3061	Valid
			S3	0.834	0.3061	Valid
Intention	0.813	Reliable	N1	0.678	0.3061	Valid
			N2	0.765	0.3061	Valid
			N3	0.869	0.3061	Valid
			N4	0.757	0.3061	Valid
			N5	0.098	0.3061	Valid

(Continued)

Table 1. (*Continued*)

Variable	Cronbach's Alpha	Reliability	Item	r-stat	r-table	Validity
Behavior	0.821	Reliable	P1	−0.208	0.3061	Valid
			P2	0.588	0.3061	Valid
			P3	0.705	0.3061	Valid
			P4	0.758	0.3061	Valid
			P5	0.600	0.3061	Invalid
			P6	0.095	0.3061	Valid
			P7	0.527	0.3061	Valid
			P8	0.677	0.3061	Valid
			P9	0.532	0.3061	Valid
			P10	0.727	0.3061	Valid

* Use IBM SPS Statistic Version 24 for Windows 64 bit.

Table 2. Operational variable.

Variable	Items Code	Measurement Items
Entertainment	H1	The advertising is entertaining.
	H2	The advertising is enjoyable.
Informativeness	K11	The advertising is a good source of product information.
	K12	The advertising supplies relevant product information.
	K13	The advertising provides timely information.
Irritation	G1	The advertising is annoying.
	G2	The advertising is irritating.
	G3	This ad is disturbing.
Credibility	KR1	The advertising is credible.
	KR2	The advertising is trustworthy.
	KR3	The advertising is believable.
Novelty	K1	The advertising is creative.
	K2	The advertising is innovative.
	K3	The advertising have novelty.
	K4	The advertising is futuristic.
	K5	The advertising is imaginative.
	K6	The advertising is inventive.
Personalization	PS1	I feel that YouTube advertising displays messages that are personalized to me.
	PS2	I feel that YouTube advertising is personalized for my usage.
	PS3	Contents in YouTube advertising are personalized.
Flow	A1	During my last visit to YouTube I was absorbed intensely in the activity.
	A2	During my last visit to YouTube my attention was focused on the activity.
	A3	During my last visit to YouTube I concentrated fully on the activity.
	A4	During my last visit to YouTube I was deeply engrossed in the activity.
Perceived Social Usefulness	SD1	I forward YouTube advertising messages I like to my friends.
	SD2	Watching YouTube advertising I can demonstrate my innovativeness to my friends.
	SD3	If I watch YouTube advertising most of the people who are important to me will regard me as clever.
Attitude	S1	Overall YouTube advertising is good.
	S2	Overall YouTube advertising is favorable.
	S3	Overall YouTube advertising can be liked.
Intention	N1	I am willing to receive one advertisement per video while watching YouTube.
	N2	I am willing to receive two advertisements per video while watching YouTube.

(*Continued*)

Table 2. (*Continued*)

Variable	Items Code	Measurement Items
	N3	I am willing to receive three advertisements per video while watching YouTube.
	N4	I am willing to receive four advertisements per video while watching YouTube.
	N5	I am unwilling to receive advertising.
Behavior	P1	If I receive advertising I will ignore or close it immediately.
	P2	If I receive advertising I will watch it occasionally.
	P3	If I receive advertising I will watch it after it has appeared too many times.
	P4	If I receive advertising I will watch it when I have time.
	P5	If I receive advertising I will watch it right away.
	P6	I don't watch YouTube advertising at all.
	P7	I watch about a quarter of the advertising that I receive.
	P8	I watch about half of the advertising that I receive.
	P9	I watch about three-quarters of the advertising that I receive.
	P10	I watch all of the advertising that I receive.

Three items had invalid results; in this case an in-depth discussion took place, after which we decide to delete the items that were already represented by other items in the related variable. From the results we obtained it can be said that the modified model in this study can be used for further research in order to measure the behavior of ad viewers on YouTube. Deeper research needs to be done for each company that wants to measure the effectiveness of advertisements made through YouTube, so the company must emphasize the factor that has the greatest strength in an advertisement on YouTube. These factors must be considered in more depth if companies want to make more effective advertising targeting potential customers.

REFERENCES

Ajzen, I. and Fishbein, M. 1977. Attitude-behavior relations: a theoretical analysis and review of empirical research. *Psychological Bulletin*. 84 (5): 888.

Bauer, H. H., Reichardt, T., Barnes, S. J., and Neumann, M. M. 2005. Driving consumer acceptance of mobile marketing: A theoretical framework and empirical study. *Journal of Electronic Commerce Research* 6(3): 181–192.

Brackett, L.K. and Carr, B.N. 2001. Cyberspace advertising vs other media: consumer vs mature student attitudes. *Journal of Advertising Research*. 41 (5): 23–32.

Ducoffe, R. H. 1996. Advertising value and advertising on the web. *Journal of Advertising Research*. 36 (5): 21–35.

Feng, Y. and Xie, Q. 2018. Measuring the content characteristic of videos featuring augmented reality advertising campaigns. *Journal of Research in Interactive Marketing* 12: 489–508.

Jones, Robert. 2017. Why do People Skip Online Video Ads? [online]. www.smartinsight.com/internet-advertising/internet-advertising-strategy/people-skip-online-video-ads.

Kotler, P., Kartajaya, H., and Setiawan, I. 2017. *Marketing 4.0: Moving from Traditional to Digital*. Hoboken, NJ: Wiley.

Koufaris, M. 2002. Applying the technology acceptance model and flow theory to online consumer behavior, *Information Systems Research*. 13 (2): 205–223.

Lee, E., Lee, S., & Yang, C. 2016. The Influences of Advertisement Attitude and Brand Attitude on Purchase Intention od Smartphone Advertising. *Industrial Management & Data System*. Vol.117 No.6: 1011–1036.

O'Brien, Clogdagh. 2017. Secrets of Super Successful Video Marketing. [online]. www.digitalmarketinginstitute.com/blog/5-secrets-super-successful-video-marketing–1.

Sheinin, D.A., Varki, S. and Ashley, C. 2011. The differential effect of ad novelty and message usefulness on Brand judgements. *Journal of Advertising*. 40 (3):5–7.

Soares, A. M. and Pinho, J. C. 2013. Advertising in online social networks: The role of perceived enjoyment and social influence. *Journal of Research in Interactive Marketing* 8(3): 245–263.

Tsang, M. M., Ho, S., and Liang, T. 2004. Consumer attitudes toward mobile advertising: An empirical study. *International Journal of Electronic Commerce* 8(3): 66–78.

Wang, Y. and Genc, E. 2018. Path to effective mobile advertising in Asian markets: Credibility, entertainment and peer influence. *Asia Pacific Journal of Marketing and Logistic* 31(1): 55–80.

We Are Social. 2019, March 3. Digital 2019: Indonesia. [online]. https://datareportal.com/reports/digital-2019-indonesia.

Webster, J., Trevino, L.K. and Ryan, L. 1994. The dimensionality and correlates of flow in human-computer interactions. *Computers in Human Behavior.* 9 (4): 411–426.

Yang, H., Liu, H. & Zhou, L. 2012. Predicting Young Chinese Consumers' Mobile Viral Attitudes, Intents and Behavior. *Asia Pacific Journal of Marketing and Logistics* 24(1): 59–77.

Yang, K., Huang, C., Yang, C. & Yang, S. Y. (2017). Consumer attitudes toward online video advertisement: YouTube as a platform. *Kybernetes* 46(5): 840–853.

Zhang, J. and Mao, E. 2008. Understanding the acceptance of mobile SMS advertising among young Chinese consumers. *Psychology & Marketing*, 25 (8): 787–805.

Measurement tool for analyzing the influence of Corporate Social Responsibility (CSR) initiatives on consumer attitude, satisfaction, and loyalty at PT. BNI Tbk

Indrawati & R.R. Padang
School of Economics and Business, Telkom University, Bandung, Indonesia

ABSTRACT: The implementation of corporate social responsibility (CSR) in activities of charity and empowerment has increased for various companies in Indonesia. One of them is PT. BNI Tbk. This research was done so that BNI can map consumer behavior for the future in terms of attitudes, satisfaction, and customer loyalty. To improve and retain customers requires understanding consumer perceptions related to CSR. This study aimed to propose a measurement tool based on the model of Paluri and Mehra (2018), which was modified for this study based on literature studies with the addition of variables. The loyalty variable has been added as a new dependent variable in this modification. Measurement tools were tested using thirty respondents. This pilot test has revealed that the measurement model fulfills the requirements of validity and reliability. Therefore, this proposed measurement material is ready for use in further study.

1 INTRODUCTION

The implementation of corporate social responsibility (CSR) is increasingly active and rampant as a method corporations use to offset business operations. The development of CSR in Indonesia became a trending issue a few years ago, when many companies, including state-owned companies and the banking industry, started CSR as a program through which to empower local communities. Of the 115 state-owned enterprises in Indonesia, 41 have implemented CSR, according to data from the Indonesia Corporate Social Responsibility Award (ICSRA) (2017) in *Indonesia-Asia Institute-Economic Review Magazine*. This means that about 35.65% of companies have implemented superior CSR programs. By contrast, in 2018, 72 companies carried out CSR and won the ICSRA. In Indonesia, the government mandates CSR under the Constitution of the Republic of Indonesia through Law No. 40 of 2007 article 74 concerning Limited Liability Companies (UUPT). BNI is one of the many banking companies in Indonesia that has carried out CSR programs.

BNI is included on the list of the best companies in the world for which to work. Forbes ranked BNI 157 out of 2,000 of the best companies in the world (DetikFinance, 2018). Besides, BNI was ranked as the fourth largest national bank in Indonesia due to its total assets of 712,213,488 million rupiahs, total capital of 97,798,040 million rupiahs, and total loans of 459,289,448 million rupiahs (Bank Performance, 2019). Based on the 2018 BNI annual report (BNI, 2019), 43,546,693 units of savings and deposit accounts were obtained through its payroll cooperation program with companies and institutions, especially BNI customers/debtors in the business banking sector (corporate, commercial, and retail), the acquisition program from school to school and campus to campus, as well as the implementation of the 2018 government program with several ministries and government institutions such as the Non-cash Social Assistance Program, the Smart Indonesia Program (PIP), and the Student Savings Program (Simple). The number of BNI account units has increased 31.7% (savings) and 8.2% (deposits) from 2017 to 2018, equivalent to 10,384,517 units in the savings category and 27,840 units in the deposit category. The achievements and successes that BNI has obtained do not make it forget CSR. According to data retrieved from the official bank website, BNI shows that the company has implemented four types of CSR programs – namely BNI Berbagi, BNI Go Green, Kampoeng BNI, and Kami Bersama BNI. BNI Berbagi is an empowerment program aiming to achieve better standards of living and improve local community conditions. The program is in the context of supporting and implementing the Ministry of State-Owned Enterprises Regulation concerning the Partnership Program and the Community Development Program. BNI Go Green is one of BNI's missions focused on increasing environmental and social care and responsibility. Kampoeng BNI is a program of local community economic empowerment, poverty alleviation, and environmental improvement. Kampoeng BNI refers to the principle of community development by displaying superior products or characteristics in an area. Kami Bersama is a program intended to provide comprehensive support to the Empowerment of Indonesian Workers (TKI) and their families, covering the period before departure (predeparture).

Implementing CSR consistently in the long run will foster acceptance of the company's presence in the community and have a positive impact on consumer attitudes, satisfaction, and loyalty (Paluri & Mehra, 2018; Perez & Rodriguez del Bosque, 2015). Customer satisfaction is an important aspect of establishing and maintaining long-term relationships with customers that are essential for the survival of the company (Siu, Zhang, & Yau, 2013). Customer satisfaction is directly related to customer loyalty (Fornell et al., 1996), which in turn results in company profitability (Bitner, Booms, & Tetreault, 1990). Based on this phenomenon, research is needed to find out what CSR initiatives impact the attitudes, satisfaction, and loyalty of BNI customers. The purpose of this research was to test the measuring instrument that henceforth can be used to test the proposed model.

2 LITERATURE REVIEW

This study employed the theory of consumer behavior, perception, CSR, attitudes, loyalty, and several previous studies. According to Kotler and Keller (2016), consumer behavior is how individuals, groups, and organizations choose, buy, and use goods, and how goods, services, ideas, or experiences satisfy the needs and desires of consumers. Solihin (2008) states that CSR is the continuing commitment of business actors to behave ethically and contribute to economic development, while at the same time improving the quality of life of their workers and their families as well as their local communities and society at large.

This study used CSR initiatives as independent variables, attitude and satisfaction as intervening variables, and loyalty as the dependent variable. The object of this research was PT. BNI Tbk. In connection with the time of implementation, this study used cross-sectional data from one period. Based on literature studies and previous research references, researchers made modifications by adding loyalty as the dependent variable where previously the dependent variable was satisfaction. The addition of variables to this modification was also based on satisfaction's very significant relationship to loyalty (Fatma, Khan, & Rahman, 2016, 2018; Park, Kim, & Kwon, 2017; Martinez & Rodriguez del Bosque, 2013; Paluri & Mehra, 2018; Perez & Rodriguez del Bosque, 2015).

The definition of each variable was as follows. The perception of the bank's involvement in CSR activities is the process and form of bank contributions to sustainable development by paying attention to economic development and improving quality of life. The preference for environment-friendly initiatives is a form of CSR that prioritizes actions focused on protecting and preserving the environment. The preference for philanthropic initiatives is a form of CSR contribution that prioritizes efforts that can be felt and are beneficial to improving the quality of life of the surrounding communities. The preference for employee-support initiatives is a form of CSR contribution that prioritizes actions that will have an impact on improving the work ethic, productivity, and welfare of employees. The preference for customer-centric activities is an effort to improve consumers' attitudes by influencing their purchasing decisions directly, through quality, innovation, compliance with standards, guarantees, and other

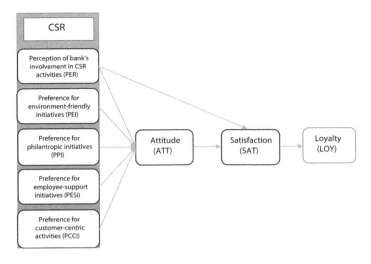

Figure 1. Modified theoretical framework.

information provided about products. Attitude is a response to stimuli and objects in a series of evaluative and effective traits. Satisfaction is a response or reaction to the performance and expectations of a product after consumption. Loyalty is a commitment consumers form after consuming a product, making repeat purchases, recommending a service or product to others, or increasing their transactions at a company. Figure 1 shows the conceptual modification of the model in this study.

3 METHODOLOGY

3.1 *Participants*

The sample examined in this pilot test research comprised thirty BNI customers in Medan, Jakarta, Bandung, Surabaya, Denpasar, and Makassar who have been customers for at least three months and are aware that BNI is involved in CSR. These areas were chosen as good representatives of a metropolitan city in Indonesia that has many branch and sub-branch offices and ATM outlets. So this shows that the city has many BNI customers.

3.2 *Measurement*

Data collection in this study used a questionnaire from thirty respondents to test for validity and reliability using IBM SPSS Statistics 23 software. Twenty-nine items from eight variables were used in this study. Testing the validity and reliability was very necessary in this pilot test. Researchers utilized several stages to test the measurement of the model. First, content validity tests aimed to ensure the items used in the questionnaire were logically suitable for measuring the variables under study, based on the definitions and indicators that have been set (Indrawati, 2015). The authors tested five academic experts in management and marketing as well as one practitioner expert in CSR. Modifications and adaptations of variables were made based on previous studies such as Abdeen and colleagues (2014), Fatma and colleagues (2018), Ilkhanizadeh and Karatepe (2017), Jean and colleagues (2014), Lombart and Louis (2014), Martinez and Rodriguez del Bosque (2013), Paluri and Mehra (2018), Park and colleagues (2017), Perez and Rodriguez del Bosque (2015, 2016), Rivera and colleagues (2016), Shabib and Ganguli (2017), Siu and colleagues (2014), Tsai and colleagues (2014), and Walsh and colleagues (2013). This was intended to ensure that all items could measure the variables examined in this study – namely the perception of the bank's involvement in CSR activities (PER), preference for environment-friendly initiatives (PEI), preference for philanthropic initiatives (PPI), preference for employee-support initiatives (PESI), preference for customer-centric activities (PCCI), attitude (ATT), satisfaction (SAT), and loyalty (LOY) of the respondents. Finally, the authors conducted a readability test with the respondents in order to ensure that the questionnaire had an adequate level of readability. Table 1 shows the items of each variable from the proposed model.

Table 1. Questionnaire items.

Item Code	Items of perception of the bank's involvement in CSR activities
PER1	BNI is involved in solving environmental problems.
PER2	BNI is involved in solving social problems.
PER3	BNI follows the laws and regulations concerning CSR.
Item Code	Items of preference for environment-friendly initiatives
PEI1	BNI participated in the use of recycled materials.
PEI2	BNI received an award for its performance in a good environment.
PEI3	BNI seeks to promote environmental sustainability.
Item Code	Items of preference for philanthropic initiatives
PPI1	BNI becomes a sports sponsor to foster young athletes.
PPI2	BNI offers free financial planning to the general public through seminars.
PPI3	BNI provides nutritious food for underprivileged children.
Item Code	Items of preference for employee-support initiatives
PESI1	BNI pays attention to the safety of its staff.
PESI2	BNI provides a sense of security to its employees.
PESI3	BNI maintains good relations with its employee unions.
Item Code	Items of preference for customer-centric activities
PCCI1	BNI staff can be relied upon.
PCCI2	BNI staff can handle complaints from customers.
PCCI3	BNI staff show positive behavior to customers.
Item Code	Items of attitude
ATT1	I like BNI.
ATT2	I enjoy visiting BNI.
ATT3	I feel good with BNI.
ATT4	I have sense of belonging to BNI.
Item Code	Items of Satisfaction
SAT1	My decision to choose BNI was right.
SAT2	I would recommend BNI to a friend.
SAT3	I feel happy with my decision to choose BNI.
SAT4	BNI is a satisfactory banking services provider.
SAT5	Overall, I am satisfied with BNI.
Item Code	Items of loyalty
LOY1	When other customers ask for my advice, I always recommend BNI.
LOY2	I am a loyal customer of BNI.
LOY3	I intend to remain a customer of BNI.
LOY4	I have a positive view of BNI when other customers ask me about it.
LOY5	I intend to contact BNI to purchase my next product.

3.3 *Data analysis*

In this study, the authors used a pilot test to ensure that the questionnaire items met validity and reliability standards. This study involved surveying thirty respondents to obtain initial sample data. The authors processed the data using IBM SPSS Statistics 23 for Windows 64-bit software with a significance level of 5%. In conducting this validity test the authors used the calculated R value > R table 0.3061 then declared the measurement valid. Meanwhile, a reliability test was performed using Cronbach's alpha. The Cronbach's alpha coefficient of at least 0.70 indicated that the questionnaire has a fairly good level of reliability.

4 RESULT AND DISCUSSION

The authors checked the validity based on the calculated R value > R table 0.3061, then declared all items valid. Next, the authors checked the reliability through obtaining the Cronbach's alpha value. Based on the calculation technique, the Cronbach's alpha value can be declared reliable if it has a value > 0.70. The study results are presented in Table 2.

As shown in Table 2, the results of the pilot study revealed that all twenty-nine items and eight variables of this measurement model fulfilled the requirements of validity and reliability.

Table 2.   Construct validity and reliability test results.

Item Code	R Table	R Value	Cronbach's Alpha
PER1	0.3061	0.832	0.713
PER2	0.3061	0.750	
PER3	0.3061	0.811	
PEI1	0.3061	0.909	0.894
PEI2	0.3061	0.882	
PEI3	0.3061	0.934	
PPI1	0.3061	0.794	0.754
PPI2	0.3061	0.774	
PPI3	0.3061	0.850	
PESI1	0.3061	0.928	0.899
PESI2	0.3061	0.923	
PESI3	0.3061	0.892	
PCCI1	0.3061	0.892	0.803
PCCI2	0.3061	0.861	
PCCI3	0.3061	0.792	
ATT1	0.3061	0.791	0.885
ATT2	0.3061	0.867	
ATT3	0.3061	0.953	
ATT4	0.3061	0.849	
SAT1	0.3061	0.917	0.943
SAT2	0.3061	0.888	
SAT3	0.3061	0.934	
SAT4	0.3061	0.876	
SAT5	0.3061	0.904	
LOY1	0.3061	0.869	0.882
LOY2	0.3061	0.757	
LOY3	0.3061	0.860	
LOY4	0.3061	0.816	
LOY5	0.3061	0.853	

* The authors used IBM SPS Statistic Version for Windows 64-bit.

5 CONCLUSION

The conclusion of this pilot test was that all eight variables and twenty-nine items under examination is this study were valid and reliable. Therefore, the questionnaire material of the proposed pilot test is ready for use in further research.

REFERENCES

Abdeen, A., Rajah, W., & Gaur, S. S. 2014. Consumers' belifs about firm's CSR initiatives and their purchase behaviour. *Marketing Intelligence & Planning* 34(1), 2–18.

BankPerformance. 2019. *Peringkat Bank Umum*. www.kinerjabank.com/peringkat_bank?bank_category=umum [March 6, 2019].

Bitner, M. J., Booms, B. H., & Tetreault, M. S. 1990. The service encounter: Diagnosing favorable and unfavorable incidents. *Journal of Marketing* 54(1), 71–84.

BNI. 2019. *Laporan Tahunan 2018*. www.bni.co.id/id-id/perusahaan/hubunganinvestor/laporanpresentasi [March 25, 2019].

DetikFinance. 2018. *4 BUMN Masuk Daftar Perusahaan Terbaik Dunia*. www.finance.detik.com/berita-ekonomi-bisnis/d-4271623/4-bumn-masuk-daftar-perusahaan-terbaik-dunia [March 6, 2019].

Fatma, M., Khan, I., & Rahman, Z. 2018. CSR and consumer behavioral responses: The role of customer-company identification. *Asia Pacific Journal of Marketing and Logistics* 30(2), 460–477.

Fatma, M., Khan, I., & Rahman, Z. 2016. The effect of CSR on consumer behavioral responses after service failure and recovery. *European Business Review* 28(5), 583–599.

Fornell, C., Johnson, M. D., Anderson, E. W., Cha, J., & Bryant, B. E. 1996. The American Customer Satisfaction Index: Nature, purpose, and findings. *Journal of Marketing* 60(4), 7–18.

Ilkhanizadeh, S., & Karatepe, O. M. 2017. An examination of the consequences of corporate social responsibility in the airline industry: Work engagement, career satisfaction, and voice behavior. *Journal of Air Transport Management* 59, 8–17.

Indrawati. 2015. *Metode Penelitian Manajemen dan Bisnis Konvergensi Teknologi Komunikasi dan Informasi*. Bandung: PT Refika Aditama.

Jean, R. B., Wang, Z., Zhao, X., & Sinkovics, R. R. 2014. Drivers and customer satisfaction outcomes of CSR in supply chains in different institutional context a comparison berween China and Taiwan. *International Marketing Review* 30(4), 514–529.

Kotler, P., & Keller, K. L. 2016. *Marketing management* (15th ed.). Upper Saddle River, NJ: Pearson Prentice Hall.

Lombart, C., & Louis, D. 2014. A study of the impact of corporate social responsibility and price image on retailer personality and consumers' reactions (satisfaction, trust and loyalty to the retailer). *Journal of Retailing and Consumer Service* 21, 63–642.

Martinez, P., & Rodriguez del Bosque, I. 2013. CSR and customer loyalty: The roles of trust, customer identification with the company and satisfaction. *International Journal of Hospitality Management* 35, 89–99.

OJK. 2019. *Undang-Undang Republik Indonesia Nomor 40 Tahun 2007 tentang Perseroan Terbatas*. www.ojk.go.id/sustainable-finance/id/peraturan/undang-undang/Pages/Undang-Undang-No.-40-tahun-2007-tentang-Perseroan-Terbatas.aspx [March 6, 2019].

Paluri, R. A., & Mehra, S. 2018. Influence of bank's corporate social responsibility (CSR) initiatives on consumer attitude and satisfaction in India. *Benchmarking: An International Journal* 25(5),1429–1446.

Park, E., Kim, K. J., & Kwon, S. J. 2017. Corporate social responsibility as a determinant of consumer loyalty: An examination of ethical standard, satisfaction, and trust. *Journal of Business Research* 76, 8–13.

Perez, A., & Rodriguez del Bosque, I. 2016. The stakeholder management theory of CSR: A multidimensional approach in understanding customer identification and satisfaction. *International Journal of Bank Marketing* 34(5),731–751.

Perez, A., & Rodriguez del Bosque, I. 2015. An integrative framework to understand how CSR affects customer loyalty through identification, emotions and satisfaction. *Journal Business Ethics* 129, 571–584.

Rivera, J. J., Bigne, E., & Perez, C. 2016. Effects of corporate social responsibility perception on consumer satisfaction with the brand. *Spanish Journal of Marketing* 20, 104–114.

Shabib, F., & Ganguli, S. 2017. Impact of CSR on consumer behavior of Bahraini women in the cosmetics industry. *World Journal of Entrepreneurship, Management, and Sustainable Development* 13(3), 174–203.

Siu, N. Y. M., Zhang, T. J. F., & Yau, K. H. Y. 2014. Effect of corporate social responsibility, customer attribution and prior expectation on post-recovery satisfaction. *International Journal of Hospitality Management* 43, 87–97.

Solihin. 2008. *Corporate social responsibility from charity to sustainability*. Jakarta: Salemba Empat.

Tsai, T. T., Lin, A. J., & Li, E. Y. 2014. The effect of philanthropic marketing on brand resonance and consumer satisfaction of CSR performance: Does media self-regulation matter?. *Chinese Management Studies* 8(3), 527–547.

Walsh, G., & Bartikowski, B. 2013. Exploring corporate ability and social responsibility associations as antecedents of customer satisfaction cross-culturally. *Journal of Business Research* 66, 989–995.

Managing Learning Organization in Industry 4.0 – Rachmawati & Hendayani (eds)
© 2020 Taylor & Francis Group, London, ISBN 978-0-367-81920-0

Effect of celebrity endorsement, EWOM and brand image on purchase decision of Nature Republic products in Indonesia

N.C. Lubis & M. Ariyanti
Telkom University, School Of Business and Economics, Bandung, Indonesia

ABSTRACT: The most favored skin care products (cosmetics) in Indonesia are not domestic products, but products from South Korea. Competition has arisen over Korean brands that have expanded in Indonesia. One of the company's brands is Nature Republic. The purpose of this research is to find out the significant effect of celebrity endorsement, electronic word of mouth, and brand image on the purchase decision of Nature Republic products In Indonesia. This type of research based on the method is quantitative research, Respondents in this study were 384 Nature Republic consumers. Data collection using a questionnaire and literature study. The sample used in this study is non-probability sampling with purposive sampling technique. Hypothesis testing uses descriptive analysis that is processed using Microsoft Excel and SEM PLS analysis. Based on the results of data processing, it can be seen that all the independent variables showed a significant effect on purchase decision.

1 INTRODUCTION

The population of Indonesia, which is around 250 million people, makes Indonesia a promising market for cosmetics companies (kemenperin.go.id, 2013). South Korean products are the most preferred in Indonesia. Data from the ZAP Beauty Index 2018, survey of 17,889 women revealed that 46.6 percent of women most liked products from the country of ginseng (South Korea). Followed by 34.1 percent who preferred products from Indonesia, then 21.1 percent chose products from Japan. Competition arises over Korean brands that have expanded in Indonesia. They came and targeted to Indonesia because of the demands and promising market potential (Kompas.com, Nurfadilah, 2018). Based on CNBC Indonesia's records, at least nearly a dozen Korean cosmetic brands are present in Indonesia, they are Etude House, The Face Shop, Laneige, Sulwhasoo, The Saem, Nature Republic, Innisfree, Tony Moly, Missha, Skinfood, and the last is Moonshot from YG Entertaiment (cnbcindonesia.com, Hasibuan, 2018).

One of them is Nature Republic, it was first established in South Korea in 2009. The founder is Jung Won-ho, until now Nature Republic already has 750 stores spread across several countries in the world, one of them is in our country Indonesia. Nature Republic products are cosmetics and skin care that originated from natural ingredients. In accordance with their motto "journey to nature", which is mean as a natural brand that discovers and shares life energy from pure nature throughout the world. Nature Republic also has 4 promises, "pure nature, for everyone, originality, and community.

Researchers take Nature Republic as the object of research in this study by seeing that Nature Republic has just expanded in Indonesia in early 2018 and continues to open branches in several cities in Indonesia. Nature Republic also had high enthusiasm and became number 1 search compared to other Korean skin care products (Google Trends, 2018). Nature Republic since 2013 until now has collaborated with EXO (boy band from one of the biggest agencies in South Korea) and was appointed as an exclusive model (Koreaboo, 2015). The spread of EXO Nature Republic digital content is happening fast enough due to the large number of EXO fans.

Referring to Jain research results, Celebrity Endorsement has an impact on sales on to a little extent and that Celebrities should not always be used to endorse Brands of various products. Agree and can't say. Although, the study has a positive inclination towards the belief that people are motivated to buy products as a result of Celebrity Endorsement. Furthermore the respondents also strongly agree that celebrities bring brand equity. The research also indicates that Celebrity Endorsement helps in brand promotion to the products. Finally he conclude that there's no harm in using celebrities for the endorsements, none the less everything has its own pros and cons (Jain, 2011:83).

According to Goldsmith & Horowitz in the journal Ismagilova et al (2017: 18), eWOM communication is dynamic and sustainable and can spread spontaneously and online. Anonymity and interactive nature allow consumers to provide and seek opinions about the experience of the product or service from people they have never met. As a result, eWOM affects the choice of brands and the sale of goods and services by consumers.

This research was conducted to analyze consumer behavior in purchasing decisions Nature Republic products. Looking at the 3 variables, namely celebrity endorsement, EWOM and brand image. This research was conducted as a reference for Nature Republic companies in Indonesia in order to create the right marketing strategy.

2 LITERATURE REVIEW

According to Kotler & Keller (2012: 498), marketing communication (marketing communication) is a tool that promotes companies, persuades, and warns consumers directly and indirectly about the products and brands sold. Electronic Word of Mouth (EWOM) is any agreement based on the experiences made by customers regarding products, services or companies that can be accessed by many people via the internet.

Messages delivered by interesting or popular sources can reach high attention and invite, discussed about advertisers who often use Celebrity Endorsement as a speaker of people (Kotler and Keller, 2012: 507). According to Mowen & Minor (in Sangadji & Sopiah, 2013: 7) [8], consumers are study units and decision-making processes involved in the acceptance, use and purchase, and also the selection of goods or services and an idea. Purchasing decisions are inseparable from how the nature of consumers (consumer behavior) so that each consumer has different habits in making purchases.

In this study, the use of celebrities was analyzed using the Vis Cap model. This Vis Cap model is used to measure the influence of EXO as a Celebrity Endorsement consisting of several factors, namely visibility, credibility, attractiveness, and strength (Kertamukti, 2015: 70) [9]. Furthermore, according to Goyette et al (2012: 14) [10] is a 3-dimensional measurement of electronic word of mouth, namely intensity, opinion and content. Kotler and Armstrong (2016: 176) [11] state that brand compartments have a strong and positive image in the eyes of consumers so the brand will always discuss and increase consumers to buy brands that correspond to very large quantities. The dimensions of the brand are divided into 3 i.e the strength of the brand association, supporting the brand association and the uniqueness of the brand association.

3 METHODOLOGY

3.1 *Participants*

Population is a whole group of people, events, objects that attract researchers to be examined which will be a limitation of the results of research that will be obtained, therefore the results obtained are only about a group of objects that have been determined by researchers (Indrawati, 2015:164). The population of this research is Nature Republic consumers who have purchased Nature Republic products at least once, know about EXO as a Celebrity

Endorsement of Nature Republic and domiciled in Indonesia. Meanwhile the sample used in this study was 384 Nature Republic consumer respondents in Indonesia which was took from purposive sampling method.

3.2 *Measurements*

To analyze the data collected, the author uses Smart PLS 3.0 software. In PLS, there are two different models of testing carried out, namely the outer and inner models. In the analysis phase of the measurement model (outer model), there are two things to be ana-lyzed, namely validity analysis (Convergent Validity, Discriminant Validity) and reliability analysis (Cronbach's Alpha, and Composite Reliability). While in the structural model analysis phase (inner model), there are two things that become testing tools namely R-square analysis (R2), Goodness of Fit (GoF) and t-statistic test both for direct effect, and indirect effect, which is obtained by using Bootstraping calculations on the Smart PLS application. The path coefficient must have a t-value of at least 1,645, each con-sidered significant at the 95% level.

4   RESULTS AND DISCUSSIONS

In this study, there are 3 independent variables and 1 dependent variable. From the processing of descriptive analysis, the results show that the variables have a high category. Descriptive analysis can be described as respondents' ratings of celebrity endorsement, electronic word of mouth and brand image of the decision to purchase Nature Republic products in Indonesia. The percentage breakdown of the variables is Celebrity Endorsement with 83%, Electronic Word Of Mouth with 77%, Brand Image with 80%, and Purchase Decision with 76%.

Furthermore, to test the research hypothesis, this study collected data from 384 valid respondents who are consumers of the Nature Republic in Indonesia. Respondents were selected using a purposive sampling technique from several cities in Indonesia (based on inter-est by sub-region from Google Trends for the keyword Nature Republic) through an online survey using a questionnaire.

Based on the convergent validity test results on PLS, it can be stated that there are three indicators that are declared invalid because the loading factor value is less than 0.500 i.e, EW 7, KP 9 and KP 10. Therefore the three indicators are eliminated. For discriminant valid-ity test results show that the factor loading in each indicator of the latent variable is proven to be greater than the relationship to the other latent variables so that it can be concluded that discriminant validity is fulfilled.

While in the structural model analysis phase (inner model), there are two things that become testing tools namely R-square analysis (R2), Goodness of Fit (GoF) and t-statistic test both for direct influence, and indirect effect, which is obtained by using Bootstraping cal-culations on the SmartPLS application. The path coefficient must have a t-value of at least 1,645, each considered significant at the 95% level. Table 1 shows the path coefficients and t-values of the model as a result of bootstrapping:

To test the overall quality of the model, Goodness of Fit is used. The calculation results are as follows:

$$GoF = \sqrt{\overline{AVE} \times \overline{R^2}} = \sqrt{0.568 \times 0.520} = 0.544$$

Table 1.   R-square analysis.

Endogenous variable	R Square (R^2)
Purchase Decision	0.520

Table 2. Statistical test results.

Hypothesis	Path coefficient	T Statistics	T Table	Hypothesis result
CE → PD	-0,078	2,187	1,645	Significant effect
EW → PD	0,344	7,324	1,645	Significant effect
BI → PD	0,502	12,235	1,645	Significant effect

5 CONCLUSIONS AND RECOMMENDATIONS

Descriptive analysis results based on responses from respondents shows that the four variables, Celebrity Endorsement (CE), EWOM (EW), Brand Image (BI) and Purchase Decision (PD) overall are in the percentage of 68-84% with high criteria. From the descriptive analysis it is known that Celebrity Endorsement (CE) variable gets an average percentage of 83%. Next, Brand Image variable (BI) the average percentage is 80%, EWOM variable (EW) with an average percentage of 77%, and the last with an average percentage of 76% on the Purchase Decision variable (PD).

From the results of testing the independent variables in this study all the independent variables showed a significant effect on purchase decision, while the details are as follows: Celebrity Endorsement Variable (CE) significant effect on the value of t arithmetic 2,187 which is bigger than t Table 1,645. Electronic Word of Mouth Variable (EW) significant effect on the value of t arithmetic 7,324 which is bigger than t Table 1,645. Brand Image Variable (BI) significant effect on the value of t arithmetic 12,235 which is bigger than t Table 1,645.

This research is expected to be able to contribute and as a consideration for companies in determining the marketing strategy of their products. especially relates to celebrity endorsement, EWOM and Brand Image. It's can be seen that celebrity endorsement variable has the lowest significance value among other variables that is equal to 2,187 and so is the indicator with the lowest percentage. The lowest indicator refers to the EXO profile and style statement according to the Nature Republic product. This is because Nature Republic still uses the same celebrity endorser for 6 years while its products continue to grow. And most consumers are female, Nature Republic can add the support of other celebrities who are female so consumers feel closer and appropriate and can imagine how the product is used by them. Because EXO is a man and not all products can suit them. For example in lip tint or lipstick products that are commonly used by women rather than men.

Based on the results of this study, the author tries to provide recommendations for further research, which is to expand or use several research objects at once. As well as using a number of additional variables that might influence the decision to purchase Nature Republic products in Indonesia, for example such as promotion, product quality and so on. Considering the other factors of this research result which were not examined are large enough (48%). In addition, the model in this study can be retested with different objects.

REFERENCES

Hasibuan, L. 2018. Sihir Kosmetik Korea Masih Ampuh, Ekspansi Makin Gencar di RI [online].

Indrawati. 2015. *Metode Penelitian Manajemen dan Bisnis Konvergensi Teknologi Komunikasi dan Informasi*. Bandung: Aditama.

Ismagilova, E., Dwivedi, Y.K., Slade, E., & Williams, M.D. 2017. *Electronic Word of Mouth (eWOM) In The Marketing Context*. Switzerland: Springer Nature.

Jaikumar, S., & Sahay, A. 2015. Celebrity Endorsements and branding strategies: event study from India. *Journal of Product & Brand Management* 24(6): 633–645.

Jain, V & Roy, S. 2016. Understanding meaning transfer in Celebrity Endorsements: a qualitative exploration", Qualitative Market Research: An International Journal. *Qualitative Market Research: An International Journal* 19(3): 266–286.

Jain, V. (2011). Celebrity Endorsement And Its Impact On Sales: A Research Analysis Carried Out In India. *Global Journal of Management and Business Research* 11(4): 68–84.

Kementrian Perindustrian Republik Indonesia. 2012. Indonesia Lahan Subur Industri Kosmetik. Jakarta: Kemenperin.

Kertamukti, R. 2015. *Strategi Kreatif dalam Periklanan*. Yogyakarta: Rajawali pers.

Koreaboo. 2015. Taeyeon and EXO renew their contract as Nature Republic's exclusive models [online]. https://www.koreaboo.com.

Kotler, P., & Keller, Kevin Lane. 2016. *Manajemen Pemasaran* (Bob Sabran, Penerjemah). Jakarta: Erlangga.

Kotler, P., and Gary, A. 2016. *Prinsip-prinsip Pemasaran* (Bob Sabran, Penerjemah). Jakarta: Erlangga.

Nurfadilah, P.S. 2018. Perempuan Indonesia Pilih Produk Kecantikan dari Korea, Bagaimana dengan Label Halal? [Online]. https://ekonomi.kompas.com.

Sangadji, E.M, & Sopiah. (2013). *Perilaku Konsumen*. (1st ed). Yogyakarta: CV. Andi Offset.

Tools for analyzing factors affecting marketplace usage by micro, small, and medium enterprises: Using a modified unified theory of acceptance and a technology 2 model in Bandung

R. Dzulfiqar & M. Ariyanti
School of Economics and Business, Telkom University, Bandung, Indonesia

ABSTRACT: Micro, small, and medium enterprises (MSMEs) play an important role in Indonesia's economy. Indonesia has a population of more than 250 million people and about 117.68 million workforces; 96.87 percent of the Indonesian people work in MSMEs. The Ministry of Cooperatives and Small and Medium Enterprises states that MSMEs accounted for up to 60.34% of the gross domestic product (GDP) in 2017. Due to the importance of MSMEs in Indonesia's economy, the application of information technology for this sector needs to be fully supported. This study aimed to propose a measurement tool based on the unified theory of acceptance and use of technology 2 (UTAUT2) model by Venkatesh, Thong, and Xu (2012). The model has been modified by adding a trust variable. The measurement tool was tested by using thirty respondents. The pilot test revealed that the measurement model fulfills the requirements of validity and reliability. Therefore, this proposed measurement material is ready for use in further study.

1 INTRODUCTION

During this digital era, technology has advanced and had a great impact on the development of the business world, especially on companies that have moved from conventional to digital markets. In this new world, small enterprises or startups can run their business activities more easily with the help of internet technology. With the Internet, businessmen can sell their products widespread at far lower cost.

The e-commerce sector in Indonesia presents a massive opportunity as well. Indonesian consumers spent more than USD 10.269 billion (around Rp140 trillion) to purchase commodities on various e-commerce platforms throughout 2017 (Hamdani, 2018). Marketplace is a third-party e-commerce platform that provides product trading services for sellers and customers on an online-based site (Putra, Nyoto, & Pratiwi, 2017).

Micro, small, and medium enterprises (MSMEs) can also use this concept to market their products. On Marketplace, sellers benefit by being able to easily promote and market their products to a wider range of potential consumers. Based on data obtained from the Ministry of Cooperatives and Small and Medium Enterprises, 3.79 million MSMEs used online platforms to market their products in 2017. This figure represents around 6% of the total MSME businessmen in Indonesia, which equals 59.2 million businessmen.

This research was carried out in Bandung City, which is considered to have a good potential to adopt internet or digital platforms and to face the Industrial Revolution 4.0, as the seventh president of Indonesia, Joko Widodo, has stated (Ashari, 2018). Bandung has become one of the biggest metropolitan cities in West Java and one of the most creative cities in Asia. Bandung has also received Natamukti Nindya awards as the city with the best development of small and medium enterprises (SMEs) in Indonesia, according to an assessment of the Indonesia Council for Small Business (ICSB) and the Ministry of Cooperatives and Small and Medium Enterprises (Ramdhani, 2016).

Based on this condition, we can conclude that the increasing number of internet users in Indonesia, as well as the huge number of people who shop online in Indonesia, have not significantly increased the number of MSMEs that use online platforms in Indonesia; it was only around 3.79% of the total MSME businessmen of 59.2 million in 2017. This could be a huge opportunity for MSME businessmen to contribute to a digital platform. Based on the data from the Ministry of Cooperatives and Small and Medium Enterprises, the contribution of MSMEs to the gross domestic product (GDP) in 2017 reached 60.34%. This could make MSMEs quite vital for the economy in Indonesia.

To see these phenomena, research is needed in order to recognize the factors that influence the use of Marketplace by MSMEs in Bandung. Two approaches were applied in this research, i.e. the unified theory of acceptance and use of technology 2 (UTAUT2) model adapted from Venkatesh and colleagues (2012) with some modification to analyze factors influencing the use of Marketplace by MSMEs in Bandung. The objective of this research was to provide a measurement tool for use as a proposed testing model.

2 LITERATURE REVIEW

Kotler and Keller (2016) state that e-commerce is something offered by a company or a site to facilitate transactions, product sales, and online services. Chaffey and Chadwick (2016) define e-commerce as anything related to financial and information exchange mediated electronically among organizations and external stakeholders.

This research analyzed Marketplace as a medium used by MSMEs in Bandung. Marketplace is a virtual medium with facilities to enable any trading between sellers and buyers (Chaffey & Chadwick, 2016). Robert and Ningrum (2017) state that Marketplace is an information system that provides data about prices, where sellers and buyers in the market communicate information about prices and products and can make transactions through electronic communication channels.

The object of this research was MSMEs in Bandung. The definition of small business, according to the Law No. 20/2008 on MSMEs in article 1, is as follows:

Micro enterprise is a productive business owned by an individual and/or private venture which meets the criteria of Micro Business as stated in this Act. Small enterprise is a productive economy business that stands alone, carried out by an individual or corporation which is not subsidiary or the branch of the company owned and ruled by, or that becomes the part both direct or indirect from the Medium or Big Business that meets the criteria of Small Enterprise as stated within this Act. Medium enterprise is a productive economy business that stands alone, carried out by an individual or corporation which is not subsidiary or the branch of the company owned and ruled by, or that becomes the part both direct or indirect from the Small or Big Business with the amount of net wealth or the annual selling yield as stated within the Act.

The approach used to recognize the factors in this research was a modified UTAUT2 model. It is the newest acceptance method and is the development or summary of the previous eight theories of technology acceptance. The eight previous model are as follows: theory of reasoned action (TRA), technology acceptance model (TAM), motivational model (MM), theory of planned behavior (TPB), Gabungan TAM 2nd TPB (C-TAMTPB), model of PC utilization (MPCU), innovation diffusion theory (IDT), and social cognitive theory (SCT). The UTAUT2 model can explain technology acceptance where the context is its usage (Venkatesh et al., 2012).

Furthermore, the researchers modified the UTAUT2 model by adding a trust variable. This additional factor is based on research contending that trust is predicted to influence the use of Marketplace. McKnight, Choudhury, and Kacmar (2002) believe that trust is important since it helps users overcome the uncertainty, risk, and "attitude related to trust" with web-based vendors, such as sharing personal information or purchasing.

Research has proven that trust can influence the use of technology, some of which was carried out by Manaf and Ariyanti (2016) about exploring key factors in the technology acceptance of mobile payment users in Indonesia. In addition, Qasim and Abu-Shanab (2016) study technology adoption in terms of the acceptance of mobile payment. Furthermore, research results from Padashetty and Kishore (2013) show that trust is one of the factors that play an important role in facilitating technology adoption for payment solutions.

Previous research used cross-sectional data or was carried out in a single time, while the variable of moderator experience was a result of the observation of the use of longitudinal or sustainable experience; therefore, the variable of experience was not used in this research model. In addition, this research modified the moderation variable by erasing the gender variable since MSMEs cannot be categorized based on gender and changed the age variable, which previously was the age of individuals, into the age of the company as suggested by Farinas and Moreno (2000), who state that a company under five years old is included in the new company category and those over five years old are included as old companies. The conceptual model used in this research is explained in Figure 1.

This research defined each original variable as adapted based on the model from Venkatesh and colleagues (2012). The definition of each variable is as follows: performance expectancy is how far someone believes that using Marketplace will benefit him/her in business activities. Effort expectancy is the level of ease of use of Marketplace. Social influence is how far the influence of social contacts, such as family and friends, influence others' attitudes when using Marketplace. Facilitating condition is how far someone believes that organizational and technical infrastructures exist to support a system. Hedonic motivation is how far users tend to use Marketplace. Price value is the benefit in using Marketplace. Habit is how far a user tends to use Marketplace automatically as it has become habit. Trust is how far MSMEs believe in using Marketplace. Behavior intention is how far MSMES will use Marketplace in the future. Use behavior is the intensity of Marketplace usage by MSMEs.

3 METHODOLOGY

3.1 *Participants*

The samples taken in this pilot test research came partly from MSMEs in Bandung that adopt Marketplace and included thirty respondents.

3.2 *Measurement*

This research used questionnaires in collecting data, and thirty respondents were tested for validity and reliability. Thirty-five items of ten variables were used in this research. The variable in the model proposed was tested using IBM SPSS Statistics version 23. Validity and reliability were highly needed on this pilot test. Therefore, the authors took steps to execute a measurement model. Modifications and variable adaptations were made based on references

Figure 1. Conceptual modified UTAUT2 model.

in national and international accredited journals, as carried out by Venkatesh and colleagues (2012), Indrawati and Sofiar (2017), Anggraini, Indrawati, and Harsono (2017), and Foon, Fah, and Fan (2011). This was to make sure that all items could measure the existing variables examined in this research – performance expectancy, effort expectancy, social influence, facilitating conditions, hedonic motivation, price value, habit, trust, behavior intention, and use behavior – from MSME respondents in Bandung. Table 1 shows the items of each variable from the proposed model.

Table 1. Questionnaire items.

Item code	Items of performance expectancy
PE1	Using Marketplace will be useful for my business.
PE2	Marketplace provides access to sell products faster.
PE3	Marketplace helps business activity become more productive.

Item code	Items of effort expectancy
EE1	It is easy for me to learn how to use Marketplace.
EE2	It is easy for me to understand Marketplace.
EE3	It is easy for me to become skillful in using Marketplace.

Item code	Items of social influence
SI1	People who are important to me say that I must use Marketplace.
SI2	People who influence me say that I must use Marketplace.
SI3	People I respect say that I must use Marketplace.
SI4	Most people around me use Marketplace.

Item code	Items of facilitating conditions
FC1	I have the device needed to use Marketplace.
FC2	I could easily acquire the right knowledge in using Marketplace.
FC3	Marketplace is compatible with other technology I use such as a smartphone.
FC4	I can easily get help from others if I find difficulties in using Marketplace.

Item code	Items of hedonic motivation
HM1	It is fun for me to use Marketplace for business needs.
HM2	I enjoy using Marketplace for my business needs.
HM3	I feel comfortable using Marketplace.
HM4	The existing features in Marketplace such as direct interaction with the customer make me feel entertained.

Item code	Items of price value
PV1	The cost to use Marketplace fits with my business needs.
PV2	Promotional costs through Marketplace make me not experience significant profit.
PV3	Premium membership in Marketplace provides more benefits and more selling.

Item code	Items of habit
H1	It has been my habit to use Marketplace.
H2	I am addicted to using Marketplace in running my business.
H3	It has been an obligation for me to use Marketplace.

Item code	Items of trust
T1	I am sure that Marketplace has a reliable payment system.
T2	I have no doubt in using Marketplace to run my business activities.
T3	I am sure of the security of my business data on Marketplace.

(Continued)

Table 1.   (*Continued*)

Item code	Items of performance expectancy
BI1	I intend to keep using Marketplace in the future.
BI2	I will keep trying to use Marketplace for my business activities.
BI3	I plan to use Marketplace more often in the future.
BI4	I will recommend more people use Marketplace.

Item code	Items of use behavior
UB1	I use Marketplace in my business activities.
UB2	I often use Marketplace in my business activities.
UB3	I use Marketplace for almost all business activities.
UB4	I prefer Marketplace to other media.

3.3   *Data analysis*

In this research, the authors conducted surveys in the form of a pilot test in order to guarantee that questionnaire items met validity and reliability standards. The test involved thirty respondents for the early data. The authors processed the data using IBM SPSS 23. In testing the validity, the authors used corrected item total correlation (CITC). According to Friedenberg and Kaplan in Indrawati (2015), the correlation coefficient > 0.3 is valid. According to Indrawati (2015), for the item reliability test, Cronbach's alpha is the most commonly used technique. The instrument could be said to have the best reliability if Cronbach's alpha > 0.70 (Hair et al., 2010; Kaplan & Saccuzzo, 1993: 126; Nunnally & Bernstein, 1994; Pedhazur & Pedhazur, 1991; cited in Indrawati, 2015).

4   RESULT

The authors examined the validity based on CITC. Furthermore, reliability was checked with Cronbach's alpha. The test result is presented in Table 2.

Table 2.   Pilot test result.

Item Code	CITC	CA
PE1	0.601	0.771
PE2	0.687	
PE3	0.616	
EE1	0.746	0.861
EE2	0.864	
EE3	0.614	
SI1	0.830	0.855
SI2	0.719	
SI3	0.845	
SI4	0.426	
FC1	0.437	0.851
FC2	0.777	
FC3	0.741	
FC4	0.876	
HM1	0.524	0.911
HM2	0.879	
HM3	0.937	
HM4	0.879	

(*Continued*)

Table 2. (*Continued*)

Item Code	CITC	CA
PV1	0.776	0.821
PV2	0.482	
PV3	0.807	
H1	0.697	0.761
H2	0.567	
H3	0.526	
T1	0.935	0.945
T2	0.897	
T3	0.837	
BI1	0.518	0.782
BI2	0.540	
BI3	0.849	
BI4	0.540	
UB1	0.403	0.761
UB2	0.613	
UB3	0.606	
UB4	0.645	

5 CONCLUSION

From the pilot test, we can conclude that the test conducted with thirty MSME respondents using Marketplace in Bandung for at least three months obtained the measurement result from ten variables with thirty-five items ready for use in further research.

REFERENCES

Anggraini, P. D., Indrawati, & Harsono, L. D. 2017. The use of modified unified theory of acceptance and use of technology 2 model to analyze factors influencing continuance intention of e-payment adoption (a case study of Go-Pay from Indonesia). *International Journal of Science and Research (IJSR)* ISSN (Online): 2319-7064 Index Copernicus Value (2016): 79.57 | Impact Factor (2015): 6.391.

Ashari, M. 2018. Jokowi Nilai Kota Bandung Siap Hadapi Tantang Revolusi Industri 4.0. www.pikiran-rakyat.com

Chaffey, D., & Chadwick, F. E. 2016. *Digital marketing*. 6th edition. Upper Saddle River, NJ: Pearson.

Farinas, C. J., & Moreno, L. 2000. Firm growth, size and age: A nonparametric approach. *Review of Industrial Organization* 17, 249–265.

Foon, S. Y., Fah, Y., & Fan, B. 2011. Internet banking adoption in Kuala Lumpur: An application of UTAUT model. *International Journal of Business and Management*. Canadian Center of Science and Education 6 (4). 161–167.

Hamdani, A. L. 2018. Melacak pemain lokal pada peta ecommerce Indonesia. www.tek.id

Indrawati. 2015. *Metode Penelitian Manajemen dan Bisnis Konvergensi Teknologi Komunikasi dan Informasi (cetakan ke-1)*. Bandung: Alfabeta.

Indrawati & Sofiar, Y. 2017. Adoption factors of online-web railway ticket reservation service (a case from Indonesia). 2017 Fifth International Conference on Information and Communication Technology (ICoICT).1–6. *Proceedings of International Conference on Information and Communication Technology (ICoICT)*. Malacca, Malaysia.

Kotler, P., & Keller, K. L. 2016. *Marketing management*. Upper Saddle River, NJ: Pearson.

Manaf, N. R., & Ariyanti, Maya. 2016. Exploring key factors on technology acceptance of mobile payment users in Indonesia using modified unified theory of acceptance and use of technology (UTAUT) model use case: ABC Easy Tap. In IIER International Conference (pp. 1–5). *Proceedings of The IIER International Conference*, Jakarta, Indonesia.

Marco, R., & Ningrum, B. T. P. 2017. Analisis Sistem Informasi E Marketplace Pada Usaha Kecil Menengah (UKM) Kerajinan Bambu Dusun Brajan. *Jurnal Ilmiah DASI* 18(2), 48–53.

McKnight, D. H., Choudhury, V., & Kacmar C. 2002. Developing and validating trust measures for e-commerce: An integrative typology. *Information Systems Research* 13(3), 334–359.

Padashetty, S., & Kishore, K. S. 2013. An empirical study on consumer adoption of mobile payments in Bangalore City: A case study. *Researchers World* 4(1), 83.

Putra, A. K., Nyoto, R. D., & Pratiwi, H. S. 2017. Rancang Bangun Aplikasi Marketplace penyedia jasa les private di kota Pontianak Berbasis Web. *Jurnal Sistem dan Teknologi Informasi* 2(5), 1–5.

Qasim, H., & Abu-Shanab, E. 2016. Drivers of mobile payment acceptance: The impact of network externalities. *Information Systems Frontiers* 18(5), 1021–1034.

Ramdhani, D. 2016. Bandung Diberi Penghargaan Kota Terbaik dalam Pengembangan UKM. https://regional.kompas.com

UU no. 20 Tahun 2008 Ayat 1. www.bi.go.id

Venkatesh V., Thong, J. Y., L., & Xu, X. 2012. Consumer acceptance and use of information technology: Extending the unified theory of acceptance and use of technology, *MIS Quarterly* 36 (1). 157-178.

The effect of Online Native Advertising on the attitude of Tokopedia consumers in Bandung city

A. Pramodhana & M. Ariyanti
School of Economics and Business, Telkom University, Bandung, Indonesia

ABSTRACT: This study aims to analyze the attitude of Tokopedia consumers in Bandung city to the new digital advertising format, namely Online Native Advertising Tokopedia which is expected to give a better impression than its former. The population in this study is domiciled in Bandung city who uses Tokopedia applications and services. The sample used in this study was 450 respondents using the Tokopedia application. The method of data collection is done through the distribution of questionnaires online such as GoogleDocs delivered to various social media to 450 respondents Tokopedia users who are domiciled in Bandung city. From the questionnaire distributed, data processing was carried out using Smart PLS 3.0. The result of the study has shown that there are four factors significantly influence the Online Native Advertisement Value, namely Entertainment, Information, Irritation (negative), and Personalization. There are three factors that significantly influence the Context Awareness Value, namely Personalization, Timing, and Location. Online Native Advertisement Value and Context Awareness Value have a significant effect on Online Native Advertising Attitude. The model in this study is proven to be able to predict Online Native Advertisement Value, Context Awareness Value, and Online Native Advertising Attitude of consumers to Online Native advertisements delivered by Tokopedia because it shows good R^2 values of 48%, 44.2%, and 56%. This research has found that the most influential factor in the model that influences the value of advertisements on Tokopedia consumer attitudes is Location. This can be a reference that Tokopedia Online Native Advertising can provide information or offer products offered based on the location of consumers with merchants who sell products in Tokopedia with the closest distance between locations.

1 INTRODUCTION

The growth of internet users in Indonesia has been increasing every year. People have been switching from shopping at the shopping center to online shopping (e-commerce). The development of e-commerce in Indonesia is also believed to be increased. The number of users has reached 147.1 million users until the beginning of the quarter of 2019 (statista.com, 2019). To compete in the increasingly fierce competition, several marketing strategies are carried out by e-commerce companies (especially marketplace businesses) to attract consumers to use their services. One of them is through a fairly new digital advertisement format, namely Online Native Advertising. It is designed to blend in the page content and display promotional contents according to the layout of the media placement. With average application visitor of 168 million visitors per month, Tokopedia is one of the biggest Indonesian marketplace businesses that implement it (iprice.co.id, 2019).

However, there is a perception that the majority of online advertisements delivered on the internet tend to be annoying or spam. Indonesia occupies the fourth position that downloads online ad-blocking applications with a total of 35 million AdBlock software which reinforces the fact that people have a bad attitude towards digital advertisements offered on the internet (pagefair.com, 2017). This study aims to analyze the attitude of Tokopedia consumers in the city of Bandung to the new digital advertising format, namely Online Native Advertising Tokopedia which is expected to give a better impression than its former.

2 LITERATURE REVIEW

2.1 *Online native advertising*

According to the Interactive Advertising Bureau, Native Advertising is a type of advertisement designed to blend in page content, consistent with the general aspects of the page and the respective media platform, from an editorial point of view (Kotler & Keller, 2016). In general, Native advertising is a term used to describe the spectrum of new forms of online advertising which is focused on minimizing the sense of disruption to the consumer experience by displaying these ads in one stream (Campbell & Marks, 2015). According to Seligman, Native Advertising is a form of sponsored ad content and is designed to be able to adjust to the editorial content where the advertisement is placed (Seligman, 2015). As for some of the techniques and formats included in Native Advertising, namely Custom Content, Content that appears in-feed, and Content that appears in a recommendation (Seligman, 2015). In accordance with the purpose of this study, one of the techniques that will be focused in this research is the Special Content (Custom Content) which can be found among the article content on a website or blog.

2.2 *Attitude*

According to Schiffman and Kanuk, attitude is a tendency that is learned in behaving in a way that is pleasant or unpleasant to a particular object (Schiffman, 2000). According to Abdullah, consumer attitudes are matters relating to the beliefs and choices of consumers towards the product brand being offered (Abdullah, 2016). The existence of a positive attitude towards a particular brand will enable consumers to make purchases of that brand.

In this study, the authors adapted the research framework of Bang Lee et al. (2017). The author has made several changes based on research background where Online Native Advertising is relatively new research and will be too complex when associated with Brand Attitude and Purchase Intention. The author intends to examine consumer behavior that is reflected by the attitudes of consumers toward relatively new ad formats such as Online Native Advertising in addition to some people's assumptions that online advertising is annoying or spam. Following the research framework adapted from the research of Bang Lee et al. (2017).

Hypothesis:

H1a: Entertainment has a positive and significant effect on Online Native Advertisement Value.

H1b: Information has a positive and significant effect on Online Native Advertisement Value.

H1c: Irritation has a negative and significant effect on Online Native Advertisement Value.

H1d: Personalization has a positive and significant effect on Online Native Advertisement Value.

H2a: Personalization has a positive and significant effect on Context Awareness Value.

H2b: Activity has a positive and significant effect on Context Awareness Value.

H2c: Timing has a positive and significant effect on Context Awareness Value.

H2d: Location has a positive and significant effect on Context Awareness Value.

H3: Online Native Advertisement Value has a positive and significant effect on Online Native Advertising Attitude.

H4: Context Awareness Value has a positive and significant effect on Online Native Advertising Attitude.

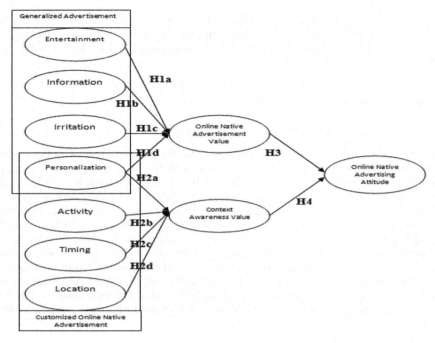

Figure 1.   Research framework.

The differentiator is, previous research classifies 4 other independent variables (Personalization, Activity, Timing, Location) as Customized Mobile Advertisement. As for this study, the author classifies these 4 variables more specifically according to related research, namely as Customized Online Native Advertisement. The difference also lies in the intervening variable where the author only takes 2 of the 4 intervening variables of previous research namely Advertisement Value (which is changed to Online Native Advertisement Value) and Context Awareness Value. The author also uses Online Native Advertising Attitude, which in previous studies was Advertising Attitude as the dependent variable of research.

3   METHODOLOGY

3.1   *Participants*

The author uses 95% confidence where the tolerance of error samples is 5% of the total. It means the estimated proportion of success and failure is 0.5. The minimum number of samples for this study is 384 which was added to the author to 450 samples (Wetzels, 2009).

Based on the results of the descriptive analysis, the sample consists of 255 males (56.6%) and 195 females (43.4%). The Majority of Respondents has a range of age between 17-24 years old (69,3%). About 91% of the total sample has Diploma/Bachelor educational background with a number of 409 respondents. The Occupation of the major respondents is a student with a percentage of 70% followed by Private Company employees with a percentage of 18% of the population.

3.2   *Measurements*

Likert scale is an instrument scale used in this study. The Likert scale will report the variables that will be indicators which then made as a starting point to make the items that will be requested. Mulyatiningsih answered to use an answer scale without a neutral choice of answers with the aim that the respondent's response more assertive (Mulyatiningsih, 2012). Each item provides four answer choices from values 1 to 4 while value 1 is the lowest means strongly agree when the highest value is 4 that strongly agrees.

3.3 *Data analysis*

There are two test models in PLS, the outer model and the inner model. Outer models are applied to latent variables or the ability level of the model indicators. Reflective indicators are tested through convergent validity, discriminant, Average Variance Extracted (AVE), and composite reliability. Meanwhile, the inner model influences one latent variable and another. The stability of this estimation is tested by the t-statistic test obtained through the bootstrapping procedure by looking at the clear percentage of variance where R^2 for the dependent latent variable is influenced by the latent variable.

4 RESULTS AND DISCUSSION

4.1 *Outer model*

There are two types to be analyzed in the analysis phase of the measurement model or outer model, namely validity analysis (Convergent Validity and Discriminant Validity) and reliability analysis (Cronbach's Alpha and Composite Reliability).

Convergent & Discriminant Validity

A measurement tool is to find the convergent validity criteria if the measurement results of items used to measure the same variable have a higher correlation when compared to the correlation of items used in different variables (Indrawati, 2015). These results can be demonstrated through the Loading Factor. The loading factor value shows how big the relationship of each latent variable is to the respective indicator and can be seen directly in the output outer setting on the results of the SmartPLS application algorithm. Based on the convergent validity test results, it shows that all indicators on each variable are declared valid which meets the standards of > 0.500. The research has other measurements to test whether or not an item meets the construct validity criteria by calculating the AVE (Average Variance Extracted) indicator. According to Indrawati (2015), AVE scores above 0.50 indicate that the items of these variables have sufficient convergent validity. Based on the data, all variables have an AVE score above 0.50 which indicate that the questionnaire in this study met the criteria of convergent validity.

One way to find out the value of Discriminant Validity is through the cross-loading value. The value of cross-loading is the result of a comparison of the magnitude of the relationship of each indicator to its variable, which is shown from the value of the loading factor, with the magnitude of the relationship of each indicator to other variables. The magnitude of the relationship of each indicator to the variable must be greater than the relationship of each indicator to the other variables in order to obtain valid results. The results obtained in this study where the factor loading of each indicator to the latent variable is proven to be greater than the relationship to other latent variables which indicate that discriminant validity has been fulfilled.

Cronbach's Alpha & Discriminant Validity

This stage is the reliability test of the measurement model after the validity test has been fulfilled. The two criteria in this test are Cronbach's Alpha & Composite Reliability which is obtained through the output overview of the algorithm results from the SmartPLS application. To meet the reliability of the measurement structure, the recommended values are above 0.6 for Cronbach's Alpha and 0.7 for Composite Reliability (Sekaran, 2003) (Hair et. al, 2012).

Based on the data, it shows that Cronbach's Alpha test results have a value of more than 0.6 and Composite Reliability has a value of more than 0.7. This indicates that the measurement model of this study has good reliability so it is worth further analysis, namely the inner model and hypothesis testing.

4.2 *Inner model*

The structural model test phase or inner model in this study is divided into 3 testing phases namely R-square (R^2) analysis, Goodness of Fit (GoF), and t-statistic test using Bootstrapping calculation on SmartPLS application.

R-square (R^2) analysis

The first structural model test is done by looking at R^2 in endogenous latent constructs to estimate the predictive power of the structural model and show how much influence the endogenous latent variables received from each exogenous variable that contributes to it. The greater the value of R^2 the greater the influence received by the exogenous variable.

The R^2 value of the Online Native Advertisement Value endogenous variable is 0.480, meaning the percentage of the Online Native Advertisement Value variable or precisely 48% can be explained by the exogenous variables related to the variable. Either with other endogenous variables, namely Context Awareness Value (44,2%) and Online Native Advertising Attitude (56%). Based on the R^2 category according to Hair et al. (2012), all of these endogenous variables categorized into the "moderate" category where all results are above 0.33.

Goodness of Fit (GOF)

The purpose of the Goodness of Fit is to test the overall quality of the model or each variable. The GoF grading classification is 0.1 (GoF) small, 0.25 (GoF) moderate, and 0.36 (GoF) large (Wetzels et. al, 2009). From the calculation results, this study has a GoF of 0.573 or included in the classification of a large GoF assessment. These results prove that this research model has a very good combined performance model, measurement and structural model.

T-Statistic and Hypothesis testing

The third test in the analysis of the structural model (inner model) is through the t-Statistics test or by looking at the value of the path coefficient and the significance value of t-Statistics to test the influence of hypotheses on exogenous latent variables on endogenous latent variables (Hair et. al, 2012). The table below shows the results of data processing with the bootstrapping process with a total sample of 450. Using a confidence level of 95% and a significance level of 5%, the t-table value was obtained as a comparison of t-statistics of 1.650.

Table 1.   T-Statistic test result.

Construct	Path Coefficient	T-Statistic	Standard	Result
ENT > ADV	0,359	7,707	1,650	Accepted
INF > ADV	0,166	3,581	1,650	Accepted
IRR > ADV	-0,097	2,059	1,650	Accepted
PER > ADV	0,238	4,906	1,650	Accepted
PER > CAV	0,333	6,209	1,650	Accepted
ACT > CAV	-0,095	1,502	1,650	Rejected
TIM > CAV	0,150	3,443	1,650	Accepted
LOC > CAV	0,398	8,529	1,650	Accepted
ADV > ADA	0,439	8,409	1,650	Accepted
CAV > ADA	0,379	7,318	1,650	Accepted

Based on data from the table above, it appears that almost all relationships or constructs have a significant effect except for one construct, namely the effect of Activity on Context-Awareness Value, where the results of t-statistics show the number 1.502 or below the standard of t table, which is 1.650. The results of the path coefficient also show negative results where Activity should had a positive influence on Context-Awareness Value.

5 CONCLUSIONS AND RECOMMENDATIONS

The model in this study uses the Bang Lee et. al (2017) and has an R-square value for Online Native Advertisement Value of 48% which shows that the variables used in this study affect 48% Online Native Advertisement Value while the remaining 52% is influenced by other variables outside this study. The R-square value for Context Awareness Value is 44.2% which means that the variables that have been used affect 44.2% Context Awareness Value while the remaining 55.8% is influenced by other variables outside this study. Online Native Advertising Attitude has an R-square value of 56% or there are 44% other variables outside this study affecting these variables. With the results of these data, it can be concluded that the adaptation model of Bang Lee et. al (2017) can be used to predict attitudes of Online Native Advertising Attitude consumers towards Online Native Advertising Tokopedia.

This research model can be developed further over time with circumstances or phenomena that are already different. In other words, researchers can then add new variables or even other relationships that are more adapted to their research. When Online Native Advertising is no longer a new concept in digital advertising and the attitude of internet users has really proven to be positive towards this advertising, researchers can develop their research further or more than just examining consumers' attitudes towards this advertising concept. In addition, this study only collected data from respondents who live in the city of Bandung. To be more representative, further research can conduct research related to a wider range of respondents in order to be more specific in providing advice in improving the delivery of a digital advertisement. The number of samples can be added more in order to better represent a population. Also, R-Square values for endogenous variables in this study (Online Native Adver-tisement Value, Context Awareness Value, Online Native Advertising Attitude) are still in the "moderate" category. This shows there are still other variables that affect endogenous variables that are not used in this study so that future research can add several other variables related to the submission of Online Native Advertising.

REFERENCES

Abdullah, Ma'ruf. 2016. *Manajemen Komunikasi Periklanan.* Yogyakarta: Aswaja Pressindo.
Bang Lee, E., Gun Lee, S. & Gyu Yang, C. 2017. The influences of advertisement attitude and brand attitude on purchase intention of smartphone advertising. *Industrial Management & Data Systems* 117(6). doi: 10.1108/IMDS-06-2016-0229
Campbell, C. & Marks, L. J. 2015. Good Native Advertising Isn't A Secret. *Business Horizons* 58(6): 599–606. Retrieved from Science Direct Education Journal Database.
Hair, J. F., Sarstedt, M., Ringle, C. M. & Mena, J. A. 2012. An assessment of the use of partial least squares structural equation modeling in marketing research. *Journal of the Academy of Marketing Science,* *40*(3). 414–433.
Indrawati. 2015. *Metodologi penelitian manajemen dan bisnis: Konvergensi teknologi komunikasi dan informasi.* Bandung: Refika Aditama.
Iprice.co.id. 2019. [online]. https://iprice.co.id/insights/mapofecommerce/en/ [22 Maret 2019]
Kotler, Philip & Kevin, Keller. 2016. *Marketing Management 15th edition.* Harlow: Pearson Education Limited.
Mulyatiningsih, Endang. 2012. *Metode Penelitian Terapan Bidang Pendidikan.* Bandung. Alfabeta.

Pagefair.com. 2017. [online]. https://pagefair.com/downloads/2017/01/PageFair-2017- Adblock- Report. pdf [2 Maret 2019]

Schiffman, L. G., Leslie, L. K. 2000. *Consumer Behavior*. Fifth Edition. Prentice-Hall Inc. New jersey.

Sekaran, Uma. 2003. *Research method for business: A skill building approach*. 4th edition. John Wiley & Sons.

Seligman, T. J. 2015. Native Advertising: The Old Is New Again. *Computer & Internet Lawyer* 32(7): 1–9.

Statista.com.2018. [online]. https://www.statista.com/outlook/243/120/ecommerce/indonesia [14 Maret 2019]

Wetzels, Martin & Odekerken, Gaby. 2009. Using PLS Path Modeling for Assessing Hierarchical Construct Models: Guidelines and Empirical Illustration. *Management Information Systems Quarterly - MISQ*. 33. 10.2307/20650284.

Managing Learning Organization in Industry 4.0 – Rachmawati & Hendayani (eds)
© 2020 Taylor & Francis Group, London, ISBN 978-0-367-81920-0

The influence of service quality on repurchase intention in the Plaza Telkomcel in Timor Leste

C. Soares & M. Ariyanti
Faculty of Economics and Business, Telkom University, Indonesia

ABSTRACT: The telecommunications industry of Timor Leste in 2018 had a market penetration among cellular service users of 132%. Telemor (575,000 mobile subscribers) (38.90%), Telkomcel (466,881 subscribers), and Timor Telecom (436,356 subscribers) are telecommunications companies in Timor Leste. Telkomcel, as the second largest company, must know that service quality is a factor in maintaining and increasing the market share. This study aimed to determine the effect of service quality dimensions – such as tangibility, empathy, assurance, responsiveness, and reliability – partially and simultaneously on customer satisfaction, and its impact on repurchase intention as the operational variable. This research was conducted using quantitative methods, with nonprobability sampling techniques. Respondents in this study were 400 Timor Leste Telkomcel customers. The data were obtained using Google Forms to distribute questionnaires, and the results of the questionnaires were statistically analyzed using the SEM-PLS method and a partial t-test formula. The results of the SEM-PLS analysis found that dimensions of service quality had a significant positive effect simultaneously on customer satisfaction of 63.9% and customer satisfaction with repurchase intention of 74.8%. Service quality had a significant positive effect on repurchase intention with customer satisfaction as the intervening variable. The t-test found that all dimensions of service quality significantly positively effect customer satisfaction. Furthermore, this study also found that the dimension of service quality that has the highest effect on customer satisfaction is reliability and the dimension that has the lowest effect on customer satisfaction is tangibility.

1 INTRODUCTION

Telkomcel is a telecommunications company in Timor Leste. Besides Telkomcel, Telemor and Timor Telecom also serve as telecommunications providers. Based on market share data in the fourth quarter from the Timor Leste Statistics Agency, Telkomcel must know that service quality can help the company to maintain and increase its market share. The market share of the telecommunications industry in Timor Leste can be seen in Table 1.

Consumers are faced with more than one choice of telecommunications providers and prefer to use products that are in accordance with the desired service quality, so Telkomcel must be able to defend and win the market in order to survive in this industry, and the company needs to know what level of service quality can satisfy its consumers and keep them loyal. Telkomcel has made various efforts to demonstrate its commitment to providing the maximum service quality, such as being the only data center service provider in Timor Leste and joining the Bridge Alliance, which is an international alliance of telecommunications operators in Asia. Previous research conducted by Al-Hashedi and Abkar (2017) explains that customer satisfaction is critical to success in the telecommunications industry and service quality can be used as a good measurement tool so as to determine the extent of customer satisfaction. The company must be able to create positive repurchase intentions in the future by providing positive satisfaction in order to retain customers.

Table 1. Market share of telecommunications in Timor Leste.

Provider	Period	Mobile subscribers	Percentage
Telemor	2018	575,000	38.90%
Telkomcel	2018	466,881	31.58%
Timor Telecom	2018	436,356	29.52%

2 LITERATURE REVIEW

Meiliani and Mustikasari (2018) explain that service quality focuses on meeting the needs and desires of customers and the provisions delivered to consumers so as to balance consumer expectations. Five main variables are arranged according to their importance: reliability, responsiveness, assurance, empathy, and tangibility (Tjiptono & Chandra, 2016).

Kotler and Keller (2012) state that customer satisfaction is a reflection of a person's assessment of a product – if a product is not in line with expectations, the customer will be disappointed. If the product is appropriate, the customer will feel satisfied, and if it exceeds expectations, the customer will feel happy. Hellier and colleagues (2003) explain that repurchase intention is a consumer's plan to buy more products or services from the same company by considering the current situation and the various possibilities. Service quality directly or indirectly influences the repurchase intention, so customer satisfaction can increase repurchase intention. This is in accordance with the formulation of the hypothesis addressed in this research. The framework of this study can be seen in Figure 1.

Based on the theory and previous research, the hypotheses of this study are as follows:

H1: Together the dimensions of service quality (tangibility, empathy, assurance, responsiveness, reliability) have a positive significant effect on customer satisfaction.
H2: Tangibility has a positive significant effect on customer satisfaction.
H3: Empathy has a positive significant effect on customer satisfaction.
H4: Assurance has a positive significant effect on customer satisfaction.
H5: Responsiveness has a positive significant effect on customer satisfaction.
H6: Reliability has a positive significant effect on customer satisfaction.
H7: Customer satisfaction has a positive significant effect on repurchase intention.
H8: Service quality has a positive significant effect on repurchase intention.
H9: Customer satisfaction has a positive significant effect in mediating the relationship between service quality and repurchase intention.

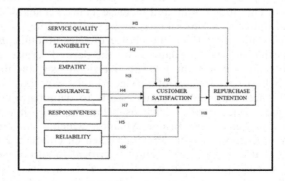

Figure 1. Framework.

3 METHODOLOGY

This research employed a quantitative method with the aim of conclusive or causal study. According to Indrawati (2015), conclusive or causal research is research conducted when researchers have seen or read previous research that discusses the relationship between variables aimed at understanding which variables are the cause and which are the effect. This study was used to determine the relationship between service quality dimensions and repurchase intention.

3.1 Participants

The researchers used nonprobability sampling with a purposive sampling method. Of its total population of 466,881 mobile subscribers, Telkomcel Timor Leste provided samples of customers who met the criteria in accordance with research needs, namely knowing Telkomcel Timor Leste products, using Telkomcel Timor Leste products, and having visited the Telkomcel plaza at least one time. The number of samples used in this study was determined using the Slovin formula with an error rate of 5% and a confidence level of 95%.

The Slovin formula is:

$$n = \frac{N}{1 + Ne^2} \tag{1}$$

where n = sum of sample, N = sum of population, and e = tolerance (5%).

Calculations using the Slovin formula obtained results of 399.66 rounded up to 400 samples. Data collection was done by distributing questionnaires online using Google Form. The questionnaire was prepared using a Likert scale, which indicates a respondent's answer with a range of values where 1 indicates strong disagreement and 4 indicates strong agreement.

3.2 Measurement

In this study service quality (X) was related to the concept of perception. Service quality has several dimensions. Baruah and colleagues (2015) define the independent variables used in measuring the dimensions of service quality as follows: tangibility (X1), empathy (X2), assurance (X3), responsiveness (X4), and reliability (X5). Customer satisfaction (M) affects repurchase intention (Choi & Kim, 2013). Previous research conducted by Astuti and Rusfian (2013) uses price, variety, information, and service quality as intervening variables, and planning to repurchase, making the product a top choice, positive reviews, and recommending to others as dependent variables (Y) (Grewal et al., cited in Astuti & Rusfian, 2013).

3.3 Data analysis

Data analysis was performed using SEM-PLS and calculations with the partial formula t-test. The relationship between service quality, customer satisfaction, and repurchase intention was analyzed using SmartPLS 3.0 software to create a path model with two steps:

1. Evaluate the measurement model (outer model) by looking at convergent validity, discriminant validity, and reliability.
2. Evaluate the structural model (inner model) by looking at the R-square value, effect size, and predictive relevance.

The hypotheses were further examined by doing bootstrapping so as to see the simultaneous effect, and through using the partial formula t-test to see the partial relationship of each dimension of service quality to customer satisfaction.

The t-test formula is:

$$t = r\sqrt{\frac{n-2}{1-r^2}} \tag{2}$$

where t = result of level significant, r = partial correlation, and n = sum of data.

4 RESULT AND DISCUSSION

4.1 Evaluation of measurement model (outer model)

The rule of thumb used for convergent validity is outer loading > 0.7, communality > 0.50, and AVE > 0.50 (Chin, cited in Abdillah & Hartono, 2015). Based on the SmartPLS results, all indicators met the criteria. Chin (cited in Ghozali & Latan, 2015) states that the use of Cronbach's alpha to test the reliability of the construct will give a lower value, so he recommends a composite reliability value > 0.70. The SmartPLS results reveal that all indicators met the criteria. These results can be seen in Table 2.

The SmartPLS results indicated that all variables have a composite reliability value and Cronbach's alpha > 0.7, meaning that the variables used in this study are reliable. These results can be seen in Table 3.

4.2 Evaluation of structural model (inner model)

The inner model looking at the R-square value was used to explain the effect of exogenous latent variables on endogenous latent variables. The construct variability of customer

Table 2. Results of convergent validity test.

Variable	Indicator	Outer loading	AVE	Result	Variable	Indicator	Outer loading	AVE	Result
Tangible	T1	0.817	0.722	Valid	Service quality	E1	0.817	0.645	Valid
	T2	0.839		Valid		E2	0.817		Valid
	T3	0.870		Valid		E3	0.882		Valid
	T4	0.870		Valid		E4	0.774		Valid
Empathy	E1	0.826	0.757	Valid		A1	0.828		Valid
	E2	0.862		Valid		A2	0.838		Valid
	E3	0.850		Valid		A3	0.803		Valid
	E4	0.728		Valid		A4	0.826		Valid
Assurance	A1	0.867	0.759	Valid		R1	0.802		Valid
	A2	0.888		Valid		R2	0.809		Valid
	A3	0.873		Valid		R3	0.842		Valid
	A4	0.856		Valid		R4	0.812		Valid
Responsiveness	R1	0.797	0.768	Valid		RE1	0.830		Valid
	R2	0.833		Valid		RE2	0.795		Valid
	R3	0.852		Valid		RE3	0.787		Valid
	R4	0.821		Valid	Customer	CS1	0.872	0.775	Valid
Reliability	RE1	0.868	0.812	Valid	satisfaction	CS2	0.903		Valid
	RE2	0.856		Valid		CS3	0.890		Valid
	RE3	0.807		Valid		CS4	0.857		Valid
Service quality	T1	0.715		Valid	Repurchase	RI1	0.897	0.801	Valid
	T2	0.753		Valid	intention	RI2	0.861		Valid
	T3	0.792		Valid		RI3	0.912		Valid
	T4	0.788		Valid		RI4	0.908		Valid

Table 3. Results of composite reliability and Cronbach's alpha.

Variable	Cronbach's alpha	Composite reliability	Variable	Cronbach's alpha	Composite reliability
Tangible	0.871	0.912	*Reliability*	0.884	0.928
Empathy	0.893	0.926	*Service quality*	0.969	0.972
Assurance	0.894	0.926	*Customer satisfaction*	0.903	0.932
Responsiveness	0.899	0.930	*Repurchase intention*	0.917	0.941

satisfaction of 0.639 (60%) and the remaining 0.4 (40%) is not explained in the hypotheses studied under this model. Furthermore, the construct variability of repurchase intention of 0.748 (70%) and the remaining 0.3 (30%) is not explained in the hypotheses studied under this model. Based on the rule of thumb, we know that the R2 value of the model is included in the category of a strong model. The path model analysis produced statistical values; all t-statistic values of each construction reached a significance level of 5%, and the one-tailed hypothesis had a t-table value of 1.64. The t-statistic value is greater than the t-table value, indicating that the exogenous variable has a significant influence on the endogenous variable. These results can be seen in Table 4.

The entire t-statistic value of each construction reached a significance level of 5%, and the one-tailed hypothesis had a t-table value of 1.96. The t-statistic value is greater than the t-table value, meaning that each variable partially effects customer satisfaction. These results can be seen in Table 5.

Service quality dimensions such as tangibility, empathy, assurance, responsiveness, and reliability have a significant partial and simultaneous effect on customer satisfaction. Service quality, customer satisfaction, and repurchase intention have a significant influence on the sales of Telkomcel products in Timor Leste. The R2 for Telkomcel customer satisfaction in Timor Leste is 63.9%, meaning that service quality (tangibility, empathy, assurance,

Table 4. Results of R-square.

Variable	Original sample (O)	Mean sample (M)	Standard deviation (STDEV)	t-Statistic (IO/ STDEV)
Service quality → Customer satisfaction	0.799	0.799	0.036	22.504
Service quality → Repurchase intention	0.508	0.513	0.070	7.241
Customer satisfaction → Repurchase intention	0.403	0.397	0.072	5.598
Service quality → Customer satisfaction → Repurchase intention	0.322	0.316	0.055	5.801

Table 5. Results of partial test.

Variable	A	β	R	t-statistic	Result
Tangible → Customer Satisfaction	3.243	0.710	0.665	17.312	Accepted
Empathy → Customer Satisfaction	3.301	0.729	0.711	20.188	Accepted
Assurance → Customer Satisfaction	2.511	0.766	0.728	21.167	Accepted
Responsiveness → Customer Satisfaction	2.388	0.793	0.768	23.936	Accepted
Reliability → Customer Satisfaction	2.251	1.076	0.790	25.688	Accepted

responsiveness, and reliability) affects customer satisfaction by 63.9%, while the remaining 36.1% is influenced by other factors. The R2 value for Telkomcel repurchase intention in Timor Leste is 74.8%, meaning that service quality and customer satisfaction affect repurchase intention by 74.8% and the remaining 25.5% is influenced by other factors. The overall dimensions of service quality influence customer satisfaction. The service quality dimension that has the most influence on customer satisfaction at Telkomcel in Timor Leste is reliability; it is hoped that Telkomcel can continue to maintain its reliability in relation to the needs of its customers going forward. The service quality dimension that has the lowest significant influence is tangibility (physical evidence). Telkomcel therefore needs to pay attention to matters related to the physical evidence they have in order to optimally improve the quality of its services. Service quality influences repurchase intention both directly and through customer satisfaction. Therefore, in the future Telkomcel is expected to be able to maintain the quality of its services so that it can influence customers' repurchase interest in Telkomcel products in Timor Leste.

REFERENCES

Abdillah, W., & Hartono, J. 2015. *Partial least square (PLS) alternative structural equation modeling (SEM)*. Dalam Penelitian Bisnis (Edisi 1). Yogyakarta: ANDI.

Al-Hashedi, A. H., & Abkar, S. A. 2017. The impact of service quality dimensions on customer satisfaction in telecom mobile companies in Yemen. *American Journal of Economics* 7(4),186–193.

Astuti, R. D., & Rusfian, E. Z. 2013. Pengaruh e- service quality terhadap repurchase intention melalui customer satisfaction (studi pada online shop gasoo galore). *Fisip Universitas Indonesia* 2013.

Beruah, D., Nath. T., & Bora, D. 2015. Impact of service quality dimension on customer satisfaction in telecom sector. *International Journal of Engineering Trends and Technology* (IJETT) 27(2), 111–117.

Choi, E. J., & Kim, S. H. 2013. The study of the impact of perceived quality and value of social enterprise on customer satisfaction and re-purchase intention. *International Journal of Smart Home* 7(1),239–252.

Ghozali, I., & Latan, H. 2015. *Partial least square, Konsep, Teknik dan Aplikasi Menggunakan Program SmartPLS 3.0 untuk penelitian empiris*. Semarang: Badan Penerbit Undip.

Hellier, P. K., Geursen., Gus, M., Carr, R. A., & Richard, J. A. 2003. Customer repurchase intention: a general structural equation model. *European Journal of Marketing* 37(11), 1762–1800.

Indrawati. 2015. *Metode Penelitian Manajemen dan Bisnis: Konvegensi Teknologi Komunikasi dan Informasi*. Bandung: PT. Refika Aditama.

Kotler, P., & Keller, K. L. 2012. Marketing management. 14th edition. Upper Saddle River, NJ: Pearson.

Meiliani, S. D., & Mustikasari, A. 2018. The influence of service quality to customer satisfaction: case study service private company at PT. Tiki jalur ekakurir (JNE), branch setrasari mall in bandung 2018. *E – proceeding of Applied Science* 4(3), 1153–1162.

Tjiptono, F., & Chandra, G. 2016. *Service, quality, and satisfaction*. 4th edition. Yogyakarta: ANDI.

Tools for measuring variables influencing customers' adoption of online tax services

Indrawati, D. Grenny & Syarifuddin
Telkom University, Bandung, Indonesia

ABSTRACT: This study aimed to provide a tool that will allow the measurement of variables that influence the intentions of users – especially Indonesian Hotel and Restaurant Association employees – of online tax services using a modification of the unified theory of acceptance and use of technology 2 (UTAUT2) model. The researchers completed five steps in order to develop and confirm the modified UTAUT2 model, from the operationalization of the variables, to content validity and a pilot study. The pilot study was completed using data from thirty respondents. The results showed that the measurement tools, which consisted of forty indicators of nine variables, are valid and reliable for use in further study, namely measuring variables that influence the behavioral intention and use behavior of Indonesian Hotel and Restaurant Association employees in using online tax services.

1   INTRODUCTION

PT. Finnet Indonesia was established on October 31, 2005, in order to provide information technology (IT) infrastructure, applications, and content. PT. Finnet also aimed to meet the information and transaction systems needs of financial institutions and the government in Indonesia through electronic payment and information and communication technologies. PT. Finnet Indonesia has collaborated with the city of Surakarta, which is well known as Solo, in implementing online tax services for the Indonesian Hotel and Restaurant Association (IHRA) starting in 2017. Solo has been selected as a smart city pioneer in Indonesia, and implementing the online tax services there represents just one of many efforts to support smart governance. Smart cities have ten dimensions: smart economy, smart buildings, smart education, smart energy, smart healthcare, smart mobility, smart water, smart safety and security, smart technology, and smart governance (Indrawati & Smart City Research Group, 2019). Smart governance is a machine that moves the other nine dimensions forward so as to achieve the objectives of a smart city. Smart governance should be implemented in three functions: public services, bureaucracy, and policy. The objectives of smart governance usually comprise effective, transparent, and accountable bureaucratic management (Ahmadjayadi, Subkhan, & Wiradinata, 2016). Hence the Solo government has collaborated with PT. Finnet Indonesia in providing online tax services to the IHRA in Solo. The implementation of these online tax services is part of smart governance efforts to make the payment of IHRA taxes transparent.

The establishment of these services requires significant investment from the Solo city government, but up to this study no research measured the factors that influence the acceptance of online tax services. It is important to find these factors so that IHRA employees' acceptance of online tax services will increase. The data show that 48% of parking area employees and 55% of entertainment area employees have not yet used online tax services. To identify the factors preventing their acceptance and use of online tax services, it is important to find tools for measuring the variables and indicators influencing customers' adoption of online tax services. The objective of this study was to provide a measurement – through employing a modified unified theory of acceptance and use of technology 2 (UTAUT2) model – of the variables and indicators that influence the intention of users, especially Indonesian Hotel and Restaurant Association employees in Solo, to utilize online tax services.

2 LITERATURE REVIEW

To achieve the objectives as described in the introduction, the researchers conducted a literature review of the theories and models related to the acceptance of technology or services based on technology. Ten main models and theories are commonly used to study consumers' acceptance of technology or services based on technology. Those models are: (1) theory of reasoned action (TRA), (2) theory of planned behavior (TPB), (3) technology acceptance model (TAM), (4) motivational model (MM), (5) combined TAM-TPB (C-TAM-TPB), (6) model of personal computer utilization (MPCU), (7) innovation diffusion theory (IDT), (8) social cognitive theory (SCT), (9) unified theory of acceptance and use of technology (UTAUT), and (10) unified theory of acceptance and use of technology 2 (UTAUT2). After reviewing all of these models, this study found that the UTAUT2 model has the highest predictive power (Venkatesh, Thong, & Xu, 2012).

The UTAUT2 model is an extension of the UTAUT model developed by Venkatesh, Morris, Davis, and Davis (2003) that combines the earlier eight theories of technology acceptance (i.e., TRA, TPB, TAM, MM, C-TA-TPB, MPCU, IDT, and SCT). The UTAUT was formulated as a new model to explain the acceptance and use of technology with the highest value of R^2 compared to previous models. The UTAUT models can predict the intention of people in organizations to adopt technology up to 70%, while the other eight models can predict only 17–53% (Venkatesh et al., 2003).

To predict the intentions of consumers using individual or consumer content, Venkatesh and colleagues (2012) redeveloped the UTAUT model into the UTAUT2 model by adding three new constructs to the UTAUT model – namely hedonic motivation, price value, and habit. This new model also involves three moderating variables such as age, gender, and experience.

The UTAUT2 model has been used to research the adoption of products or services in various studies. Thomas, Singh, and Gaffar's (2013) study of the adoption of mobile learning at the higher education level in Guyana; Dhulla and Mathur's (2014) analysis of the willingness of tertiary-level students in Mumbai to adopt cloud computing; Escobar-Rodríguez and Carvajal-Trujillo's (2014) research into the determinants of purchasing flights from low-cost carrier (LCC) websites by 1,096 Spanish consumers; Attuquayefio and Addo's (2014) study of tertiary-level Ghanaian students' acceptance of information and communication technology (ICT); and Nasri and Abbas's (2015) examination of the attitude of Kuwaiti citizens toward e-government applications. The UTAUT2 model has also been used in several studies in Indonesia, such as research by Indrawati and Marhaeni (2015) that studied the use of a modified UTAUT2 model to predict behavioral intention toward instant messenger applications; by Indrawati and Ariwiati (2015) that studied the use of a modified UTAUT2 model to predict behavioral intention toward e-commerce of micro, small, and middle enterprises (SME) in Indonesia; by Indrawati and Adicipta (2017) on the adoption of internet banking in Indonesia; by Indrawati and Utama (2018) on Indonesians' adoption of 4G; and by Indrawati and colleagues (2019) on the adoption of teleconsultation technology among employees of Telkom Indonesia.

This research eliminated the independent price value variable from the original UTAUT2 model since users pay no cost when using the application under examination here. The moderating variable of experience was also eliminated from the original UTAUT2 model, due to the fact that this study just took the data one time. A new variable of trust was added since some of the respondents who participated in preliminary data gathering thought they might use trusted online tax services.

3 METHODOLOGY

In order to develop a questionnaire through which to measure the model of this study, the researchers completed five steps. The first step was to determine the operationalized variable through a process of breaking the variables contained in the problem of research into their smallest parts so they could be known by classification, making it easier to get the data needed to solve the research problem at hand (Indrawati, 2015). The second step involved ensuring content validity by checking that each item used on the questionnaire was logical to measure items from definition and that the indicator was applied (Indrawati, 2015: 147). The authors adopted and modified questionnaire items from previous studies, either international or national journals related to this research. Questionnaire items were adapted and modified from previous study by Venkatesh and colleagues (2003), Venkatesh and colleagues (2012), Indrawati (2015), and Masri and Tarhini (2017). The third step was testing the questionnaire items with three experts in the fields of marketing and technology adoption. The objective of this test was to obtain suggestions or recommendations from the experts in order to improve the questionnaire items and to better meet the needs of this research. The fourth step was readability test, which was done with prospective respondents in order to make sure that each item in the questionnaire was understandable. The respondents found no difficulties in understanding the questionnaire. The last step was a pilot study to test that each item in the questionnaire was valid and reliable enough to conduct this research. The pilot test survey items were answered by thirty respondents and those data were used to perform validity and reliability tests. The authors used IBM SPSS Statistic 25 as a tool and a corrected item–total correlation (CITC) value was used as a validity reference score. A CITC greater than the r-table with a value of 0.316 meant that an item was valid.

4 RESULTS AND DISCUSSION

Based on the aforementioned processes, the results of the tests are shown in Table 1.

Table 1. Validity test results.

Indicator	Code	CITC score
IHRA online tax services shorten the time needed to create IHRA tax reports.	PE1	0.433
IHRA online tax services help in monitoring IHRA tax every time.	PE2	0.756
IHRA online tax services cause no device damage.	PE3	0.613
Overall, IHRA online tax services help tax work become more efficient and effective.	PE4	0.510
It is easy to learn how to operate IHRA online tax services.	EE1	0.613
It is clear how to use IHRA online tax services.	EE2	0.543
It is easy to understand how to operate IHRA online tax services.	EE3	0.637
IHRA online tax services are easy to use.	EE4	0.595
IHRA online tax services are easy to move around.	EE5	0.378
People who are important to me suggest IHRA online tax services.	SI1	0.672
People who are influential to me suggest IHRA online tax services.	SI2	0.509
People whose opinions I respect suggest IHRA online tax services.	SI3	0.587
People who are close to me recommend IHRA online tax services.	SI4	0.688
The regional government suggests IHRA online tax services.	SI5	0.383
	FC1	0.566

(*Continued*)

Table 1. (*Continued*)

Indicator	Code	CITC score
Equipment/tools/facilities needed to operate IHRA online tax services are adequately available.		
I have knowledge of how to use IHRA online tax services.	FC2	0.739
IHRA online tax services provide the most suitable system to record, calculate, and make a report of IHRA taxes.	FC3	0.619
If trouble occurs while using IHRA online tax services, a person or other parties will help me.	FC4	0.723
IHRA online tax services are operable on the device I have currently.	FC5	0.636
I feel happy when I use IHRA online tax services.	HM1	0.632
I feel satisfied when I use IHRA online tax services.	HM2	0.737
I feel comfortable when I use IHRA online tax services.	HM3	0.679
I used to use IHRA online tax services.	HB1	0.736
I used to record IHRA tax transactions using IHRA online tax services.	HB2	0.840
I used to calculate IHRA taxes using IHRA online tax services.	HB3	0.674
I used to create IHRA tax reports using IHRA online tax services.	HB4	0.719
I used to monitor IHRA taxes using IHRA online tax services.	HB5	0.773
I believe and have faith in using IHRA online tax services.	TR1	0.855
I believe and have faith in the security of IHRA online tax services.	TR2	0.801
I don't hesitate to use IHRA online tax services.	TR3	0.821
I believe and have faith in the guarantee of IHRA online tax services.	TR4	0.835
I believe and have faith that errors while using IHRA online tax services are very minimal.	TR5	0.829
I have interest in using IHRA online tax services in the future	BI1	0.482
I used to use IHRA online tax services to calculate and report IHRA taxes.	BI2	0.719
I used to operate IHRA online tax services to monitor IHRA taxes.	BI3	0.633
I plan to use IHRA online tax services in the future.	BI4	0.616
I choose to use IHRA online tax services rather than conventional or manually inputted tax services.	BI5	0.678
I check taxes through IHRA online tax services when I want to pay IHRA taxes.	UB1	0.529
In the future, I plan to keep using IHRA online tax services.	UB2	0.294
I use IHRA online tax services for every tax payment transaction.	UB3	0.348
I use IHRA online tax services to create tax reports.	UB4	0.726

Based on the result as shown in Table 1, nine variables and forty indicators or items are valid except for item UB2, which is considered invalid due to its CITC value of less than 0.316.

5 CONCLUSIONS AND RECOMMENDATIONS

The results show that the measurement tool, which consists of forty indicators of nine variables, is valid and reliable to be used for further study – namely measuring variables that influence the intention and use behavior of IHRA employees utilizing online tax services.

It is suggested to use this measuring tool for further study not more than six months after the validity test.

ACKNOWLEDGMENT

The authors express their gratitude to the Ministry of Research, Technology and Higher Education of Indonesia for its financial support in carrying out this research.

REFERENCES

Ahmadjayadi, C., Subkhan, F., & Wiradinata, M. 2016. *Melesat atau Kandas? New Indonesia*. Jakarta: PT. Elex Media Komputindo.

Attuquayefio, S. N., & Addo, H. 2014. Using the UTAUT model to analyze students' ICT adoption. *International Journal of Education and Development Using Information and Communication Technology* 10(3), 75–86.

Dhulla TV, Mathur SK. Adoption of cloud computing by tertiary level students–a study. *Journal of Exclusive Management Science*. 2014; 3(3):1–15.

Escobar-Rodríguez, T., & Carvajal-Trujillo, E. 2014. Online purchasing tickets for low cost carriers: An application of the unified theory of acceptance and use of technology (UTAUT) model. *Tourism Management* 43, 70–88.

Indrawati. 2015. *Metodologi Penelitian Manajemen dan Bisnis Konvergensi Teknologi Komunikasi dan Informasi*. Bandung: PT. Refika Aditama.

Indrawati & Adicipta, S. R. M. 2017. Factors influencing internet banking acceptance (a case study of ABC internet banking in Bandung, Indonesia). *Journal of Engineering and Applied Science* 12(7),1705–1709. doi: 10.3923/jeasci.2017.1705.1709

Indrawati & Ariwiati. 2015. Factors affecting e-commerce adoption by micro, small and medium-sized enterprises in Indonesia. Proceedings of the International Conference on E-Commerce and Digital Marketing. July 21–23, 2015, Las Palmas de Gran Canaria, Spain.

Indrawati & Marhaeni, G. A. M. M. 2015. Measurement for analysing instant messenger application adoption by using unified theory of acceptance and use of technology 2 (UTAUT2). *International Business Management* 9(4), 391–396.

Indrawati & Smart City Research Group. 2019. *Inilah Cara Mengukur Kesiapan Kota Pintar*. Bandung: Intrans Publishing.

Indrawati & Utama, K. P. 2018. Analyzing 4G adoption in Indonesia using a modified unified theory of acceptance and use of technology 2. 2018 6th International Conference on Information and Communication Technology (ICoICT). Institute of Electrical and Electronics Engineers. doi: 10.1109/ICoICT.2018.8528744

Indrawati, U. M., & Yogi, O. 2019. Predicting behavior intention to adopt teleconsultation technology (a perspective from the implementation of Udoctor at Telkom Indonesia). *Journal of Advanced Research in Dynamical and Control Systems* 11(5) (Special Issue), 352–357.

Masri, E., & Tarhini, A. 2017. Factors affecting the adoption of e-learning systems in Qatar and USA: Extending the unified theory of acceptance and use of technology 2 (UTAUT2). *Education Tech Research Dev*. 65, 743–763. https://doi.org/10.1007/s11423-016-9508-8 1–21.

Nasri, W., & Abbas, H. A. 2015, May. Determinants influencing citizens' intention to use e-Gov in the state of Kuwait: Application of UTAUT. *International Journal of Economics, Commerce and Management (IJECM)* 3(5), 517–540.

Thomas, T. D., Singh, L., & Gaffar, K. 2013. The utility of the UTAUT model in explaining mobile learning adoption in higher education in Guyana. *International Journal of Education and Development Using Information and Communication Technology* 9(3), 71–85.

Venkatesh, V., Morris, M. G., Davis, G. B., & Davis, F. D. 2003. User acceptance of information technology: Toward a unified view. *MIS Quarterly* 27(3), 425–478.

Venkatesh, V., Thong, J. Y. L,& Xu, X. 2012. Consumer acceptance and use of information technology: Extending the unified theory of acceptance and use of technology. *MIS Quarterly* 36(1), 157–178.

Measuring the smart office index: A case study from Telkomsel smart office

Indrawati, A.A. Sekarini & A. Husni
Telkom University, Bandung, Indonesia

ABSTRACT: In realizing a sustainable smart city, a smart building concept is needed that can support energy efficiency and maintain the security and comfort level of residents/employees. One of the building forms that use the most energy is office buildings; therefore, office buildings must be managed intelligently or converted into smart offices. Previous research on smart building has found variables and indicators that builders can use to measure smart building. However, this model has not been tested to measure building a smart office. Required data were collected by conducting in-depth interviews with thirty-eight respondents. The result revealed that the Telkomsel Smart Office (TSO) building has an index of 85.92, which is in the good category. Even though it was considered good, the authors suggest that the building owner can improve the smart elevator and HVAC systems in the building since these two indicators received poor scores.

1   INTRODUCTION

According to the United Nations' report on world urbanization prospects (UN, 2018), 68% of Asia's population will live in urban cities by 2050, which shows the need for strategically developing urban areas around the world into smart cities with smart buildings that minimize energy consumption as urban areas contribute to around 40% of the total annual energy consumption. The concept of smart buildings is aimed at achieving not only savings in energy consumption but also increased convenience and benefits to various stakeholder – namely, building users, owners, tenants, and smart service providers – as "smartness" adds, creates, and generates value to buildings.

The benefits of smart building include increased safety and security, increased comfort and convenience, reduction in CO_2 emissions, and reduction in operational costs (heating, cooling, and lighting). A study published in the *Harvard Business Review* (*HBR*) (2016) shows that smart buildings save up to 30% of water usage, 50% of natural gas usage, and 40% of energy usage, hence smart buildings are considered critical and a precondition for transforming a city into a smart city (Indrawati & Smart City Research Group. 2019; *HBR*, 2016).

In modern society, people spend most of their time in their office. According to Li (2014), there is no doubt that the office environment directly influences work efficiency, so comfort is needed in the office environment. On the other hand, the current energy crisis and the dilemma of environmental contamination that is developing throughout the world, especially in developing countries, make energy conservation a new trend in office buildings. The smart office system emerged in response to these complicated problems.

The present urbanization rate in Indonesia is around 4.1% per year, which is the fastest among the Asian countries (World Bank, 2016), indicating the need for developing urban cities into smart cities by constructing smart buildings, including smart offices, in Indonesia. The Telkomsel Smart Office (TSO) was built at a very high cost and through development that involved many parties in the installation of technology.

In its planning, TSO adapted the latest technology to create an office with a high level of intelligence, also called a smart office. TSO carries the concept of a seamless, wireless, paperless, and cashless office that supports a mobile and digital lifestyle. The present study focused on TSO in order to explore the possibility of improving its status as a smart office to a higher level in the future by concentrating on various variables and indicators used for measuring smart buildings identified in earlier research (Indrawati & Amani, 2017) for developing TSO's smart office index (SOI). The two research questions (RQ) the present paper tries to answer are: RQ1: What is the right model for measuring the SOI? RQ2: What is the TSO's SOI?

2 LITERATURE REVIEW

Muñoz and colleagues (2018) argue that a smart office is an environment that supports and influences employees in their daily tasks. This system uses information collected by various sensors and trigger actions that adjust the environment to the needs of users through actuators.

Smart offices must be aligned with the company's business objectives and must enable a productive environment that maximizes employee satisfaction and company performance. Thus, smart offices must manage the Internet of Things (IoT) infrastructure used in the workplace efficiently and proactively. In addition, according to Furdik and colleagues (2013), smart offices must be integrated with smartphones and help employees to unite personal communication with professional activities.

In addition, Li (2014) proposes a smart office system design that involves the control of heating, lighting, and ventilation, and the reconfiguration of offices and meeting rooms. Smart office systems typically consist of embedded automation systems, information technology, and automation technology. Some objects are controlled and respond according to sensors. These systems must adjust to users' wishes and analyze them, finally reacting at that moment. The comprehensive smart office system concentrates on door access, lighting, ventilation, heating, and reconfiguration designed to save energy and increase employee satisfaction.

3 METHODOLOGY

The entire process of the present study was carried out in six stages: first, reviewing and confirming the variables and indicators of smart office buildings; second, identifying best practices data for smart office implementation from around the world; third, finding the data for smart office building implementation indicators for TSO; fourth, conducting in-depth interviews and collecting qualitative and quantitative data; fifth, calculating the individual scores of each variable and indicator to ensure validity for inclusion in the final model; and, finally, calculating the SOI for TSO.

Both qualitative and quantitative data were used in the study. Qualitative data were used for assessing and confirming smart building status based on various variables and indicators. An in-depth survey was carried out with business players, experts or researchers, and users or society in order to identify the variables and indicators used for measuring smart buildings and confirming the validity of the identified variables and indicators. Quantitative data were used for calculating the SOI, which was based on best practice data from other countries that have already implemented a smart building index (Creswell, 2009; Indrawati, 2015). The data were collected using a nonprobability sampling method – namely, purposive sampling (Indrawati, 2015; Zikmund et al., 2010) from three categories of respondents, the details of which are given in Table 1.

Table 1. List of respondents.

Category	Respondents	Number
Expert/ researchers	Experts in smart office building operations	4
Business players	Smart office building management companies engaged in building procurement or smart office building	6
Users	People who know about the technology of smart buildings	28

4 VALIDATION OF SMART OFFICE INDICATORS AND VARIABLES

Seven variables to measure smart building are identified in an earlier work by Indrawati and Amani (2017) – namely, a building automation system, building control system, energy management system, safety and security management system, enterprise management system, IT network connectivity, and green building construction. These seven variables have twenty-four indicators, which were used in the present study after validating the same before calculating the SOI. The details of all twenty-four indicators/constructs of the seven variables with their validity test results are shown in Table 2.

Validation of the variables and their relevant indicators was carried out based on the in-depth survey data collected from the respondents with the aim of exploring the possibility of adding or accommodating a new indicator(s) and variable(s). Respondents were asked to score (ranging from 1 to 100) each of the twenty-four indicators of the seven

Table 2. Validity test results of smart building variables and indicators.

#Variables	#Indicators/Constructs	R	Validity
1 Building automation system	1 Sensors	0.232	Not Valid
	2 Smart elevator	0.725	Valid
	3 HVAC system	0.616	Valid
2 Building control system	4 Remote monitoring	0.845	Valid
	5 Real-time monitoring	0.928	Valid
	6 Software connected to existing equipment	0.870	Valid
3 Energy management	7 Power consumption and system monitoring	0.913	Valid
	8 Energy-efficient electrical appliances	0.873	Valid
	9 Backup energy	0.890	Valid
4 Safety and security	10 Threat detection and response	0.660	Valid
	11 Control of access to the facility	0.798	Valid
	12 Securing lives and assets	0.730	Valid
	13 Security framework and cybersecurity	0.863	Valid
	14 Safety and privacy policy	0.815	Valid
5 Enterprise management system	15 Data management framework	0.892	Valid
	16 System information management	0.931	Valid
	17 Data analytics	0.945	Valid
6 IT network connectivity	18 Wireless communication	0.872	Valid
	19 Devices connected with wireless communication	0.823	Valid
	20 Available and reliable network	0.906	Valid
7 Green building construction	21 Green building architecture	0.611	Valid
	22 Low impact on the environment	0.881	Valid
	23 Resource efficiency	0.817	Valid
	24 Healthy environment	0.784	Valid

Source: Spearman rank test of quantitative survey of thirty respondents

variables, which allows the indicators to be grouped into five categories – namely, scores from 1 to 60 indicating a worst performance scenario that needs a lot of improvement, scores from 61 to 70 indicating a bad performance scenario that needs a lot of improvement, scores from 71 to 80 indicating a good or good enough performance scenario that still needs a lot of improvement, scores from 81 to 90 indicating a good or satisfying performance scenario that needs little improvement, and scores from 91 to 100 indicating a very good performance scenario where no improvement is required.

The scores collected from the respondents were then tested for their validity using the Pearson product moment correlation, and the resulting value of at least 0.364 was considered valid with an error rate of 0.05. The result of the validity test shown in Table 2 indicates that except one construct – namely "sensors implementation" (indicator number 1 of variable 1) – all the constructs are valid. This indicator was removed while calculating the SOI. Subsequently, Cronbach's alpha was also calculated so as to ensure reliability and found to be above 0.70 for all seven variables and twenty-four indicators.

5 LEVEL OF TSO'S SOI AND PUBLIC PERCEPTION OF SMART OFFICES

Based on the secondary data from the best practices in smart cities from the 2018 Global Cities Index Ranking (Peña et al., 2018), TSO data were collected as explained in the previous section based on the twenty-four indicators of the seven variables that determine the status of a building as a smart building. The calculation of SOI was carried out in two stages – namely, first the calculation of the "Indicator Index" and then the "Smart Office Index."

$$Indicator\ Index = \Sigma x / \Sigma x$$

Σx = Total score of an indicator given by respondents
Σy = Total respondents who gave a score for an indicator

$$Smart\ Office\ Index = \Sigma Ii / \Sigma Ti$$

ΣIi = Total average score of all indicators given by respondents
ΣTi = Total indicators

The result of the "Indicator Index" and "Smart Office Index," after removing the one indicator, is given in Table 3. The scenario of smart office building implementation given in the last column clearly indicates that TSO is rated at good level; out of the twenty-three remaining indicators, one fell into the "bad scenario" category, one was in the "fair scenario," and twenty-one came under the "good scenario." Only "building automation system" came under "fair scenario" and "bad scenario," whereas the rest of the indicators were in the "good scenario." The main reason for such a bad scenario in TSO is the fact that the number of elevators in the building is inadequate to meet the needs of employees, consultants, and building visitors. Builders plan to convert TSO into a digitally integrated smart office; the preparations are already done, so that the technology adopted is the latest. However, implementation has changed over time and has had limited funds so that not all technologies are as complex as originally planned.

Table 3. Indicator scores and SOI.

#Variables	#Indicators/Constructs	Score	Scenario
1 Building automation system	Sensors	removed	
	1 Smart elevators	64.47	Bad
	2 HVAC system	79.77	Fair
2 Building control system	3 Remote monitoring	85.93	Good
	4 Real-time monitoring	87.07	Good
	5 Software connected to existing equipment	86.60	Good
3 Energy management	6 Power consumption and system monitoring	86.93	Good
	7 Energy-efficient electrical appliances	87.60	Good
	8 Backup energy	89.90	Good
4 Safety and security	9 Threat detection and response	86.90	Good
	10 Control of access to the facility	88.90	Good
	11 Securing lives and assets	87.63	Good
	12 Security framework and cybersecurity	88.83	Good
	13 Safety and privacy policy	97.80	Good
5 Enterprise management system	14 Data management framework	86.97	Good
	15 System information management	87.03	Good
	16 Data analytic	85.97	Good
6 IT network connectivity	17 Wireless communication	88.80	Good
	18 Devices connected with wireless communication	88.03	Good
	19 Available and reliable network	89.03	Good
7 Green building construction	20 Green building architecture	97.43	Good
	21 Low impact on the environment	86.00	Good
	22 Resource efficiency	86.80	Good
	23 Healthy environment	87.03	Good

Source: Scoring of in-depth interview with thirty respondents

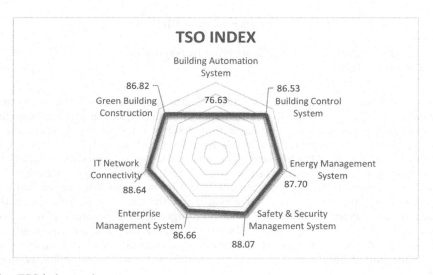

Figure 1. TSO index result.

6 CONCLUSION

The present study identified seven variables and twenty-three indicators for measuring TSO's SOI, which resulted in a score of only 85.92, leading to the conclusion that TSO has achieved very good and satisfying performance, aligned with its mission to have a smart futuristic office where all Telkomsel digital lifestyle solutions are experienced along with a comfortable, cozy, and collaborative working environment.

Among the indicators, the best indicator is "available and reliable network" with a score of 89.03, and the worst indicator is "smart elevators" with a score of 64.50. Only the "automation system" variable has one indicator falling under the "fair scenario" category and one indicator in the "bad scenario," whereas the rest of the indicators were in the "good scenario." The main reason for such a bad scenario in TSO is the fact that the number of elevators in the building is fewer than what employees, consultants, and building visitors need. Builders plan to convert TSO into a digitally integrated smart office; the preparations are already done, so that the technology adopted is the latest. However, implementation has changed over time and due to limited funds so that not all technologies are as complex as originally planned. TSO must also focus on improving the smart elevator and HVAC systems by providing the required facilities based on the high occupancy of the building. A well-planned implementation program must be developed by including various stakeholders, such as management, general service, building management, and educational institutions. Well-educated and well-informed employees will help TSO attain the status of a sustainable smart office in the coming years.

ACKNOWLEDGMENT

The authors express their gratitude to the Ministry of Research, Technology and Higher Education of Indonesia for its financial support in carrying out the research.

REFERENCES

Angdini, D. N. 2019. *Indeks Kesiapan Smart Building Kota Bandung*. Bandung: Universitas Telkom.

Creswell, J. W. 2009. *Research Design: Quantitative, Qualitative and Mixed Method*. London: Sage.

Furdik, K., Lukac, G., Sabol, T., & Kostelnik. P. 2013, November. The network architecture designed for an adaptable IoT-based smart office solution. *International Journal of Computer Networks and Communications Security* 1(6), 216–224. Available at www.ijcncs.org.ISSN 2308-9830.

Harvard Business Review (HBR). 2016. Smart cities start with smart buildings. Available at https://hbr.org/sponsored/2016/01/smart-cities-start-with-smart-buildings.

Indrawati. 2015. *Metode Penelitian Manajemen dan Bisnis*. Bandung: PT. Refika Aditama.

Indrawati, Yuliastri, R., & Amani, H. 2017. Indicators to measure a smart building: An Indonesian perspective. *International Journal of Computer Theory and Engineering* 9(6), 406–411. doi: 10.7763/IJCTE. 2017.V9.1176

Indrawati & Smart City Research Group. 2019. *Inilah Cara Mengukur Kesiapan Kota Pintar*. Bandung: Intrans Publishing.

Li, H. 2014. A novel design for a comprehensive smart automation system for the office environment. Proceedings of the 2014 IEEE Emerging Technology and Factory Automation (ETFA). doi: 10.1109/etfa.2014.7005267.

Muñoz, S., Araque, O., Sánchez-Rada, J. F., & Iglesias, C. A. 2018. An emotion aware task automation architecture based on semantic technologies for smart offices. *Sensors* 18(5), E1499. doi: 10.3390/s18051499

Peña, A. M., Hales, M., Peterson, E. R., & Dessibourg, N. 2018. 2018 global cities report: Learning from the East: Insights from China's Urban Success. Available at www.atkearney.com/2018-global-cities-report.

United Nations Department of Economic and Social Affairs, Population Division. 2018. World urbanization prospects: The 2018 revision, methodology. Working Paper No. ESA/P/WP.252. New York: United Nations.

World Bank. 2016. Kisah Urbanisasi Indonesia. Available at www.worldbank.org/in/news/feature/2016/ 06/14/indonesia-urban-story.
Zikmund, W. G., Babin, B. J., Carr, J. C., & Griffin, M. 2010. *Business research methods.* 8th edition. Singapore: Cengage Learning.

Measuring smart office index as part of smart building: A case from Telkom Landmark Tower

Indrawati, M.R.M. Siahaan & H. Amani
Faculty of Economics and Business, Telkom University, Bandung, Indonesia

ABSTRACT: The Telkom Landmark Tower (TLT) building is owned by PT. Telkom Indonesia and used for office space. The TLT was designed as a smart office intended to keep the building comfortable and safe, as well as to make operating costs and energy consumption more efficient. No study has measured if the objective of converting the TLT into a smart building has been achieved. Therefore, this research explored the variables and indicators used for measuring a smart office, which are then used to measure the TLT. To achieve this objective, researcher performed a literature review and in-depth interviews with thirty-seven experts and users in building management and office building operations. This study found seven variables – namely, building automation system, building control system, energy management system, safety and security management system, enterprise management system, IT network connectivity, and green building construction – with twenty-three indicators to measure the index of the TLT as a smart office. The result showed that the index of the TLT is 86.68, which is in the good or satisfactory category.

1 INTRODUCTION

Buildings consume a lot of energy to regulate air conditioning and lighting, as well as lifts or escalators. One way to control energy consumption in office buildings and commercial and public facilities is to implement smart building solutions. In Jakarta, the governor's Regulation No. 38 of 2012 mandates that building managers conserve energy resources. The goal of implementing a smart building management system is to be more cost-efficient, save energy consumption, increase building safety, and provide comfort (Yudis, 2014).

Telkom supports converting Jakarta into a smart city, hence Telkom has built an office in the Telkom Landmark Tower (TLT) building. The TLT is strategically positioned on the main highway from Jl. Jend Gatot Subroto in the central business district (Telkom Landmark Tower, 2016). The official social media account Telkom @ Telkom Indonesia posted on April 15, 2017, that the TLT is designed as a smart office with complete facilities and high technology that is environmentally friendly (Telkom Indonesia, 2017). If the smart office is implemented, it will help the Jakarta city government in structuring and building the city of Jakarta so that it is comfortable and safe, as well as more energy efficient.

The TLT is used for offices that have been declared smart offices by Telkom. No study has measured if the objective of converting the TLT into a smart building has been achieved. Therefore, this research explored the variables and indicators used for measuring a smart office, which were then used to measure the TLT. This study aimed (a) to find out what variables and indicators are appropriate for measuring the level of "smartness" of the TLT; (b) to find out how well respondents' assessments of variables and indicators measure the level of "smartness" of the TLT.

2 LITERATURE REVIEW

Marketing is about identifying and meeting human and social needs. One of the shortest good definitions of marketing is meeting need profitably (Kotler & Keller, 2012). Referring to this definition, we can conclude that the importance of marketing is meeting the needs of customers and of society at large.

A smart city is system integration that involves humans and social interactions by using technology that aims to efficiently achieve sustainability and resilience for the development and quality of life of the city (Monzon, 2015).

Smart building can be interpreted as construction that uses a technology system to enable service and operation to be used by residents and management (Vattano, 2014).

A smart office is an environment that supports and influences employees in their daily tasks. This system uses information collected by various sensors and triggers actions that adjust the environment to the needs of users through actuators (Muñoz, 2018).

3 METHODOLOGY

3.1 *Measurements*

As explained in the introduction, this study attempted to answer two research questions; hence researchers created map and matrix designs. The stages of research were as follows: conduct a literature review to determine the variables and indicators of smart offices, confirm the variables and indicators reached from the literature review, search for best practice data and data from TLT, interview experts in smart office variables and indicators, and ask the experts to score each indicator and variable.

The experts (thirty-seven altogether) in this study were divided into three groups: building management, office building operations, and users.

Based on the literature review, this study found seven research variables – namely, building automation systems, building control systems, energy management systems, safety and security management systems, enterprise management systems, IT network connectivity, and green building construction (Indrawati, Yuliastri, & Amani, 2017).

The literature review result found indicators as follows: implementation of sensors, implementation of remote monitoring of building and occupancy conditions, implementation of real-time monitoring, implementation of software that can connect all devices, both new and old, with various brands, implementation of monitoring and control of power consumption, implementation of energy-efficient electrical appliances, implementation of energy backup, implementation of detection and response to hazards, implementation of access control to facilities, implementation of personal and asset security, implementation of security and cybersecurity frameworks, issuance of safety and security regulations, data management frameworks, information system management, implementation of analytic data, implementation of wired/wireless communication, all devices connected with various communication services, availability and reliability network, environmentally friendly building construction, low impact on the environment, healthy environment, and resource efficiency (Indrawati et al., 2017).

3.2 *Data analysis*

This research employed a mixed-method approach, which combines and associates qualitative and quantitative forms (Creswell, 2016). Qualitative research involves data analysis in the form of descriptions when the data cannot be directly quantified; quantitative data are studied by giving codes or categories (Indrawati, 2015).

Index is a value resulting from the measurement of a concept related to data collection in the field (Zikmund, Babin, Carr, & Griffin, 2010). This assessment range was obtained from the Ministry of Research and Technology in 2018 and used to assess national research assessors/reviewers. The agreed rating ranges were as follows. Scores of 0–60 were rated very poor and indicated a need for a lot of improvement. Scores 61–70 were considered not good; there were still many shortcomings, but within reason. Scores of 71–80 were considered sufficient and satisfactory, but some things are still lacking. Scores of 81–90 were considered good, satisfactory, in line with expectations, and slightly lacking. Scores of 91–100 were considered very good, very satisfactory, and according to participants' expectations; there were almost no shortcomings (Angdini, 2019).

The researcher used several data validity techniques: (a) The researcher used data obtained from interviews, observation, and documentation (triangulation). Triangulation in this study was divided into three categories – namely, method triangulation, triangulation between researchers, and triangulation of data sources. (b) Peer debriefing involved conducting discussions and adding information with teammates who are also considered knowledgeable about the characteristics of the subject while the results were discussed with the supervisor. (c) Quantitative validation was performed in order to validate the variables and indicators resulting from the in-depth interviews; this study used the Spearman rank test. The data reliability test was also carried out by using the Cronbach's alpha test.

4 RESULT AND DISCUSSION

Based on the statements obtained from experts or informants, labeling was done of each variable and indicator. The informants were also asked to rank their agreement with variables and indicators. Some of the agreement results revealed that more than 60% of the informants agreed with seven variables and twenty-two indicators. However, at the same time during the data-collection process, researchers found several indicators that they added as important for measuring a smart building. Hence, this study included two indicators confirmed by more than 60% of the informants. These indicators were smart elevators and HVAC implementation. The proposed model based on a qualitative approach consisted of seven variables and twenty-four indicators.

At the time of the interview, the informants underwent a process of evaluating variables and indicators; besides being asked about their perceptions per indicator, they were also asked for their perceptions per variable. This was intended as a way for researchers to perform inferential statistics. The first hypothesis examined the relationship between indicators and variables, and the second addressed the relationship between variables of smart offices. In this study, researchers performed calculations using the Spearman rank correlation. Data processing was completed using IBM SPSS Statistics 17 software. The result of the calculation is shown in Table 1.

The results show that the security framework and cybersecurity indicators do not have a positive correlation, which means they are invalid; the indicator was removed. The remaining seven variables and twenty-three indicators are valid and reliable. Hence, the proposed model for measuring the TLT as a smart building consists of seven variables and twenty-three indicators, as shown in Figure 1.

Table 2 shows the score of each variable based on the perception of informants.

Most of the indicators are considered good and very good; only one is considered bad – namely, smart elevator implementation. Figure 2 shows the scores for each variable.

Figure 2 shows that the building automation system variable has a score of 78.18, building control system has a score of 85.56, energy management system has a score of 86.04, safety and security management system has a score of 86.04, enterprise management system has a score of 88.48, IT network connectivity has a score of 89.02, and green building construction has a score of 89.79. The calculation shows that the smart office index of Telkom Landmark Tower is in the good category (with an average score of 86.68), which means it is satisfactory.

Table 1. Validity test result of smart office variables and indicators.

#Variables	#Indicators/constructs	r	Validity
1 Building automation system	1 Sensors	0.593	Valid
	2 Smart elevator	0.878	Valid
	3 HVAC system	0.812	Valid
2 Building control system	4 Remote monitoring implementation	0.702	Valid
	5 Real-time monitoring implementation	0.796	Valid
	6 Software connected to existing equipment	0.603	Valid
3 Energy management	7 Power consumption and system monitoring control	0.561	Valid
	8 Energy-efficient electrical appliances	0.516	Valid
	9 Implementation of backup energy	0.700	Valid
4 Safety and security	10 Threat detection and response	0.552	Valid
	11 Controlling access to the facility	0.389	Valid
	12 Securing lives and assets	0.459	Valid
	13 Security framework and cybersecurity	0.313	Valid
	14 Safety and privacy policy	0.629	Valid
5 Enterprise management system	15 Data management framework	0.661	Valid
	16 System information management	0.617	Valid
	17 Data analytics	0.877	Valid
6 IT network connectivity	18 Wireless communication	0.690	Valid
	19 Devices connected with wireless communication	0.654	Valid
	20 Availability and reliability of network	0.829	Valid
7 Green building construction	21 Green building architecture	0.606	Valid
	22 Low impact on the environment	0.934	Valid
	23 Resource efficiency	0.791	Valid
	24 Healthy environment	0.795	Valid

Source: Spearman rank test of quantitative survey of thirty respondents

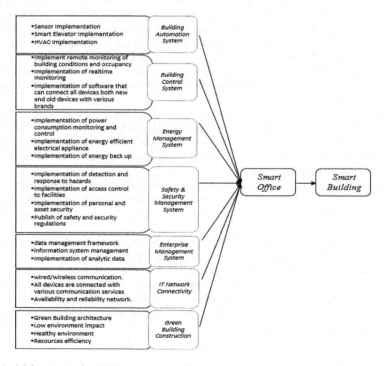

Figure 1. Model for measuring TLT as a smart office.

62

Table 2.   Indicator scores and smart building index.

##Variables	#Indicators/Constructs	Score	Scenario
1 Building automation system	1 Sensors	83.16	Good
	2 Smart elevator	69.80	Not Good
	3 HVAC system	81.56	Good
2 Building control system	4 Remote monitoring	85.50	Good
	5 Real-time monitoring	86.10	Good
	6 Software connected to existing equipment	85.06	Good
3 Energy management	7 Power consumption and system monitoring control	85.83	Good
	8 Energy-efficient electrical appliances	84.86	Good
	9 Backup energy	87.43	Good
4 Safety and security	10 Threat detection and response	88.40	Good
	11 Controlling access to the facility	91.93	Good
	12 Securing lives and assets	90.66	Very Good
	13 Safety and privacy policy	87.86	Very Good
5 Enterprise management system	14 Data management framework	87.63	Good
	15 System information management	89.43	Good
	16 Data analytics	87.90	Good
6 IT network connectivity	17 Wireless communication	89.43	Good
	18 Devices connected with wireless communication	90.20	Very Good
	19 Availability and reliability of network	87.43	Good
7 Green building construction	20 Green building architecture	90.40	Very Good
	21 Low impact on the environment	89.60	Good
	22 Resource efficiency	89.43	Good
	23 Healthy environment	89.73	Good

Source: Scores from in-depth interviews with thirty informants

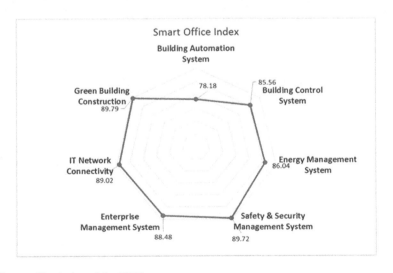

Figure 2.   Smart office index of the TLT.

5 CONCLUSIONS AND RECOMMENDATIONS

The present study finds seven variables and twenty-three indicators for measuring the TLT's smart office index, which resulted in a score of 86.68, leading to the conclusion that the TLT has very good and satisfying performance.

This study suggests that the TLT should maintain and improve user experience; we can see from the users' assessment that some indicators need to be improved, especially the indicators for smart elevators, which fall into the unfavorable category.

ACKNOWLEDGMENT

The authors express their gratitude to the Ministry of Research, Technology and Higher Education of Indonesia for its financial support in carrying out this research.

REFERENCES

Angdini. 2019. *Indeks Kesiapan Smart Building Kota Bandung*. Bandung: Universitas Telkom.

Creswell, J. W. 2016. *Research Design: Qualitative, Quantitative, and Mixed Methods Approaches*. 4th edition. London: Sage.

Indrawati. 2015. *Metode Penelitian Manajemen dan Bisnis*. Bandung: PT. Refika Aditama.

Indrawati, Yuliastri, R., & Amani, H. 2017. Indicators to measure a smart building: An Indonesian perspective. *International Journal of Computer Theory and Engineering* 9(6), 406–411). doi: 10.7763/IJCTE.2017.V9.1176

Kotler. P., & Keller, K. L. 2012. *Marketing management*. 14th edition. Upper Saddle River, NJ: Pearson.

Monzon, A. 2015. *Smart cities concept and challenges: Bases for the assessment of smart city projects*. Lisbon, Portugal: IEEE Publications.

Muñoz, S., Araque, O., Sánchez-Rada, J. F., & Iglesias, C. A. 2018. An emotion aware task automation architecture based on semantic technologies for smart offices. *Sensors* 18, E1499. doi: 10.3390/s18051499

Telkom Indonesia. 2017. Tweet smart office. https://twitter.com/telkomindonesia/

Telkom Landmark Tower. 2016. Buildings. http://tlt.co.id/

Vattano, S. 2014. Smart buildings for a sustainable development. *Economics World* 2(5), 310–324.

Yudis. 2014. Smart building Sangat Mendesak Diterapkan. http://housingestate.id/read/2014/03/16/smart-building-sangat-mendesak-diterapkan

Zikmund, W. G., Babin, B. J., Carr, J. C., & Griffin., M. 2010. *Business research methods*. Boston: South-Western Cengage Learning.

Planned behavior of millennial women

H. Rachma & N. Sobari
University of Indonesia, Indonesia

ABSTRACT: This study sought to investigate consumer intentions in purchasing herbal cosmetic products from non-halal-certified manufacturers. This study was broadened with the inclusion of religious knowledge related to herbal cosmetics. The researchers employed planned behavior theory by including the dimensions of perceived value (health, safety, and environmental) and religious knowledge, purchase intentions, subjective norms, and perceived behavioral control. This empirical study aimed to develop and examine structural models through which to investigate purchase intentions using a combination of planned behavior theory, perceived value, and halal knowledge. The context of this research was unique because it helps in predicting the behavior of Muslim communities in Muslim-majority countries toward cosmetic products that have herbal associations but are not certified halal, in the context of cosmetic products used by millennials. It shows an intention to purchase products with herbal associations that comply with sharia.

1 BACKGROUND STUDY

Herbal cosmetics are made from active ingredients (derived from natural resources) and cosmetics agents. They consist of facial treatments and makeup such as facial soap, moisturizer, toner, serum, powder, lipstick, etc. A large number of cosmetic formulations and toiletries have been developed modern production combined with natural resources and chemical agents (Joshi & Pawar, 2015). There is an increased emphasis on safer, natural products. This is because the effects of chemicals on health have been scientifically proven. The fact is that consumers want healthy living. Based on research, herbal cosmetics are shown to be skin-friendly and have reduced side effects, although this market segment is smaller than the natural food segment. One of the big constraints in developing herbal cosmetics is that consumers lack information about the benefits of the products. In cosmetics, consumers may be confused by this term because they are not given the right definition. If they want to know, then they must read labels and avoid dangerous things (Petrescu, 2013). Another constraint is that Indonesia has a small GDP. People's purchasing power is low. Indonesian consumers are very price sensitive. This explains why people still consider natural consumption. This is said to be a luxury product; only the middle and upper classes can access this sector.

Religion influences consumer lifestyles, which affects consumers in decision-making (Briliana & Mursito, 2017). Religion can increase or weaken choice. Several studies have shown that religion can influence consumer attitudes and behavior (Briliana & Mursito, 2017). Indonesia's population is 87% Muslim. This made Indonesia the country with the largest number of Muslims in 2018 (Inclusive & Economy, 2018). The halal sector is dominated by small and medium-sized companies faced with challenges, especially from multinational companies. They have a large marketing budget and a strong retail presence. Halal cosmetics and cosmetics care brands also face challenges and opportunities from the growing segment of natural, herbal, and, vegan cosmetics. Consumers around the world are increasingly worried about the ingredients in the products they use, giving rise to trends for herbal cosmetic products made with natural ingredients that are produced ethically and do not endanger the environment. This trend is potentially congruent with halal cosmetics (Wilson & Liu, 2010).

Previous studies have shown that attitude toward halal products influences purchase intentions. But how about overseas manufacturers that are not certified? The fastest strategy, in this case, is to assume that their products are halal. This can be done in various ways, namely by taking certification and associating their products with halal. By associating their products with halal, they can convince consumers in strengthening purchasing decisions (Inclusive & Economy, 2018). Companies are now trying to associate their products with things that create positive impressions and that are in line with sharia such as being made from nature, cruelty free, and fair trade (Mukhtar & Butt, 2012).

2 THEORY OF PLANNED BEHAVIOR

In 1980, Ajzen proved a relationship between attitude, subjective norm, and perceived behavior control and intention. This influence will direct people into a specific action in certain situations. This theory developed previously from theory reason action. Intention is a motivational factor that can influence a person's behavior and their intention to try, and an indication of how much they try to do a behavior. The stronger the intention to participate in a certain behavior, the greater the performance (Ajzen, 2009). The theory of planned behavior provides a conceptual framework for dealing with the complexity of human social behavior.

This theory explains social behavior and defines concepts to predict and understand certain behaviors in a context (Ajzen, 2009). Attitude toward behavior, subjective norms, and perceived control can predict behavioral intentions with a high degree of accuracy (Ajzen, 2009; Ghazali et al., 2017). This theory has been widely used to predict behavioral intentions and behavior. The theory of planned behavior implemented previously aims to predict the behavior of green consumers. The robustness of the theory has been confirmed (Ajzen, 2009; Van Loo et al., 2013). This can explain the antecedents of intention to buy green or natural products.

According to the theory of planned behavior, intention is influenced by three determinants: attitude, perceived behavioral control, and subjective norms. This forms the intention to purchase. This research broadens this model by examining the potential antecedents of subjective value attitudes and knowledge about halal products, then lists three predictors of intention to purchase, subjective norm, and attitude.

2.1 *Perceived value*

In this research, researchers used a multidimensional approach so as to develop a perceived value concept. The multidimensional approach is needed in order to calculate two things, namely functional value and affective value (Lee et al., 2019). The functional dimension relates to the quality of products made by consumers. The affective dimension encapsulates feelings and emotions that result from the product. The affective dimension considered is the social impact of the experience of consumer use. This study proposed three values important to consumers, namely health, safety, and environmental value, a dimension that has previously been used (Ghazali et al., 2017).

Perceived value is objective knowledge. Objective knowledge is the actual content stored in memory. This includes attributes and their evaluation, terminology, brand facts, and decision-making procedures. The more types of knowledge stored in one's memory, the greater the level of objective knowledge (Dodd et al., 2005).

2.1.1 *Health value*
Health value means the health value an individual perceives in product (Ghazali et al., 2017). Health and herbal consumption are associated with minimal use of chemicals. Many consumers feel that decreasing the use of chemicals can reduce health risks. (Ghazali and

colleagues 2017) show that concerns about personal and family health are important factors that can influence consumer attitudes toward products with healthy values (Kim & Chung, 2011).

Health awareness shows readiness for healthy behavior (Michaelidou & Hassan, 2008). Health-aware consumers care about their welfare and are motivated to improve and maintain their health and quality of life. Based on previous research, health value can be used to predict attitude (Michaelidou & Hassan, 2008). It is one predictor of intention to purchase natural foods. Improving health is one aspect of consumption of natural products, and health improvement is associated with a product when consumers perceive a product as able to make people healthy (Kim & Chung, 2011).

Fishbein's attitude theory states that attitude builds behavior. Ghazali and colleagues (2017) argue that the perception/perceived value of a product builds consumers' attitude toward the product. This attitude is then used to build confidence in the attributes of a product. In 1984, Michell (cited in Ajzen, 2009) found that confidence in the attributes of a product can strengthen the function of the product, thereby attaching an attitude.

H1: Health value has a positive influence on attitude toward herbal cosmetics.

2.1.2 *Safety value*
Safety value means that products are safe for the environment. Safety also means avoiding ingredients that are harmful to the skin. Buyers of herbal products care about physical risks. Cosmetic products call for safe treatment and should not cause damage to human health when used on a normal and reasonable scale (EU, 1976; ASEAN, 2008; NPCB, 2009a, all cited in (Jihan, Hashim, & Musa, 2014). Products made from nature can be perceived as safer than those manufactured through conventional production. Previous studies revealed that 67.5% of respondents chose cosmetics with natural ingredients because they felt they did not understand the contents of the ingredients in manufactured products and felt safer using natural products (Ghazali et al., 2017; Yin et al., 2010). Safety is represented as consumer care for residues related to artificial materials and preservatives (Michaelidou & Hassan, 2008).

(Yeung and Morris (2001) conceptualize a positive correlation between risk perception related to safety and natural cosmetics purchases. The more one perceives risk in a purchase, the lower the purchase intention. When consumers feel that a product is not safe, they reduce, delay, and change the purchase pattern of the product. Conversely, when they feel that the product is safe, the attitude will be good and the purchase intention will be high.

H2: Safety value has a positive influence on attitude toward herbal cosmetics.

2.1.3 *Environmental value*
Environmental value is defined as value perception that consists of a set of values built with an awareness of a better environment (Zhu & Hennings, 2019). Previous studies have explained that some consumers consider environmental benefits when purchasing products (Ghazali et al., 2017; Kim & Chung, 2011). Dembkowski, Lloyd, and Hanmer (2010) contend that consumers feel free from environmental responsibility when buying natural cosmetics. The generation that lives on earth today has a personal responsibility to protect the environment. They feel that they may face poor quality of life in the future, so they desire to make the world a better place. When they have that awareness, they instinctively want to consume products that are friendly to the environment.

H3: Environmental value has a positive influence on attitude toward herbal cosmetics.

2.2 *Religious knowledge*

Religious knowledge induces a moral obligation to perform or reject certain behaviors (Abd Rahman, Asrarhaghighi, & Ab Rahman, 2015). Religious knowledge influences intention through attitude toward products perceived as halal. Salehudin and Mukhlish (2012) define religious knowledge about halal as the ability to distinguish between halal products and services. Religious knowledge is an important factor for Muslims in deciding to consume a product or service. They will go through a process of knowledge, persuasion, decision-making, and confirmation (Antara, Musa, & Hassan, 2016). Religious knowledge is the ability to combine knowledge, awareness, and expertise so as to differentiate between halal and haram products and services based on sharia. Knowledge increases confidence. This can happen because there is a congruence between the value of natural products and the value of knowledge in matters of Islamic religion such as being environmentally friendly, cruelty-free, and fair trade.

H4: Religious knowledge has a positive influence on attitude toward herbal cosmetics.

2.3 *Attitude*

Attitude is an evaluation of good or bad action. Attitudes are influenced by values (Kim & Chung, 2011). Herbal products promote a healthy and sustainable lifestyle. People who want to maintain their appearance and care about their health will look for personal care products with minimal chemicals (Kim & Chung, 2011). Herbal cosmetic care products are made with minimum chemicals and are less harsh, assuming that consumer awareness of appearance is positively related to attitudes toward purchasing herbal products (Kim & Chung, 2011).

Ajzen (2009) shows that the more a consumer has a good perception, the greater the purchase intention (Van Loo et al., 2013) This is based on positive evaluations that drive intentions to purchase. Van Loo and colleagues (2013) also demonstrate a positive relationship between attitude and purchase frequency. An attitude built through values will create a good perceived attitude. Awareness of the benefits of a product can educate consumers. When the belief in those attributes is awakened, a good attitude is awakened. This contributes to the effectiveness of the intentions of consumers. Consumer education can strengthen consumers' positive attitude toward herbal products.

H5: Attitude has a positive influence on purchase intention.

2.4 *Subjective norm*

A subjective norm is a belief that can determine a person's thoughts related to the acceptance of behavior in one particular group (Lada et al., 2009). Subjective norms encourage someone to think about whether they have behaved in a certain way that conforms to the behavior of their group. This concept is a personal assessment of environmental pressures. When an individual feels that people around him consume certain products, then he will follow their example.

Muslim communities generally interact and live in groups. Subjective norms can influence purchasing decisions. When a behavior is considered important, it can influence the intention to purchase through referencing family, friends, and colleagues encouraging a product (salient referent). Word of mouth can encourage someone to make a purchase (Elseidi, 2018).

H6: Subjective norms have a positive influence on purchase intention.

2.5 *Perceived behavioral control*

Perceived behavioral control refers to the perception of ease or difficulty to conduct behavior (Ajzen, 2009; Jaffar & Musa, 2014). This concept shows the ability to perform certain behaviors, where the perception experiences fluctuations and depends on self-confidence in eliciting a person's behavior. Ajzen (2009) explains that the resources and opportunities available to someone must more or less determine the possibility of attaining behavior. Someone's behavior is greatly influenced by their belief in their ability to do it (perceived behavioral control) (Ajzen, 2009).

Internal factors, namely belief in perceived behavioral control, refers to the presence or absence of resources and opportunities. When consumers are convinced of the risks and benefits of a product or service, this will affect the degree of control. Confidence in control is based on the past, information, or experiences of acquaintances, friends, and other parties that can increase or reduce the perceived difficulties in performing certain behaviors. The more resources and trust in opportunities, the fewer obstacles consumers anticipate, so this increases perceived behavioral control. Salient control beliefs can produce perceptions of behavioral control. Ajzen (2009) and Yin and colleagues (2010) argue that the higher the perceived behavior control, the greater the intention to buy. However, when it is difficult to identify a product, the intention to purchase is lower. Ajzen (2009) reveals that consumer behavior, the ease of accessing products, and consumers' self-control in avoiding excess purchases affect the intention to purchase.

2.6 *Purchase intention*

Intention itself is a person's motivation. Purchase intention is the antecedent that drives the purchase of consumer products and it serves as an alternative to measuring consumer purchases (Nurhayati & Hendar, 2019). Intention motivates consumers and in turn affects their behavior. This is also interpreted as a desire to try, as well as the amount of effort they wish to invest. That probability determines that individuals will carry out certain behaviors depending on the strength of their intentions. If the intention to perform certain behaviors is strong, there is a greater likelihood that every behavior will occur. Purchase intention is an effective tool used in predicting the buying process (Al Khalaileh, Alshraideh, & Bond, 2012; Nurhayati & Hendar, 2019).

H7: Perceived behavior control has a positive influence on purchase intention.

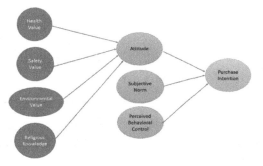

Figure 1.   Theoretical framework.

3   METHOD

This research used a conclusive descriptive research design. Malhotra (2007) explains that the purpose of conclusive research is to measure phenomena that are clearly defined and have clear hypotheses to measure the relationship between variables. Primary data are used to meet these research needs. Primary data in this study were obtained through survey research data gleaned through a questionnaire. This study used a self-administered questionnaire. The questionnaire was disseminated with non-probability purposive sampling using Google Form as an online method because this study had certain requirements for respondents: female, Muslim, routinely use cosmetics but have not used herbal cosmetics.

The study consisted of a wording test, a pretest, CB-SEM analysis, and frequency distribution analysis. The CB-SEM analyzed measurement and structural models and was chosen because the theoretical model consists of seven reflective constructs, two used as an endogenous variable, and five as exogenous variables (Wijayanto, 2008).

3.1 Characteristic of the sample

Respondents were young women aged between eighteen and thirty-five years, commonly referred to as millennials. This generation is the generation born between 1980 and 2000 (Wilson, 2010). Based on research conducted by Ishak and colleagues (2019), female millennial consumers have several considerations before buying cosmetic products. They have good information and technology search capabilities and easily access product information. This includes searching for information on ingredients, halal status, safety and health, and benefits. Young female millennial consumers desire to pay more for a reputable brand. This cohort displays different buying behavior from that of other generations. The number of samples taken on a representative basis (Hair et al., 2006) is the number of samples multiplied by five so that the total sample needed was 36 indicators x 5 = 180 respondents. Furthermore, 200 surveys were collected.

Table 1.   Demographics.

		Frequency	%
Age	18–20	132	66
	21–25	65	32.5
	26–30	2	1.0
	31–35	1	0.5
Education level	Junior high school	1	0.5
	High school	160	80
	S1	38	19
	S2	1	0.5
Income	< Rp.3.500.000	175	87.5
	3.500.000–4.000.000	19	9.5
	4.500.000–5.500.000	1	0.5
	5.500.000	5	2.5

Table 2.   Top three jobs.

Job	Frequency	%
University student	156	78
Private employee	26	13
Public employee	3	1.5

Table 3.   Top three popular cosmetics brands.

Brand name	Frequency
Garnier	52
Ponds	41
Wardah	50

Table 4.   Top three popular cosmetics brands.

Brand name	Frequency
Nature Republic	172
The Body Shop	136
Innisfree	122

3.2 Measurement model

Table 1 summarizes the indicator used in each variable, referenced study, and the number of items. This study used a seven-point Likert scale ranging from 1 (strongly disagree) to 7 (strongly agree). The measurement model assessed reliability and validity for the constructs. Validity was assessed by a standard loading factor ≥ 0.5. Thirty-six indicators were analyzed. Internal consistency was measured using composite reliability. Convergent validity was measured using AVE. All values were reliable and valid (details in Table 5).

Table 5. Model construct, loading, AVE, and CR.

Model construct	Measurement item	Loading	AVE	CR
Healthy value	I am sure that natural cosmetics can make my life healthy.	0.73	0.93	0.76
	I see that using natural cosmetics can encourage healthy living.	0.79		
	I think there is a relationship between natural cosmetics and healthy living awareness.	0.97		
	I think the use of natural cosmetics can improve health.	0.96		
Safety value	I believe that natural cosmetics are free of harmful chemical.	0.75	0.89	0.52
	I believe that natural cosmetics have security for the skin.	0.73		
	I believe that natural cosmetics are safer than nonnatural cosmetics.	0.57		
	I have an opinion that natural cosmetics are not contaminated by harmful ingredients.	0.81		
Environment value	I believe that natural cosmetic products are environmentally friendly.	0.82	0.89	0.68
	I believe that natural cosmetic products are produced responsibly.	0.76		
	I feel that natural cosmetics have environmental value.	0.87		
	Natural cosmetic products are produced in an environmentally friendly manner.	0.83		
Religious knowledge	I understand the rules in Islam regarding halal and haram food and drink.	0.87	0.86	0.56
	I have knowledge about what foods and drinks are forbidden in Islam.	0.86		
	I can distinguish which foods should and should not be eaten.	0.80		
	I understand the issue of ingredients in cosmetics that are not allowed according to Islamic sharia.	0.55		
	I can distinguish between halal certification and other certifications.	0.58		
Attitude	Natural cosmetics are good.	0.79	0.94	0.76
	Natural cosmetics are favorable.	0.82		
	Natural cosmetics are desirable.	0.82		
	Natural cosmetics are wise.	0.85		
	Natural cosmetics are fun.	0.85		
Subjective norms	My family believes that buying natural cosmetics is a good idea.	0.77	0.83	0.55
	My friend thinks that buying natural cosmetics is a good idea.	0.72		
	In general, people who are important to me buy natural cosmetic products.	0.75		
	People I listen to and who influence me buy natural cosmetics.	0.73		

(*Continued*)

Table 5. (*Continued*)

Model construct	Measurement item	Loading	AVE	CR
Perceived behav-ioral control	I have complete freedom to buy natural cosmetics or not.	0.61	0.85	0.54
	I have the opinion that nothing prevents me from buying natural cosmetics.	0.74		
	I have complete control over buying natural cosmetics.	0.75		
	I have the resources to buy natural cosmetics.	0.78		
	I am sure I can buy natural cosmetics.	0.79		
Purchase intentions	I am willing to pay more for natural cosmetic products.	0.78	0.91	0.67
	I am willing to wait for natural cosmetic products when products are unavailable.	0.80		
	I am willing to look for natural cosmetic products.	0.90		
	I will make more effort to buy natural cosmetic products.	0.86		
	I intend to buy natural cosmetics in the future.	0.75		

3.2.1 *Goodness of fit*

The analysis used measurement and structural models in order to evaluate the measurement and structural models. This is the appropriate tool to evaluate fitness between the covariance matrix and the estimated population covariance matrix. Ten statistics measured the chi-square statistic, RMSEA, NFI, NNFI, CFI, IFI, RFI, SRMR, GFI, and AGFI. Most of the statistics met recommended guidelines only: chi-square statistics, GFI, and AGFI, which is low between the standard level of the match both for measurement and according to the structural model (Wijayanto, 2008).

Table 6. Goodness of fit measurement model.

Goodness of fit	Measurement value	Structural value	Explanation
Absolute Measurement			
Statistic Chi-Square (P ≥ 0.05)	0.000	0.000	Marginal fit
Goodness of Fit Index (GFI) (GFI ≥ 0.90)	0.75	0.74	Marginal fit
Standardized Root Means Square Residual (SRMR ≤ 0.50)	0.077	0.073	Good fit
Root Mean Square Error of Approximation (RMSEA ≤ 0.080)	0.076	0.073	Good fit
Incremental Measurement Goodness of Fit			
Adjusted Goodness of Fit Index (AGFI ≥ 0.90)	0.71	0.70	Marginal fit
Non-normed Fit Index (NNFI ≥ 0.90)	0.98	0.96	Good fit
Normed Fit Index (NFI ≥ 0.90)	0.96	0.93	Good fit
Relative Fit Index (RFI ≥ 0.90)	0.96	0.93	Good fit
Incremental Fit Index (IFI ≥ 0.90)	0.98	0.96	Good fit
Comparative Fit Index (CFI ≥ 0.90)	0.98	0.96	Good fit

3.3 *Structural model*

The assessment of the significance and relevance of the structural model relationship is shown in Table 2. All of the t-value show > 1.96; this means that seven hypotheses are significant.

Table 7. Hypotheses results.

No.	Structural relationship	T-values	Conclusion
1.	Health value → Attitude	2.89	Significance
2.	Safety value → Attitude	3.00	Significance
3.	Environmental value → Attitude	5.67	Significance
4.	Religious knowledge → Attitude	2.63	Significance
5.	Attitude → Purchase intention	4.58	Significance
6.	Subjective norms → Purchase intention	3.96	Significance
7.	Perceived behavioral control → Purchase intention	2.51	Significance

3.4 Hypotheses analysis

Hypothesis 1 examined the effect of health value on attitude. Health value is significantly related to attitude with t-value of 2.89. This study has found that a product that has a natural claim means the concept of health value fit with what consumers believe. Consumers judge that the product has health value and this can convince consumers to evaluate it with a good attitude.

Consumers who aware of health value judge whether products have health value (Michaelidou & Hassan, 2008). This is encouraged to improve and maintain their welfare. Health value can be used to predict attitude. This is consistent with research conducted by Ghazali and colleagues (2017). Perceived health value can be used to build confidence in a product. Belief in attributes can strengthen the function of products in herbal cosmetics and build good attitudes toward them.

This study support H2 with t-value of 3.00. Consumers feel that natural cosmetic products are safe and can be consumed daily. With safety attributes, this can reduce the perception of unwanted risk in purchasing decisions. The product meet expectations. This reinforces the confidence that is built on product attributes (Ghazali et al., 2017). In the context of natural cosmetics, consumers need to believe that product is safe. This means that natural products can be promoted as having health value

Millennial Muslim consumers form a market segment that cares about the content of ingredients. This is in line with Ghazali and colleagues (2017). Consumers feel safe using a product, so they will evaluate the it positively.

Yeung and Morris (2001) state that when a consumer feels that a product is safe, then he will tend to evaluate it well (good attitude); this will affect the intention to purchase. Manufacturers of herbal products must maintain their reputation for protecting products from contamination by packaging the products properly and securely.

Hypothesis 3 studied the effect of environmental value on attitude, with a t-value of 5.67. Consumers consider the environmental benefits of purchasing herbal products. This explains why consumers feel they have no responsibility for the environment when they buy herbal cosmetics. They feel that by consuming herbal cosmetic products, they can reduce damage to the environment. This impulse exists because they are threatened with poor quality of life in the future. Confidence in this attribute is present because they believe that such products are produced an environmentally friendly manner.

This attribute, according to (Ajzen, 2009), forms attitudes with the information cognitive process approach. The objective value is an environmental value stored in a consumer's memory. Consumers automatically appreciate this attribute; this shows that, generally, humans tend to support the attitude that is believed to have the consequences they want. It is this expectation of consequences that shapes attitude.

This study is in line with previous studies by indicated a significant hypothesis; the t-value for H4 was 2.63. Religious knowledge is a perception based on subjective interpretations. When consumers have good knowledge, this has a positive effect on self-confidence (Ajzen, 2009). Increased knowledge means increased confidence. This can happen because there is a congruence between the value of natural products and the value of knowledge in matters of Islamic religion such as being environmental friendly, cruelty free, and fair trade.

Decisions are evaluated based on perceptions of individuals with knowledge. This indicates that a product's attributes follow expectations and that consumers can trust the product. confidence in the product is good and the attitude toward the product becomes positive. This attitude determines perceptions and preferences that are influenced by factual and subjective information. This shows that the more knowledge a person has, the more positive consequences and benefits. This supports research by Bang and colleagues (2000) finding that consumers with qualified knowledge tend to engage in behavior that is consistent with their beliefs or knowledge. Awareness of the knowledge gained from experience or learning will increase attitudes on a subject. Knowledge of religion can increase attitudes toward cosmetics that are herbal and close to nature (Abd Rahman et al., 2015).

Behavior itself, according to (Ajzen, 2009), is the result of salient belief, subjective evaluation, and attributes. When consumers are confronted with a new product, before purchasing it they will consider whether the product violate religion. In general, people who consume halal products are consideredloyal (Zailani, 2018).

This study proved H5, that attitude has a significant effect on purchase intention, as indicated with a t-value of 4.58 (Fishbein, 1980, cited in Ajzen, 2009). (Ghazali and colleagues, (2017) explain that when someone determines that a product is profitable, they tend to be more involved. The more confidence or perception is built, the greater the purchase intention (Ajzen, 2009). Awareness of the benefits of a product is a foundation on which consumers realize that the product can improve their quality of life.

The research indicates that beliefs are derived from the values of health, safety, environment, and religious knowledge. This belief provides a positive evaluation and influences positive behaviors with intentions. The attributes of a product give a strong and positive value congruent with the value of the consumers. Thus, consumers determine the attitude to buy products. This relationship is influenced by cognitive processes related to affective processes so that confidence arises in the product and raises the purchase intention (Ajzen, 2009).

In H6, subjective norms have a significant effect on purchase intention, with a t-value of 3.96. Mukhtar and Butt (2012) explain that humans live in groups and consider references and suggestions from others concerning behavior. Recommendations from close people are an important factor in making a purchase. This aspect directs someone to buy products that are suitable according to their group norms. Wilson and Liu (2010) explain that the Muslim community generally lives in groups. People in this group assume they are part of the group. To be part of the group they must meet a set of norms. When a behavior is considered important, then this can influence the intention to purchase through subjective aspects such as referencing family, friends, or colleagues who encourage the use of a product (salient referent belief). Word of mouth can encourage someone to make a purchase (Elseidi, 2018).

This study proved H7, that a significant relationship exists between perceived behavioral control and purchase intention. Perceived behavioral control contains two aspects, internal and external. External factors comprise time, availability, and presence. Internal factors mean confidence in the existence of resources and opportunities. When consumers feel they can control the risks and benefits received, it will increase perceived behavioral control. This belief in control is based on the past, information, and the experiences of friends and relatives. When consumers feel they have the resources to realize an action, they will see an opportunity. Thus, they feel that fewer obstacles can be anticipated. This increases perceived behavioral control and purchase intention. This is in line with research conducted by Ajzen (2009). Yin and colleagues (2010) reveal that the control of the individual, ease of accessing products, and no obstacle to making a purchase will affect purchase intentions.

4 DISCUSSION AND CONCLUSION

The purpose of this study was achieved by understanding what factors influence intentions to purchase herbal cosmetics. This confirmed the theory of planned behavior by adding religious knowledge variables. Purchase intention is built on attitude, perceived behavioral control, and subjective norms. Attitude is built on product attributes such as health value, safety value, and environmental value. Subjective and religious knowledge also play a role.

Muslim millennial consumers can be attracted by associating products with positive impressions and promoting them as in line with sharia, such as being made from nature, cruelty free, and fair trade. This can increase purchase intentions and create positioning for a product that is suitable for consumption. This will reinforce consumers' positive attitude toward herbal cosmetic products. Based on this study, young Muslim consumers will have a purchase intention when the product is close to the values that match with their Islamic beliefs.

Positioning as herbal cosmetics is beneficial with the ability to reach both environmentalist and Muslim consumers. Based on this study environmentalist and Muslim consumers have similar preferences. In cosmetics products, the issue of halal certification is not the main issue; the main issue is consumer confidence that a product does not use prohibited ingredients, is produced and processed properly, and is safe for the environment. Producers and marketers can highlight the attributes of health, safety, and the environment. These can be a competitive advantage. The premise of delivering attributes can be conveyed through packaging and promotion. Halal cosmetics players can reinforce their positive attitudes with herbal products. Herbal cosmetic products have a positive response, but currently, the market is still dominated by international brands. For local players, this provides insight that there are opportunities that can be exploited. Based on research, the Muslim segment positively accepts herbal products. They see a congruence between Islamic sharia and the value of herbal products.

REFERENCES

Abd Rahman, A., Asrarhaghighi, E., & Ab Rahman, S. 2015. Consumers and halal cosmetic products: Knowledge, religiosity, attitude and intention. Journal of Islamic Marketing, 6(1), 148–163. https://doi.org/10.1108/JIMA-09-2013-0068

Ajzen, I. (2009). Theory of planned behavior measure. Change: Journal of Health Psychology, 12(1), 1–8. https://doi.org/10.1037/t15668-000

Al Khalaileh, M., Alshraideh J. A., & Bond, E. 2012. Jordanian nurses' perceptions of their preparedness for disaster management. International Emergency Nursing, 20(1), 14–23. https://doi.org/10.1016/j.ienj.2011.01.001

Antara, P. M., Musa, R., & Hassan, F. (2016) Bridging Islamic financial literacy and halal literacy: The way forward in the halal ecosystem. Procedia Economics and Finance, 37(16), 196–202. https://doi.org/10.1016/s2212-5671(16)30113-7

Bang, H., Ellinger, A. E., Hadjimarcou, J., & Traichal, P. A. (2000) Consumer concern, knowledge, and attitude toward renewable energy449–468.

Briliana, V., & Mursito, N. (2017). Exploring antecedents and consequences of Indonesian Muslim youths' attitude towards halal cosmetic products: A case study in Jakarta. Asia Pacific Management Review, 22(4), 176–184. https://doi.org/10.1016/j.apmrv.2017.07.012

Dembkowski, S., Lloyd, S. H., & Hanmer, S. (2010). The environmental value attitude system model: A framework to guide the understanding of environmentally conscious consumer behaviour: Dembkowski and the environmental value-attitude-system model: A framework to guide the understanding of the environment (September 2013), 37–41. https://doi.org/10.1080/0267257X.1994.9964307

Dodd, T. H., Laverie, D. A., Wilcox, J. F., & Duhan, D. F. (2005). Differential effects of experience, subjective knowledge, and objective knowledge on sources of information used by consumers. 29(1), 3–19. https://doi.org/10.1177/1096348004267518

Elseidi, R. I. 2018. Determinants of halal purchasing intentions: Evidences from UK. Journal of Islamic Marketing, 9(1), 167–190. https://doi.org/10.1108/JIMA-02-2016-0013

Ghazali, E., Soon, P. C., Mutum, D. S., & Nguyen, B. (2017) Health and cosmetics: Investigating consumers' values for buying organic personal care products. Journal of Retailing and Consumer Services, 39(August), 154–163. https://doi.org/10.1016/j.jretconser.2017.08.002

Hair, J. R., Black, W. C., Babin, B. J., & Anderson, R. E. 2006. Multivariate data analysis. 7th edition (pp. 1–761). Upper Saddle River, NJ: Pearson.

Inclusive, A., & Economy, E. (2018). Your gateway into the Islamic economy.

Ishak, S., Che Omar, A. R., Khalid, K., Ab. Ghafar, I. S., & Hussain, M. Y. (2019). Cosmetics purchase behavior of educated millennial Muslim females. Journal of Islamic Marketing. https://doi.org/10.1108/jima-01-2019-0014

Jaffar, M. A., & Musa, R. (2014). Determinants of atitude towards Islamic financing among halal-certified micro and SMEs: A preliminary investigation. Procedia: Social and Behavioral Sciences, 130, 135–144. https://doi.org/10.1016/j.sbspro.2014.04.017

Jihan, A., Hashim, M., & Musa, R. (2014). Factors influencing attitude towards halal cosmetics among young adult urban Muslim women: A focus group analysis. Procedia: Social and Behavioral Sciences, 130, 129–134. https://doi.org/10.1016/j.sbspro.2014.04.016

Joshi, L. S., & Pawar, H. A. (2015). Natural products chemistry & research herbal cosmetics and cosmeceuticals: An overview. 3(2). https://doi.org/10.4172/2329-6836.1000170

Kim, H. Y., & Chung, J. E. 2011). Consumer purchase intention for organic personal care products. Journal of Consumer Marketing, 28(1), 40–47. https://doi.org/10.1108/07363761111101930

Lada, S., Harvey Tanakinjal, G., & Amin, H. (2009). Predicting intention to choose halal products using theory of reasoned action. International Journal of Islamic and Middle Eastern Finance and Management, 2 (1), 66– 76. https://doi.org/10.1108/17538390910946276

Lee, S., Sung, B., Phau, I., & Lim, A. (2019). Communicating authenticity in packaging of Korean cosmetics. Journal of Retailing and Consumer Services, 48(February), 202–214. https://doi.org/10.1016/j.jretconser.2019.02.011

Malhotra. (2007). An applied approach. Updated 2nd European edition.

Michaelidou, N., & Hassan, L. M. (2008). The role of health consciousness, food safety concern and ethical identity on attitudes and intentions towards organic food. International Journal of Consumer Studies, 32(2), 163–170. https://doi.org/10.1111/j.1470-6431.2007.00619.x

Mukhtar, A., & Butt, M. M. (2012). Intention to choose halal products: The role of religiosity. Journal of Islamic Marketing, 3(2), 108–120. https://doi.org/10.1108/17590831211232519

Nurhayati, T., & Hendar, H. (2019). Personal intrinsic religiosity and product knowledge on halal product purchase intention: Role of halal product awareness. Journal of Islamic Marketing. https://doi.org/10.1108/JIMA-11-2018-0220

Petrescu, D. 2013. Consumer behaviour towards organic, natural and conventional skin care products: A pilot study. Advances in Environmental Sciences: International Journal of the Bioflux Society 5(3), 274–286.

Salehudin, I., & Mukhlish, B. M. (2012). Pemasaran Halal: Konsep, Implikasi dan Temuan di Lapangan. SSRN Electronic Journal. https://doi.org/10.2139/ssrn.1752567

Van Loo, E. J., Diem, M. N. H., Pieniak, Z., & Verbeke, W. (2013). Consumer attitudes, knowledge, and consumption of organic yogurt. Journal of Dairy Science, 96(4), 2118–2129. https://doi.org/10.3168/jds.2012-6262

Wijayanto, S. (2008). Structural Equation Modeling dengan Lisrel 8.8: Konsep dan Tutorial.

Wilson, J. A. J., & Liu, J. (2010). Shaping the halal into a brand? Journal of Islamic Marketing, 1(2), 107–123. https://doi.org/10.1108/17590831011055851

Yeung, R. M. W., & Morris, J. (2001). Food safety risk. British Food Journal, 103(3), 170–187. https://doi.org/10.1108/00070700110386728

Yin, S., Wu, L., Du, L., & Chen, M. (2010). Consumers' purchase intention of organic food in China. Journal of the Science of Food and Agriculture, 90(8), 1361–1367. https://doi.org/10.1002/jsfa.3936

Zailani. 2018. Halal logistics service quality: Conceptual model and empirical evidence. https://doi.org/10.1108/BFJ-07-2017-0412

Zhu, Y., & Hennings, M. (2019). Emerging mid-career transformation in Japan. International Journal of Human Resource Management, 0(0), 1–36. https://doi.org/10.1080/09585192.2019.1651375

Understanding the effect of celebrity endorsers and electronic word of mouth on e-commerce in Indonesia

M.E. Saputri, T.G. Sarawati & F. Oktafani
School of Communication and Business, Telkom University, Bandung, Indonesia

ABSTRACT: The development of the Internet that has penetrated all aspects of human life, including economic matters, has led to the birth of e-business or e-commerce. This e-business or e-commerce was created for business transactions that utilize the Internet and the Web, in order to make it easier for people to conduct business transactions with less energy and shorter time required to search for items they need. These online transactions have allowed the buying and selling process to not be obstructed by distance and time. Among the factors that can affect a company's income and profits are celebrity endorsers and electronic word of mouth. Good celebrity endorsers and electronic word of mouth will increase consumers' purchase of a product. This study aimed to determine the effect of celebrity endorsers and electronic word of mouth on the purchase decisions of Tokopedia customers. This research used both quantitative and descriptive analytic methods. The data analysis technique used was multiple linear regression. The results of the study have indicated that the responses from respondents to the celebrity endorser variable belong to the excellent category, responses to the electronic word of mouth variable to the good category, and responses to the purchase decision process variable to the excellent category. The conclusion of this study was that celebrity endorsers and electronic word of mouth affect the purchase decisions of Tokopedia customers with a percentage of 52.6%, while the remaining 47.4% is influenced by other variables not examined in this study. Thus, it is deemed that the better the celebrity endorsers and electronic word of mouth of Tokopedia, the better the purchase decision of consumers.

1 INTRODUCTION

The growth of digital commerce in Indonesia is increasingly promising. Based on McKinsey's predictions, e-commerce in Indonesia has increased eightfold, from total online spending of USD 8 billion in 2017 to USD 55–65 billion in 2020. McKinsey also predicts that Indonesian online shopping penetration will increase to 83% of total Internet users, an increase of around 9% compared to 2017 (www.mckinsey.com).

Indonesia as a potential market for e-commerce attracts a lot of players such as Tokopedia, Shopee, Bukalapak, Lazada, JD.id, Blibli.com, Zalora, AliExpres, Zilingo Shopping, and Amazon. According to data released by iPrice, Tokopedia is the e-commerce platform with the highest monthly active users (MAU) in Indonesia. The site is visited by around 137 million users every month. This company has consistently occupied the top position for the average number of monthly visitors since the second quarter of 2018. Another e-commerce platform from Indonesia, Bukalapak, occupies the second position with an average number of 115 million visitors per month. Shopee, Lazada, and Blibli take the third, fourth, and fifth positions, respectively. Although Tokopedia still had the highest MAU in Q1 2019, the average number of Tokopedia website visitors has decreased significantly compared to Q4 2018. At the end of 2019, Tokopedia's MAU touched 168 million, a decrease of around 18% at the beginning of 2020 (https://id.techinasia.com).

In the midst of fierce competition, the company uses a variety of strategies to attract customers. Companies are required to be more sensitive to consumer desires and to communicate about products in a good and efficient manner. The marketing strategy will greatly affect sales, especially in terms of promotion. To market products, producers use various strategies to attract the attention of consumers; creativity is needed in making an advertisement. One such strategy is to use a figure who has charisma and the ability to win the hearts of consumers. In the business world, this figure is called a celebrity endorser.

A celebrity endorser is a famous personality whom the general public recognizes by name and face (Schimmelpfennig, 2018). Celebrity endorsers as digital celebrities are typical individuals who became famous through online blogging, vlogging, or social networking sites (SNS). Digital celebrities have perceived social influence owing to their large number of followers (Hwang & Zhang, 2018; Jin & Phua, 2014). Celebrity endorsers have become so important for companies because endorser credibility has an indirect impact on brand equity when this relationship is mediated by brand credibility (Spry, Pappu, & Cornwell, 2011). Endorsers have a positive influence on young adults' product-switching behavior, complaint behavior, positive word of mouth, and brand loyalty (Dix, Phau, & Pougnet, 2010). Previous research shows companies can use celebrity endorsers as a marketing strategy.

As the number of users of social media has been increasing exponentially, social media is regarded as among the most prominent tools for digital advertisement. The emergence of social media allows one person to communicate with thousands of others about a company and the products and services it provides. This has a significant impact on corporate communications with consumers. Social media is a hybrid element of the promotion mix because in the traditional sense it allows companies to talk to their customers, while in the nontraditional sense it allows customers to talk directly with each other. The content, time, and frequency of social media-based conversations between consumers is under the direct control of outside managers (Mangold & Faulds, 2009).

Tokopedia implements various strategies to compete with its competitors, including using celebrity endorsers to influence potential customers, especially on Instagram. Instagram is one of the most popular social networks worldwide. Due to its visual and high user engagement rate, Instagram is also a valuable media marketing tool. That's why most companies use this platform to generate more customers and earn greater revenues. Examples of celebrity endorsers Tokopedia uses to influence purchasing decisions include Arief Muhammad and Cinta Laura Kiehl. Figure 1 presents an example of their posts on Instagram. The use of online media as a means of promotion and sales can foster positive and negative statements from consumers, known as electronic word of mouth (e-WoM).

Tokopedia is not focused on just one or two celebrity endorsers. Since 2016 Tokopedia has used several artists as celebrity endorsers such as Rachel Venya, Sandra Dewi, and Ricky Harun. Tokopedia chose these people based on criteria of celebrity endorsers such as visibility, credibility, attraction, and power (Shimp, 2014: 258), and has verified their social media

Figure 1.   Celebrity endorsers of Tokopedia.

profiles. This study aimed to answer the following research question: "How does celebrity endorsement and e-WoM influence the purchase decisions of e-commerce's customers?" The structure of this paper is described hereafter. First, a literature review is reported along with the hypotheses proposed in this research, and a conceptual model is drawn. Second, the methodology used for data collection and measurement of variables is explained. The latter is followed by an empirical results section where multiple regression is conveyed. Last, a concluding section discusses the findings.

2 LITERATURE REVIEW

2.1 *Celebrity endorser*

Marketing strategy is the tactics business units use to reach predetermined markets, including decisions regarding target markets, product placement, marketing mix, and marketing costs needed to achieve demand (Sunyoto, 2015: 98). Nowadays, many companies use celebrity endorsers as a way of communicating and engaging with customers. Moreover, celebrity support provides an excellent vehicle for achieving the main objectives of marketing – communication and differentiation of a company's brands and products from those of its competitors (Erdogan, Baker, & Tagg, 2001). Companies use this phenomenon to expand their businesses. They offer their products to celebrities who are quite well known. In Indonesia, especially on Instagram, many people, especially young people, use the word *celebgram* as a synonym for *celebrity endorser*. The companies and their celebrity endorsers mutually increase their social media presence and attract the attention of active social media users so as to get a large following. The use of the term *celebgram* is aimed at social media artists who were originally noncelebrities. The use of artists or celebrities as endorsers is frequent and the strategy is preferred by brand owners and advertising agencies because consumers think that the brands celebrities carry have interesting properties similar to those of the celebrities themselves (Kertamukti, 2015: 74). Many studies have confirmed the effectiveness of celebrity endorsers as a new way to increase customer engagement, as shown in Table 1.

Table 1.

Author and year	Subject of study	Method and sample size	Results (related to the usage and nature of celebrities and anonymous models in advertising)
Belch and Belch (2013)	Usage of celebrities in advertisements in different types of magazines, of different types of products/ services, and bases for their use	Content analysis of advertisements in high-circulation magazines in the United States; 2,358 ads in total	9.5% of the advertisements used celebrities; celebrities are used in advertising for all product categories except "financial services"; 34% of the celebrities featured are actresses/actors, 27% athletes, 18% fashion models, 16% entertainers, 2% business executives, 1% news personalities, and 2% other types of celebrities. 68% of the advertisements contain one or more persons.
Hwang and Zhan (2018)	Influence of parasocial relationship between digital celebrities and their followers on followers'	A survey was administered to 389 Chinese SNS users following digital celebrities; valid data were analyzed	Parasocial relationships positively affected followers' purchase and e-WoM intentions. The second

(Continued)

Table 1. (*Continued*)

Author and year	Subject of study	Method and sample size	Results (related to the usage and nature of celebrities and anonymous models in advertising)
	purchase and e-WoM intentions, persuasion, and knowledge	using structural equation modeling.	model investigated the negative effects of persuasion on followers' purchase and e-WoM intentions. The results indicate that parasocial relationships moderate the paths between followers' persuasion and purchase intentions and between followers' persuasion and e-WoM intentions.
Schimmelpfennig (2018)	The prevalence of (different types of) celebrity endorsers and anonymous models in contemporary advertising	As Kassarjian notes: "content analysis is a scientific, objective, systematic, quantitative, and generalizable description of communications content."	In total, 2,877 advertising units were identified in the 80 magazines considered in this study. Of those advertisements, 121 (4.2%) were classified as celebrity endorsements.
Phua, J., Lin, J.-S. (Elaine), & Lim, D. J. (2018)	The effects of product–celebrity image congruence, consumer-celebrity risk-oriented image congruence, and parasocial identification on consumer engagement with celebrity-endorsed e-cigarette advertising on Instagram	A total of 600 participants (Mage = 28.9) were recruited through a Qualtrics panel (about 10% response rate) for the main study. Inclusion criteria were active Instagram users between the ages of 18 and 34 (considered "millennials"). A total of 300 male and 300 female participants were recruited.	Results indicated that high product–celebrity image congruence, and high consumer–celebrity risk-seeking image congruence led to significantly more positive ad attitude and greater intention to spread e-WoM and use of the product.
McCormick (2016)	The purpose of this study was to investigate if the presence of a congruent product–endorser match helped influence purchase intent of millennial consumers or resulted in unfavorable attitudes toward the advertisement.	A pretest was conducted to pick the celebrity endorsements to use for this study. Researchers have used various methods when selecting celebrity endorsers; therefore, this study employed a previously used method. First, an exhaustive search was done to find celebrity endorsements of consumer products. Then those endorsements were shown to several undergraduate classes where students rated familiarity and likeability of celebrities endorsing various consumer products on a seven-point semantic differential scale.	Millennials evaluated an unfamiliar celebrity endorsement where they indicated they had little intent to purchase the product endorsed by the unfamiliar celebrity, but the unfamiliar celebrity did lead to favorable evaluations of the advertisement.

Based on Rossiter (Kertamukti, 2015: 70), the criteria for celebrity endorsers consist of four dimensions, namely:

1. Visibility: A celebrity's popularity. Popularity can be measured through how many followers a celebrity endorser has and how often he or she appears in public.
2. Credibility: Two important characteristics of credibility are expertise that refers to the knowledge, experience, and skills a celebrity endorser possesses related to the topic of the ad, and trustworthiness which refers to honesty, integrity, and whether a celebrity endorser can be trusted as a source. The more reliable the source is considered, the more messages the source communicates to customers.
3. Attraction: The attraction of the star. More than just physical attractiveness, this include a number of positive characteristics that consumers can see in an endorser (Shimp, 2014: 261), namely the level of likeability and the similarity between the star's personality and the qualities desired by a product's users.
4. Power: Celebrities' ability to convince consumers to buy products.

2.2 E-WoM

Electronic word of mouth (e-WoM) is every positive or negative statement made by a potential customer or by a customer who has already purchased a product/service regarding a company's product that is made available to many people and companies via the Internet (Hennig et al., 2004, cited in Panwar & Rathore, 2015: 32). According to Park and colleagues (2011), e-WoM is basically traditional word of mouth on the Internet. Unlike word of mouth, e-WoM happens online. According to Syafaruddin and colleagues (2016: 65), differences exist between e-WoM and traditional word of mouth. Word of mouth and e-WoM can be distinguished based on the media used:

a) Traditional word of mouth is usually face to face.
b) Electronic word of mouth (e-WoM) is online via cyberspace.

Electronic word of mouth can also be interpreted as a form of consumer willingness to voluntarily provide recommendations to others to buy or use products from a company through internet media. Electronic word of mouth is a communication process that provides recommendations both individually and in groups and that aims to provide personal information online. According to Cheung and Lee (2014: 219), e-WoM is communication and information exchange between old and new consumers, using technological developments such as online discussion forums, electronic bulletin boards, newsgroups, blogs, review sites, and social networking sites. Ekawati and colleagues (2014: 2) argue that word of mouth has undergone a paradigm shift to e-WoM. In the past, word-of-mouth communication was done face to face with people who were already well known, but now word of mouth can be done in cyberspace with a wider scope – that is, other people can read what we share within seconds.

Rathore and Panwar (2015) contend that the channels used for e-WoM can include opinion platforms, discussion groups and forums, emails, blogs, and social networks such as Facebook, Instagram, and Twitter.

Lin, Wu, and Chen (2013) divide E-WoM into three dimensions, namely:

1. E-WoM quality: E-WoM quality refers to the persuasive power of comments embedded in information messages. Purchase decisions can be based on several criteria or requirements that meet consumers' needs, and their willingness to buy will be based on the perceived quality of the information they receive. Therefore, it is important to determine consumers' perceptions of information quality as an element through which to assess their potential purchase decisions.
2. E-WoM quantity: E-WOM quantity refers to the number of posted comments. Product popularity is determined by the quantity of online comments because it represents the market performance of a product. Consumers also need references to strengthen their confidence and to reduce the feeling of making a mistake or taking a risk when shopping, and the quantity of comments online represents a product's popularity and importance.

3. Sender's expertise: Sender's expertise is required training in a specific experience or domain. Expertise can be seen as authority or competence. A sender's expertise when he or she posts consumer reviews will convince users to trust the information the sender provides and encourage them to make a decision to buy.

Electronic word of mouth has now developed rapidly in the form of review sites where customers can share experiences and exchange information about products and services. The impact of e-WoM on the market is its ability to share experiences that affect individuals looking for information.

3 METHODOLOGY

3.1 Participants

In this study, the authors used a quantitative approach. The quantitative method is a research method based on the philosophy of positivism that is used to examine populations or certain samples, using research instruments for data collection, and analyzing statistical data, in order to test the hypotheses under study (Sugiyono, 2015). The online survey ran through the Google Form application. Using the screening question items ("Are you a follower of Tokopedia on Instagram?" and "Do you recognize that Arief Muhamad or Cinta Laura are endorsers of Tokopedia?"), the survey was administered only to respondents who are followers of celebrity endorsers of Tokopedia. The questionnaire items were adapted from relevant literature in order to ensure the content validity of the constructs. A total of 100 responses was obtained and used for data analysis.

3.2 Research framework

In this study, we use linear multiple regression analysis to analyze the relationship between the independent variable and the dependent variable; the model used for the analysis was as follows:

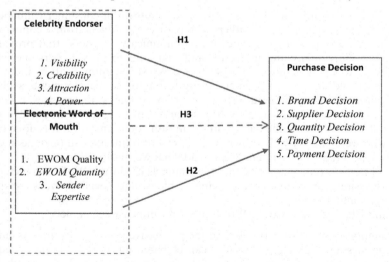

The hypotheses studied in this research were:

H1: Celebrity endorsers influence purchase decisions.
H2: Electronic word of mouth influences purchase decisions.
H3: Celebrity endorsers and e-WoM influence purchase decisions simultaneously.

In this study, the authors wanted to examine the influence of celebrity endorsers and e-WoM on purchase decisions. The use of artists or celebrities is frequent and is preferred by brand owners or advertising agencies because consumers think that the brands celebrities carry have the same qualities as the celebrities themselves (Kertamukti, 2015: 74).

Research conducted by Jalilvand and Samiei (2012: 15) reveals that e-WoM has a positive effect on brand image and purchase interest. Consumer reviews posted on the Internet are the most important form of e-WoM; consumers tend to look for product reviews online with the aim of getting information about a particular product before they ultimately make a purchase.

According to Jalilvand and Samiei (2012: 466), consumers' reasons for the strong influence of WoM on purchasing decisions include:

a) I collect information from consumer reviews online before I buy a product/brand.
b) If I can't check the reviews from online consumers, I'm worried about my decision.
c) When I buy a product/brand, online reviews make me confident.

Based on previous research, we can conclude that e-WoM influences purchasing decisions.

4 RESULTS AND DISCUSSION

As presented in Table 2, the sample had more female (56%) than male (44%) respondents, and 65% of the respondents were aged twenty-one to twenty-five years. Most respondents were university students (65%) with incomes ranging from 1.5 million to 3 million Rupiah. The demographic analysis result matches the data from Hootsuite and We Are Social in January 2019 and indicates that social media use is dominated by young people eighteen to thirty-four years old. These data also show that male users (55.5%) are more active on social media than female users (44.5%), which is opposite to our survey results. As the current study focuses on consumers who are influenced by digital celebrities and e-WoM, our data may have more female than male respondents.

After assessing the validity, the overall fit of the research model was tested. The result of multiple regression is shown in Table 3.

Table 2.   Demographic analysis.

Measure	Items	Frequency	Percentage
Gender	Female	56	56%
	Male	44	44%
Age	15–20	30	30%
	21–25	65	65%
	26–30	5	5%
Occupation	University student	87	87%
	Employee	9	9%
	Entrepreneur	2	2%
	Housewife	2	2%
Income	< 1,500,000	35	35%
	1,500,000–3,000,000	45	45%
	3,000,000–6,000,000	16	16%
	> 6,000,000	4	4%

Table 3.   Multiple regression.

Model	Unstandardized coefficients		Standardized coefficients		
	B.	Std. error	Beta	t	Sig.
(Constant)	7.493	2.542		2.947	0.004
Celebrity endorser	0.302	0.0105	0.229	2.865	0.005
Electronic word of mouth	8.859	0.12	0.572	7.151	0.000

Table 4. T-test.

Model	Sum of Squares	df	Mean Square	F	Sig
Regression	2,314.95	2	1,157.48	59.319	00.000
Residual	2,087.86	107	19.513		
Total	4,402.81	109			

a. Predictors : (Constant), Electronic Word of Mouth, Celebrity Endorse
b. Dependent Variable. Purchase Decision

Table 5. F-test.

Model	Unstandardized coefficients B.	Std. error	Standardized coefficients Beta	t	Sig.	Correlations Zero Order	Partial	Part
(Constant)	7.493	2.542		2.947	0.004			
Celebrity endorser	0.302	0.0105	0.229	2.865	0.005	0.547	0.267	0.191
Electronic word of mouth	8.859	0.12	0.572	7.151	0	0.7	0.569	0.476

a. Dependent Variable. Purchase Decision

Based on the table, we can conclude that the equation for multiple regression – Y = 7.493 + 0.302X1 + 0,859X2 – means that using celebrity endorsers and e-WoM to promote Tokopedia will increase the purchase decisions of Tokopedia's customers. To determine the significance of the independent variables on the dependent variables, a t-test (partial) and an f-test (simultaneous) were conducted. The test results based on SPSS are presented in Tables 4 and 5.

Table 3 shows that the t-value of the celebrity endorser variable is 2.865 (to) and the t-value of the electronic word of mouth variable is 7.151 while the t-table value is 1.658 (tα). The following is a description of these data:

1. Celebrity endorser variable: 2.865 (to) > 1.658 (tα) = H1 accepted.
2. Electronic word of mouth variable: 7.151 (to) > 1.658 (tα) = H2 accepted.

From the results of the hypothesis test, we can conclude that a celebrity endorser and e-WoM partially influence the purchase decisions of Tokopedia customers. The f-test presented in Table 4 shows the calculated f-value of 59.319, and the f-table value is 3.09. The f-value is 59.319 > the f-table of 3.09. The results of the f-test support the hypothesis that celebrity endorsers and e-WoM simultaneously influence the purchase decisions of Tokopedia customers.

Table 6. Coefficient of determination.

Variable	Variable standardized coefficients	Correlations	Coefficient of determination	%
Celebrity endorser	0.229	0.547	0.126	12.60%
e-WoM	0.572	0.7	0.4	40%

Based on Table 5, we can see that the influence of celebrity endorsers on purchase decisions partially is 12.6%. The effect of e-WoM on purchase decisions partially is 40%. The total effect of the celebrity endorser and e-WoM variables on the purchase decision simultaneously is 52.6%. This can also be seen from the results of the coefficient of determination.

5 CONCLUSIONS AND RECOMMENDATIONS

Online business decisions in this Industry 4.0 era must use a current renewal strategy. Such a strategy, like used celebrity endorsers and e-WoM, usually has a good impact on consumers. Based on multiple regression analysis, the promotion of celebrity endorsers and e-WoM on Tokopedia will increase the purchase decisions of Tokopedia's customers. This study shows that Tokopedia's customers are young people who connect to technology. The t-test and f-test show that celebrity endorsers and e-WoM partially and simultaneously influence purchase decisions. This study can be used not only for Tokopedia but also for other e-commerce platforms because they have similar customers and business processes.

REFERENCES

Belch, George & Belch, Michael. (2014). A content analysis study of the use of celebrity endorsers in magazine advertising. International Journal of Advertising. 32. 369. 10.2501/IJA-32-3-369-389.

Cheung, Christy & Lee, Matthew. (2008). Online Consumer Reviews: Does Negative Electronic Word-of-Mouth Hurt More?. 14th Americas Conference on Information Systems, AMCIS 2008. 5. 143.

Das, K., Gryseels, M., Sudhir, P., & Tan, K .T. 2016, October. Unlocking Indonesia's digital opportunity. McKinsey.com

Dix, S., Phau, I., & Pougnet, S. 2010. "Bend it like Beckham": The influence of sports celebrities on young adult consumers. *Young Consumers* 11(1), 36–46.

Ekawati, Mustika. 2014. Pengaruh Electronic Word Of Mouth Terhadap Pengetahuan Konsumenserta Dampaknya Pada Keputusan Pembelian (Survei Pada Followers Account Twitter @Wrpdiet). Jurnal Administrasi Bisnis. VOL 14, NO 2 (2014).

Erdogan, Z. B., Baker, M. J., & Tagg, S. 2001. Selecting celebrity endorsers: The practitioner's perspective. *Journal of Advertising Research* 41(3), 39–48.

Hennig-Thurau, T., Gwinner, K. P., Walsh, G., & Gremler, D. D. 2004. Electronic word of mouth via consumer-opinion platforms: What motivates consumers to articulate themselves on the Internet? *Journal of Interactive Marketing*. 38–52. ww.marketingcenter.de/lmm/research/publications/download/I11_Hennig-Thurau_el_al_JIM_2004_Electronic_WoM.pdf

Hwang, K., & Zhang, Q. (2018). Influence of parasocial relationship between digital celebrities and their followers on followers' purchase and electronic word-of-mouth intentions, and persuasion knowledge. Comput. Hum. Behav., 87, 155–173.

Indrawati. 2015. *Metode Penelitian Manajemen dan Bisnis Konvergensi Tekonlogi Komunikasi dan Informasi*. Bandung: PT Refika Aditama.

Jalilvand, Mohammad Reza & Samiei, Neda. (2012). The effect of electronic word of mouth on brand image and purchase intention: An empirical study in the automobile industry in Iran. Marketing Intelligence & Planning. 30. 460–476. 10.1108/02634501211231946.

Jin, S. A., & Phua, J. 2014. Following celebrities' tweets about brands: The impact of Twitter-based electronic word-of-mouth on consumers' source credibility perception, buying intention, and social identification with celebrities. *Journal of Advertising* 43(2), 181–195. doi: 10.1080/00913367.2013.827606

Kassarjian, H. H. 1977. Content analysis in consumer research. Journal of Consumer Research 4 (1): 8–18.

Kertamukti, Rama. (2015). Strategi Kreatif dalam Periklanan (Konsep Pesan, Branding, Anggaran), Jakarta: PT Raja Grafindo.

Lin, Chinho & Wu, Yi-Shuang & Chen, Jeng-Chung Victor. 2013. "Electronic Word-of-Mouth: The Moderating Roles of Product Involvement and Brand Image," Diversity, Technology, and Innovation for Operational Competitiveness: Proceedings of the 2013 International Conference on Technology Innovation and Industrial Management, ToKnowPress.

Mangold, W. G., & Faulds, D. J. 2009. Social media: The new hybrid element of the promotion mix. *Business Horizons* 52(4), 357–365. doi: 10.1016/j.bushor.2009.03.002

McCormick, Karla. (2016). Celebrity endorsements: Influence of a product-endorser match on Millennials attitudes and purchase intentions. Journal of Retailing and Consumer Services. 32. 39–45. 10.1016/j.jretconser.2016.05.012.

Park, C., Wang, Y., Yao, Y, & Kang, Y. R. 2011. Factors influencing EWOM effects: Using experience, credibility, and susceptibility. *International Journal of Social Science and Humanity* 74–79. www.ijssh.org/papers/13-H10193.pdf

Phua, J., Lin, J.-S., & Lim, D. J. 2018. Understanding consumer engagement with celebrity-endorsed E-cigarette advertising on Instagram. *Computers in Human Behavior* 84, 93–102.

Rathore, D., & Panwar. A. 2015. Developing a research framework to assess online consumer behavior using netnography in India: A review of related research. In *Capturing, analyzing, and managing word of mouth in the digital marketplace.* 154–167.

Schimmelpfennig, C. 2018. Who is the celebrity endorser? A content analysis of celebrity endorsements. *Journal of International Consumer Marketing* 30(4), 220–234. doi: 10.1080/08961530.2018.1446679

Shimp, T. A. 2014. *Komunikasi Pemasaran Terpadu dalam Periklanan dan Promosi.* Jakarta Selatan: Penerbit Salemba Empat.

Spry, A., Pappu, R., & Cornwell, B. T. 2011. Celebrity endorsement, brand credibility and brand equity. *European Journal of Marketing* 45(6), 882–909.

Sugiyono. 2015. *Metode Penelitian & Pengembangan Research and Development.* Bandung: Alfabeta.

Sunyoto, D. (2015). Perilaku Konsumen dan Pemasaran. Yogyakarta: CAPS.

Syaffarudin, et al. 2016. Pengaruh Komunikasi Electronicword Of Mouth Terhadap Kepercayaan (Trust) Dan Niat Beli (Purchase Intention) Serta Dampaknya Pada Keputusan Pembelian Jurnal Bisnis dan Manajemen Vol. 3 No.1, Januari 2016.

Zheng, X., Cheung, C., Lee, M., & Liang, L. 2015. Building brand loyalty through user engagement in online brand communities in social networking sites. *Information Technology & People* 28(1), 90–106. https://doi.org/10.1108/ITP-08-2013-0144

Analysis of customer satisfaction as an intervening variable on the effect of retail service quality on customer loyalty at Uniqlo Indonesia

T.G. Saraswati & M.E. Saputri
Telkom University, Indonesia

ABSTRACT: This study aimed to analyze customer satisfaction as an intervening variable on the relationship between the influence of retail service quality and customer satisfaction at Uniqlo Indonesia. Consumer loyalty is an important aspect to consider in the clothing retail business because consumer loyalty will directly influence the business continuity. This is a major concern for Uniqlo, a clothing business from Japan, in reaching the Indonesian market share. Uniqlo consumers show a great deal of loyalty, and whether their loyalty could be influenced by the retail service quality offered by Uniqlo is an issue that needs further testing. This research included a descriptive study with an associative method using path analysis involving 100 consumers who have shopped at Uniqlo Indonesia as respondents, with purposive sampling techniques. The results of the study stated that retail service quality has an influence on customer loyalty at Uniqlo. Likewise, when tested using customer satisfaction as an intervening variable, retail service quality displayed a greater effect on customer loyalty. For that reason, Uniqlo Indonesia should pay attention to its retail service quality and customer satisfaction in order to gain more customer loyalty.

1 INTRODUCTION

Uniqlo is a clothing retail company originating from Japan that began to enter the Indonesian market in 2013. Under the management of PT Fast Retailing Indonesia, Uniqlo successfully captured the hearts of the Indonesian people. Indonesia is not the first country in Southeast Asia to be visited by Uniqlo. Previously, Uniqlo was present in Singapore, Malaysia, Thailand, and the Philippines. Tadashi Yanai, the chief executive officer of Fast Retailing Co., Ltd., claims sales in the four countries have great potential. At present Uniqlo has more than 1,100 stores spread across thirteen countries – namely the United States, Britain, France, Russia, South Korea, China, Hong Kong, Taiwan, Thailand, Singapore, Malaysia, the Philippines, and Japan itself. In 2018, Uniqlo recorded a sales record where for the first time sales figures outside Japan were far higher than sales in Japan itself. Uniqlo's parent company, Fast Retailing, announced on January 11, 2019, that Uniqlo's international revenue for the previous quarter was 258.2 billion yen (around US $2.3 billion). The amount skyrocketed over Uniqlo's revenue in Japan of 257 billion yen, marking a milestone for the retailer.

In Indonesia, Uniqlo sales also show a pretty good number. This number is also marked by the repurchases of consumers loyal to the products Uniqlo offers. In a survey of fifty Indonesian Uniqlo consumers, 90% stated that they had purchased Uniqlo products more than once. The reasons they gave included the quality of Uniqlo products, the strategic locations of the stores (located in large malls), and varied choices. The remaining 10% did not make repeat purchases due to price.

Consumer loyalty that occurs at Uniqlo is allegedly due to retail service quality and good customer satisfaction. This needs to be reviewed and examined further in order to prove it. Therefore, the author conducted preliminary research regarding these two variables. Unfortunately, behind the success of Uniqlo sales, a number of complaints have emerged regarding Uniqlo services that can affect customer satisfaction. This can be seen from several websites and social media platforms that mention Uniqlo's bad service in the aspect of transaction services at the cashier. Transactions can take too long, making the queue quite long. The difficulty of finding employees in the store is also an issue for customers.

This research is in accordance with previous research by Wantara (2015) stating that service quality and customer satisfaction are significantly and positively related to customer loyalty. Customer satisfaction has been found to be an important mediator between service quality and customer loyalty. Research done by Nuchsarapringviriya (2015) found that service quality has a direct effect on customer satisfaction and customer loyalty. However, the relationship between service quality and customer loyalty is partly mediated by customer satisfaction. Therefore, the authors are interested in looking at the relationship between these variables and the phenomena that have been explained before.

2 LITERATURE REVIEW

2.1 Relationship of retail service quality and customer loyalty

Siat (1997: 4) states that customer satisfaction is the basic building block for companies to increase profitability and to earn customer loyalty. Basically, the elements of the service marketing mix will create satisfaction for customers. To understand the concept of customer loyalty, we must begin with customer satisfaction. Loverlook (2012), quoted in Adam (2015: 60), argues that the basis for true loyalty is customer satisfaction and service quality, which are key inputs in the service process. Customer satisfaction and customer loyalty have an inseparable relationship even though satisfied customers sometimes do not engage in loyal behavior. Oliver (1999), also cited in Adam (2015: 60), makes the same contention, although it is understood that customer loyalty and customer satisfaction have a very close but asymmetrical relationship. Although loyal customers are usually satisfied, that satisfaction does not necessarily translate into loyalty. Even so, satisfaction is an important step in the formation of loyalty and becomes less significant when other mechanisms influence it. The relationship between retail service quality and customer loyalty will be through customer satisfaction when viewed from the flow depicted in Figure 1. Customer loyalty occurs when customers feel satisfied with the retail service quality provided by a company.

2.2 Relationship of retail service quality and customer satisfaction

According to Cronin and Taylor (1992), as quoted in Tjiptono and Chandra (2017: 219), satisfaction helps customers revise their perception of service quality.

Their rationale includes:

1) If a consumer has no prior experience with a company, his perception of the quality of the company's services will be based on his expectations.

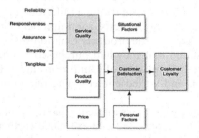

Figure 1. Customer perception of quality and satisfaction.

2) The next interaction (service encounter) with the company will cause the consumer to enter the disconfirmation process and revise their perception of service quality.
3) Any additional interactions with the company will strengthen or otherwise change the customer's perception of its service quality.
4) Perception of service quality that has been revised modifies consumers' interest in purchasing from the company in the future.

Teas (1993), as cited in Tjiptono and Chandra (2017: 219), proposes that the causal relationship between customer satisfaction and service quality can be integrated by determining two perceived quality concepts – namely transaction-specific quality and relationship quality. Perceived transaction-specific quality is treated as a component of the specific performance of certain transactions in the contemporary customer satisfaction model. This implies that transaction-specific satisfaction is a function of perceived transaction-specific performance quality. In contrast, perceived relationship quality is assumed as an assessment or global attitude regarding service superiority. Consequently, transaction-specific satisfaction is a predictor of perceived long-term relationship quality.

2.3   *Relationship of customer satisfaction and customer loyalty*

Band (2001) states that customer satisfaction is a situation where the needs, desires, and expectations of consumers are met, which results in a repeat purchase and continues to loyalty. Oliver (1999), referenced in Tjiptono and Chandra (2017: 203), summarizes six relationships between customer satisfaction and customer loyalty as often found in the marketing literature:

1) Perspective (A): Satisfaction and loyalty are basically the same construct.
2) Perspective (B): Satisfaction and loyalty are the core concepts for loyalty. Without satisfaction, loyalty will not exist.
3) Perspective (C): Satisfaction is only one component of loyalty.
4) Perspective (D) confirms the existence of a superordinate concept – namely, ultimate loyalty, which includes "simple" satisfaction and loyalty.
5) Perspective (E): Overlapping occurs between satisfaction and loyalty, but the percentage of overlap is small compared to the content of each construct.
6) Perspective (F): Satisfaction is the starting point of the transition phase, which culminates in a separate loyalty level.

2.4   *Research framework*

Tjiptono and Chandra (2017: 169) state that in the model they have developed, the quality of retail services is evaluated at three different levels: the dimension level, the overall level, and the sub-dimensional level. The quality of retail services includes five main factors: physical aspects, reliability, personal interaction, problem solving, and policy.

According to Tjiptono (2014: 368), the dimensions of customer satisfaction are overall customer satisfaction, confirmation of expectations, repurchase intention, willingness to recommend, and customer dissatisfaction. Saladin (2013: 153) argues that customer loyalty consists of loyalty to product purchases (repeat purchase), resistance to negative influences (retention), and referrals.

Based on the indicators for each variable, researchers developed a framework for use as a thinking path or reference flow for researchers conducting research with related problems; this framework can be seen in Figure 2.

3   RESEARCH METHODOLOGY

This research used a quantitative approach. Quantitative research is scientific research that examines a problem or a phenomenon and looks at the possible relationships between its variables. The linkage used in this research was the causal relationship. According to Indrawati (2015: 117), causal research is conducted when the researcher wants to describe the cause of a problem (carried

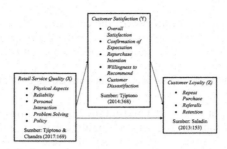

Figure 2. Research framework.

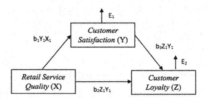

Figure 3. Path Analysis Model

out experimentally or non-experimentally). Atif (2014: 51) defines causality as a relationship between variables where changes in one variable cause changes in other variables without the possibility of the opposite.

Data analysis techniques in quantitative research use statistics. Two kinds of statistics can be used for quantitative research analysis – namely, descriptive and inferential statistics. In this study, researchers used descriptive statistical data analysis techniques. Indrawati (2015: 116) states that descriptive statistics analyze data by describing data collected without intent to make conclusions or generalizations that apply to the public. The purpose of descriptive research is to describe the characteristics of a group, to estimate the percentage of the unit analyzed, and to determine the perceptions of users of a product. Descriptive statistics present data in tabular and graphical forms.

The population in this study comprised Uniqlo consumers who had made purchases directly at Uniqlo Indonesia at least twice whose numbers were not known with certainty. Based on calculations using the Bernoulli formula, a sample of 96 was obtained and rounded to 100 respondents.

Path analysis was used to examine the effect of the retail service quality variable on customer loyalty by involving the customer satisfaction intervening variable. According to Ghozali (2013), path analysis is an extension of multiple linear regression analysis, or the use of regression analysis to estimate causal relationships between variables that have been predetermined based on theory. The path analysis model can be seen in Figure 3.

4 RESULTS AND DISCUSSION

Based on descriptive analysis, the retail service quality variable resulted in a value of 83% in the good category, with respondents agreeing that Uniqlo provides services appropriately from the start. Although the number of Uniqlo employees was small, employees paid individual attention to customers, and Uniqlo was willing to handle returns and exchanges. Based on descriptive analysis, the average score of customer satisfaction is 75.58%, which is in the good category, because respondents were satisfied after shopping at Uniqlo compared to other clothing retail companies, service met expectations, and customers have no complaints about Uniqlo.

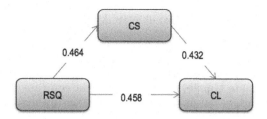

Figure 4. Hypothesis testing of retail service quality on customer loyalty through customer satisfaction

Based on descriptive analysis, the average score of customer loyalty was 79.7%, which is in the good category, because respondents agreed to shop again at Uniqlo and to recommend Uniqlo to other people. Even though negative issues occurred, customers would still shop at Uniqlo.

Based on the results of the path analysis, the effect of retail service quality on customer satisfaction is 21.5%. The higher the retail service quality, the higher the customer satisfaction. This means that customer satisfaction is created when the level of quality of products and services received meets consumers' needs and is in accordance with their expectations. Satisfied consumers feel they get more value from quality services; satisfaction occurs if consumers' desires are met with quality service. So retail service quality is an important variable consumers consider to increase their satisfaction with the performance of Uniqlo Indonesia.

Based on the results of the path analysis, which can be seen in Figure 4, the effect of customer satisfaction on customer loyalty is 18.7%. The higher the customer satisfaction, the higher the customer loyalty. This means that customer loyalty occurs when consumers feel satisfied with the service given by Uniqlo Indonesia. Satisfaction is obtained after consumers use Uniqlo products, then consumers compare expectations and reality to service quality. If consumers are satisfied with the quality of service, customer loyalty will increase and consumers will not switch to another company. So customer satisfaction is an important variable consumers consider to increase their loyalty.

Based on the results of the path analysis, which can be seen in Figure 4, the direct effect of retail service quality on customer loyalty is 21%. The higher the retail service quality, the higher the customer loyalty. Service quality is one of the important factors influencing consumer loyalty. Providing direct evidence of service, reliability of employees in providing services, accuracy of employees' ability to respond to consumer complaints, providing service guarantees, and paying attention to consumer complaints will encourage consumers to make repeated purchases of products, stay loyal to Uniqlo, and recommend Uniqlo to other people. So retail service quality is an important variable consumers consider to increase their loyalty to Uniqlo.

Based on the results of the path analysis, the indirect effect of retail service quality on customer loyalty through customer satisfaction has an intervening variable of 20%. The higher the retail service quality, the higher the customer satisfaction, and the higher the customer satisfaction, the higher the customer loyalty. This is proven by employee reliability, store layouts that make it easier for consumers to find goods, willingness to handle returns and exchanges, and ability to meet consumers' needs and expectations.

5 CONCLUSION

The higher the customer satisfaction, the higher the customer loyalty. This means that customer loyalty occurs because of the customer loyalty to the company and it happens when consumers feel satisfied with the service given by Uniqlo Indonesia. Satisfaction is obtained after consumers use Uniqlo products, then consumers compare expectations and reality to service quality. If consumers are satisfied with the quality of service, customer loyalty will increase and consumers will not switch to another company. So customer satisfaction is an important variable consumers consider to increase Uniqlo customer loyalty. The higher the retail service quality, the higher the customer loyalty. Service quality is one of the important factors influencing consumer loyalty. Direct evidence of service, reliability of employees in providing services, employees' ability to respond to consumer complaints, service guarantees, and paying attention to consumer complaints will encourage consumers to make repeated purchases of products, be loyal to Uniqlo, and recommend the company to other people. So retail service quality is an important variable consumers consider to increase consumer loyalty to Uniqlo. The higher the retail service quality, the higher the customer satisfaction, and the higher the customer satisfaction, the higher the customer loyalty. This is proven by employee reliability, store layouts that make it easier for consumers to find goods, willingness to handle returns and exchanges, and the ability to meet consumers' needs and expectations, and it also has an impact on customers repurchasing products from Uniqlo.

REFERENCES

Atif, N. F. 2014. *Metode Penelitian: Kuantitatif, kualitatif, dan campuran untuk manajemen, pembangunan, dan pendidikan* (p. 280). Bandung: PT. Rafika Aditama.
Band, W. A. 2001. Creating value for customers. New York: Wiley.
Ghozali, I. 2013. *Aplikasi Analisis Multivariate dengan Program SPSS, Edisi Ketujuh*. Semarang: Badan Penerbit Universitas Diponegoro.
Indrawati. 2015. *Metode Penelitian Manajemen dan Bisnis Konvergensi Teknologi Komunikasi dan Informasi*. Bandung: PT. Refika Aditama.
Nuchsarapringviriya, F., & Ismail, S. 2015. Service quality, customer satisfaction and customer loyalty in Thailand's audit firms. *International Journal of Management and Applied Science* 1(5), 34–40.
Saladin, D. (2010). *Manajemen Pemasaran (Edisi Pertama)*. Bandung: CV. Linda Karya.
Siat, J. 1997. *Relationship Marketing*. Swasembada 3 XXVI.
Sunyoto, D. 2011. *Metodologi Penelitian*. Yogyakarta: Center of Academic Publishing Service.
Tjiptono, F. 2000. *Prinsip & Dinamika Pemasaran*. Edisi Pertama. Yogyakarta: J&J Learning.
Tjiptono, F. 2004. *Manajemen Pemasaran Edisi Pertama*. Yogyakarta: Andi Offset.
Tjiptono, F. 2014. *Pemasaran Jasa: Prinsip, Penerapan, dan Penelitian*. Yogyakarta: Andi.
Tjiptono, F., & Chandra, G. 2017. *Service, quality & satisfaction* Edisi 4. Yogyakarta: CV. ANDI.
Tjiptono, F., & Diana, A. 2015. *Pelanggan Puas? Tak cukup!* (p. 254). Yogyakarta: CV. ANDI.
Wantara, P. 2015. The relationships among service quality, customer satisfaction, and customer loyalty in library services. *International Journal of Economic and Financial Issues* 5(1), 264–269.

Measurement of the green building index: Case study of Bandung City

Indrawati, Harvianto & D.Tricahyono
Faculty of Economics and Business, Telkom University, Bandung, Indonesia

ABSTRACT: Increased awareness of healthy lifestyles and the need for a comfortable environment has motivated cities to implement green building for their citizens. Green building is construction that meets measurable performance in saving energy, water, and other resources. Bandung is one of the cities trying to implement the smart building concept in several areas. The objectives of this study were to determine the indicators and variables needed to measure the green building index of Bandung. The method applied in this research was mix-method research, qualitative methods were used to validate the variables and indicators, and a quantitative method was used to determine the value or score so as to get the index. Thirty persons from government, business, academia, and society who were selected using purposive sampling techniques served as the resources of data for this study.

The study found six variables – namely, energy efficiency, water efficiency, site management, indoor environmental quality, building management system, and awareness and education – and twenty-six indicators to measure the index of green building in Bandung. The resulting index is 79, which can be interpreted as sufficient and satisfactory with some things still lacking.

1 INTRODUCTION

Bandung is the capital of the West Java province of Indonesia, which has an area of 16,729 acres with the population reached 2.5 million in 2018. Bandung is surrounded by mountains, hence if viewed from its highest part, Bandung is like a lake (Hadiani, 2018).

Bandung's economic growth is the highest among cities in West Java; this has become a special attraction for the surrounding population and for travelers from all regions in Indonesia, making Bandung a destination city for doing business and settling.

Increased awareness of healthy lifestyles and the need for a comfortable environment drives the Bandung city government to implement the green city concept as one of the answer to these problems. In order to achieve green city status in Bandung, several attributes need to be considered: green planning and design, green communities, green open spaces, green buildings, green energy, green transportation, green water, and green waste. This study concentrated on green building. Green building is one of the solutions to make life better and meet the needs of the next generation.

The aims of this research were as follows:

a. To find out the variables and indicators that can be used to measure green building in Bandung.
b. To measure the green building index of Bandung.

2 LITERATURE REVIEW

A green city is a city built without sacrificing the city's natural assets, while continuously cultivating all resources such as people, the environment, and the infrastructure (Konsep pengembangan kota hijau, 2016).

Green building is a concept for sustainable building, and it has certain requirements – i.e., location, planning system and design, renovation, and operation – that adhere to the principle of saving energy and that must have a positive impact on the environment, economy, and society (Sudarwani, 2012).

Benefits to the environment if the concept of green buildings is implemented include:

1. The building is more durable and stronger, with minimal maintenance.
2. Energy efficiency makes routine financing more effective.
3. The building is more comfortable to occupy.
4. Occupants see a healthier quality of life.
5. The building planners have participated in environmental awareness (Triwidiastuti, 2017).

The green building strategy can be achieved by five stages of Go Green:

1. reduces resources consumption (energy and water);
2. reduces waste and makes recycling efforts;
3. building materials should be safe for the ozone layer; and
4. the environment inside the building such as air quality, room temperature, and air conditioning should be maintained. (Armstrong, 2008)

3 METHODOLOGY

3.1 Steps of the research

As explained in the introduction, this research had two objectives. The research was designed to answer the research questions through the following stages of research: a literature review of variables and indicators, confirmation of variables and indicators from literature review results, best practice data search, and data processing. Interviews with respondents provided green building variables and indicators, qualitative and quantitative data analyses, presentation of the research data, and conclusions and suggestions.

3.2 Measurements

Based on the literature review, eight research variables were confirmed: energy licensing, water efficiency, land management, indoor environmental quality, building materials and resources, building management systems, awareness and education, and innovation by conducting interviews with experts.

At the interview stage, researchers conducted in-depth interviews so as to obtain in-depth information and to confirm related variables and indicators from the literature review results with the experts from the quadruple helix (government, business, academia, and civil society) chosen through purposive sampling.

3.3 Data analysis

The researchers applied a mixed-method approach; qualitative methods were used to validate the variables and the indicators, and the quantitative method was used to determine the value or score so as to get the index. Qualitative research methods involve data analysis in the form of descriptions, and these data cannot be directly quantified. Quantification of qualitative data is done by giving a code or category. Quantitative research methods try to make accurate measurements of behavior, knowledge, opinions, or attitudes (Indrawati, 2015). Quantitative methods are widely used in various studies because of their suitability for testing models or hypotheses.

The method of statistical data analysis in this study was the Spearman Rank correlation method. The Spearman Rank correlation is useful to measure the level of closeness between two variables – namely the independent variable and the dependent variable – which has an ordinal scale, knows the level of compatibility of the two variables with the same group, obtains the empirical validity of the data collection tools, and finds out the reliability of the data collection tools (Riduwan & Sunarto, 2014).

The range of ratings agreed by the author was as follows:

1. Scores of 0–60 are considered very poor and need a lot of improvement.
2. Scores of 61–70 are considered bad; there are still many shortcomings, but within reason.
3. Scores of 71–80 are considered sufficient and satisfactory, but some things are still lacking.
4. Scores of 81–90 are considered good and satisfactory, in line with expectations, and slightly lacking.
5. Scores of 91–100 are considered very good and very satisfactory according to the participants' expectations, there were almost no shortcomings.

This range was used to assess the research of national research assessors/reviewers from the Ministry of Research and Technology in 2018 (Angdini, 2019; Indrawati & Smart City Research Group, 2019).

4 RESULTS AND DISCUSSION

This study found six variables and twenty-six indicators that are valid and reliable to be used for measuring the green building index, as shown in Figure 1.

Using the model as shown in Figure 1, the scores obtained based on the perception and experience of green building experts are shown in Table 1.

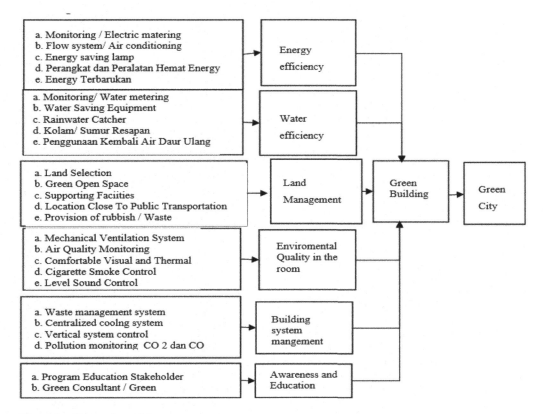

Figure 1.   Research model.

Table 1. Indicators' scores.

Variables	Indicator	Average	Category
Energy efficiency	Electricity meters	83.50	GOOD
	Air flow/conditioning system	82.00	GOOD
	Light bulb energy	79.17	ADEQUATE
	Renewable energy	80.50	GOOD
Water efficiency	Monitor/water meter	77.33	ADEQUATE
	Water-saving equipment	79.17	ADEQUATE
	Rainwater collection	75.83	ADEQUATE
	Infiltration well/pond	77.00	ADEQUATE
	Reuse of recycled water	80.17	GOOD
Land management	Land selection	80.67	GOOD
	Green open space	84.17	GOOD
	Support facilities	72.50	ADEQUATE
	Proximity of location to facilities and public transportation	71.33	ADEQUATE
	Provision of waste/waste management	82.67	GOOD
Environmental quality	Mechanical ventilation system	76.17	ADEQUATE
	Air quality monitoring	74.17	ADEQUATE
	Visual and thermal comfort	78.83	ADEQUATE
	Cigarette smoke control	78.67	ADEQUATE
	Noise level control	66.33	BAD
Building management system	Waste management system	86.67	GOOD
	Centralized cooling system	74.67	ADEQUATE
	Vertical transportation system	70.50	ADEQUATE
	CO_2 and CO pollutant monitoring system	75.00	ADEQUATE
Awareness and education	Educational program for stakeholders	89.00	GOOD
	Green consultant	80.50	GOOD
INDEKS		79.00	ADEQUATE

Based on Table 1, we can conclude that the lowest indicator of green building is "noise level control" with a value of 66.33. According to the results of interviews with building analysts of the Spatial Planning Office, noise indicators depend on specific locations such as industrial estates. For office areas, the level of noise is quite good or in the range of acceptance. The highest indicator of green building is the "stakeholder education program" with a value of 89.00. This indicates that the information related to the green building program of the city is already distributed well and accommodated in several regulations, as stated by a resource person from the Spatial Planning and Green Building Council of Indonesia. He said that the regulation outlines several requirements, such as that the building should set aside 30% of its total land area as green open space. This regulation makes the stakeholder aware of green building. The green building index value achieved by the city of Bandung is 79.00, which is considered sufficient and satisfactory, but a number of things are still lacking.

5 CONCLUSIONS AND RECOMMENDATIONS

This study found six variables and twenty-six indicators through which to measure the green building index of Bandung. The six variables that affect green building are energy efficiency, water efficiency, land management, quality in space, land management systems, and awareness and education. The green building index of Bandung is 79.00. The resulting index value can be interpreted as sufficient and satisfactory, but things are still lacking. This is in line with the results of interviews with several speakers from the Spatial Planning Office, who stated that there is currently no provision from the Board of Trustees that requires buildings in the city of Bandung to implement the green building concept. The requirements are only related to the use of land, and they state that 30% of the total land area should be green open space and that buildings should use energy-saving lamps and other requirements as stated in Bandung City Guard No. 1023 of 2016.

Table 2. Program recommendations.

	Subject	Action	Program	Target achievements	Timeline	Responsibilities
1	Support the mayor's Regulation No. 1023 of 2016 concerning green building	Provide socialization to entrepreneurs related to water use that is managed by the local government	Give rewards and punishments to entrepreneurs who have followed the rules	Efficiency in the use of water used to anticipate drought/lack of water	Q4 2019–Q2 2020	Government
2	Integrated public transportation	Provide a platform for integrated public transportation systems	Campaign for Tuesdays and Thursdays without private vehicles	Use of public vehicles to reduce congestion and carbon emissions	Q1 2020	Government
3	The use of renewable energy	Require building entrepreneurs to use solar cells as a renewable energy source	Provide faster IMB manufacturing services (SLA 7 x 24 hours)	Energy efficiency in electricity usage and savings in electricity usage costs up to 20% for building users	Q1 2020	Government
4	Air quality monitoring	Install air quality monitoring tools	1 Company 1 Air monitoring	Assess the air quality around the building and the road	Q2 2020	Government and business actors
5	Waste sorting	Waste segregation starts from the household	Provide a trash dump in every local environment	Maintain cleanliness and provide additional income by sorting waste that still has economic value	Q3 2019–Q4 2020	Civil society
6	Publication of research results	Publication of research results on social and electronic media	1 publication 1 month	Provide insight and awareness at all levels of society related to the results of research	Q4 2019	Researchers/ experts
7	Use of light bulbs	Replace lamps with light bulbs gradually	Use light bulbs in rooms with more frequent traffic	Electricity cost savings from using lights	Q3 2019–Q4 2020	Civil society
8	CSR program	Provision of CSR in accordance with the priority scale	CSR of providing clean water and making toilet facilities	Provide education related to clean and healthy lifestyles	Q4 2019	Businesspeople
9	Double skin design	Double skin design on the outside of the building	Renovation or addition of building's outer layer for design and comfort	Reducing the heat load in the room, so that air conditioner use can lowered by up to 30%	Q4 2019	Businesspeople

The lowest index value is in the indicator of "noise level control" with an average value of 66.33, which is considered bad; there are still many shortcomings but within reason. According to the results of interviews with building analysts of the Spatial Planning Office, noise indicators depend on specific locations. The level of noise is high for industrial estates, but for office areas, the level is acceptable. Based on the findings of this research, the suggested programs for Bandung are shown in Table 2.

ACKNOWLEDGMENT

The authors express their gratitude to the Ministry of Research, Technology and Higher Education of Indonesia for its financial support in carrying out this research.

REFERENCES

Angdini. 2019. *Indeks Kesiapan Smart Building Kota Bandung*. Bandung: Universitas Telkom.
Armstrong, B. 2008. Green building strategies. *Manitoba Business* 30(2), ABI/INFORM Complete, 14.
Hadiani, V. 2018. Asal muasal dan sejarah Bandung. Diambil dari. http://disdik.jabarprov.go.id
Indrawati. 2015. *Metode Penelitian Manajemen dan Bisnis: Konvergensi Teknologi Komunikasi dan Informasi*. Bandung. Refika Aditama.
Indrawati & Smart City Research Group. 2019. *Inilah Cara Mengukur Kesiapan Kota Pintar*. Bandung: Intrans Publishing.
Konsep pengembangan kota hijau. 2016. http://sim.ciptakarya.pu.go.id
Manfaat ruang terbuka hijau. 2016. http://sim.ciptakarya.pu.go.id
Peraturan Walikota Bandung No. 1023 Tahun. 2016 Tentang Bangunan Gedung Hijau.
Riduwan & Sunarto, H. 2014. *Pengantar Statistika Untuk Penelitian: Pendidikan, Sosial, Komunikasi, Ekonomi, dan Bisnis*. Bandung: Alfabeta.
Sudarwani, M. M. 2012. *Penerapan Green Architecture dan Green Building Sebagai Upaya Pencapaian Sustainable Architecture, Page 6*. Majalah Ilmiah Universitas Pandanaran.
Triwidiastuti. 2017. *Optimalisasi Peran Sains dan Teknologi untuk Mewujudkan Smart City*, page 143–144. Banten: Universitas Terbuka.

Analysis of makeup product reviews using the LDA-based topic modeling method (Case study: Cushion Pixy Make It Glow)

N.A. Lukman & N. Trianasari
Faculty of Economics and Business, Telkom University, Bandung, Indonesia

ABSTRACT: Indonesia is a promising market for the beauty industry. Based on the 2015–2035 National Industrial Development Master Plan (RIPIN), the cosmetics industry has a role as a prime mover in the future economy. PIXY is a local cosmetic manufacturer in Indonesia that has just repositioned its products with the aim of becoming the one and only local choice for Indonesian people. PIXY won the Best Beauty Products 2018 cushion series Make It Glow from Femaledaily based on user reviews. User reviews are also referred to as user-generated content (UGC) and can be utilized for the first stage of product development – namely, consumer analysis. Topic modeling was used to find out which topics consumers often discuss. The topics formed were classified as satisfied and unsatisfied. A total of 817 data points were obtained. The topic discussed as a whole is an expression of praise with the results of this product. Suggestions for PIXY indicated that overall results are good while maintaining the formulation of glowing and wet, mild, upfront effects to improve product formulation. Therefore the product does not oxidize so that it makes the skin gray.

1 INTRODUCTION

Indonesia is a potential market for the beauty industry. Based on Government Regulation No. 14 of 2015 concerning the National Industrial Development Master Plan (RIPIN) for 2015–2035, the cosmetics industry is one of Indonesia's mainstay industries, meaning priority industries that act as prime movers for the future economy.. Data from the Food and Drug Supervisory Agency (BPOM) regarding products in circulation in 2019 indicate that as much as 60.9% are cosmetic products with a total of 47,494 products.According to the Ministry of Industry, trade increased by 6.35% in 2017 and by 7.33% in 2018. In 2019, the industry was projected to grow by 9%.

In 2017, 85% of skincare products and 53% of local brand cosmetics still dominated the market among Indonesian consumers. In 2018, Indonesia had five players in the cosmetics industry that had listed their shares on the Indonesian Stock Exchange, one of which was PT. Mandom Indonesia Tbk with PIXY brand cosmetic products (sigmarearch, 2017). In 2018, PIXY repositioned its products with the objective of becoming the "one and only" local product in Indonesia. In its planning, PIXY will continue to develop better products in order to remain the choice of the people of Indonesia. PIXY focuses on product development in the base makeup and decorative makeup categories, which are also accompanied by a rebranding process with a more chic and modern concept. At the end of 2018, PIXY won the 2018 Best of Beauty Award with the Best Cushion Compact organized by Femaledaily. Product development is a series of activities that start from the analysis of market perceptions and opportunities, then end with the production, sale, and delivery of products (Rini, 2013: 31).

Femaledaily has become one of the most popular websites containing reviews of beauty products. Product reviews appear on the Femaledaily website; approximately 13,500 products from 1,000 brands and more than 100,000 reviews are featured. Users of Femaledaily write reviews of products and give a rating ranging from 1 to 5 (Rosi, Fauzi, & Perdana, 2018: 1992).

Reviews on Femaledaily can be referred to as user-generated content (UGC). UGC can be interpreted as a product review in the form of media content produced by general consumers (not paid professionals) who buy and have experience with a product or service. Topic modeling is one way to identify the topic of a text or document in social media. Topic modeling represents each document as a complex combination of several topics and each topic as a complex combination of several words. It is used as a word miner to classify documents based on the results of topic inference (Alamsyah et al., 2018: 255).

2 LITERATURE REVIEW

2.1 *Big data*

A collection of information that cannot be processed and analyzed through traditional methods, big data involves making data, storing data, extracting information, and analysis that stands out in terms of volume, variety, velocity, and veracity. In general, big data can be interpreted as a collection of very large data (volume), present in various forms or formats (variety), that has speed (velocity), and comes from a strong source (veracity) (Pujianto, Mulyati, & Novaria, 2018: 129).

2.2 *Text mining*

Text mining is the process of obtaining high-quality information from text. Usually it involves structuring text input, finding patterns within that structure, and evaluating the output in order to interpret the results. Typical text mining occurs in the form of categorization, text clustering, document summarization, keyword extraction, and others (Tong & Chang, 2016: 201–202).

2.3 *Topic modeling*

A topic consists of certain words that make up the subject of discussion, and one document has the possibility of consisting of several topics with their respective probabilities. The distribution of documented topics is a hidden structure. The purpose of topic modeling is to find topics and words contained in those topics (Blei, cited in Putra & Kusumawardani, 2017: A312).

2.4 *Latent Dirichlet allocation*

Latent Dirichlet allocation (LDA) is a generative probability model of a collection of writings called a corpus. The basic idea proposed by the LDA method is that each document is represented as a random mix of hidden topics, where each topic has a character determined based on the distribution of words contained in it (Blei, cited in Putra & Kusumawardani, 2017).

2.5 *User-generated content*

User-generated content is online content that consists of reviews of a product or service that are uploaded onto social media in a certain form by nonprofessionals or general consumers (Onny & Kusumawati, 2019: 190).

2.6 *Product development*

Product development is a series of activities that begin from the analysis of perceptions, market opportunities, production, sale, and delivery of products (Ulrich & Steven, cited in Rini, 2013).

2.7 *Rating*

Rating involves customer opinion at a certain scale. A popular ranking scheme for rating on online stores is by giving stars. The more stars are given, the better the seller's rating becomes as a representation of consumer opinion on a specific scale (Lackermair, cited in Auliya, Umam, & Prastiwi, 2017: 92).

3   METHODOLOGY

3.1   *Participants*

The data used in this project comprised product reviews of the PIXY Make It Glow cushion from the Femaledaily website during the period September 16, 2018, to September 25, 2019; data were taken using the Parsehub application and preprocessing data came from the Rstudio application. A total of 817 data points was collected.

3.2   *Data analysis techniques*

1. Data collection was done by scrapping using the Parsehub application. Data retrieval was done by taking three attributes on the Femaledaily website, linking usernames with ratings and reviews, which were then cleansed, and preprocessing data.
2. The data attributes taken were the username, rating, and reviews that were successfully collected so as to carry out cleansing and preprocessing the data. The steps of preprocessing data were as follows:
   a. Tokenizing is decomposing a description that originally consisted of sentences into words.
   b. Filtering is taking important words from the results of the token process.
   c. Stemming is mapping and decomposing various forms (variants) of a word into its basic word form (stem).
3. To be able to know each topic from consumer data, a review must be grouped into two parts – satisfied and dissatisfied – using the median of the serial scale.

Femaledaily uses ratings of 5 = very satisfied, 4 = not satisfied, 3 = quite satisfied, 2 = not satisfied, 1 = very dissatisfied. The median value of the rating value is 3, then the data grouping limit grouped satisfied ratings into the rating values of 5, 4, 3 and dissatisfied ratings into 2 and 1.

4. Data that were processed using the Rstudio application were repeated on each of the different attributes to determine directly the number of modeling topics to be analyzed later. Topic modeling is related to finding structures or topics in a set of documents that come from the appearance of words in the document.

4   RESULTS AND DISCUSSION

4.1   *Results of LDA-based topic modeling*

By using topic modeling and LDA with the data in the process using Rstudio, the data held will be seen as the whole topic discussed by PIXY series Make It Glow users on the Femaledaily website as follows:

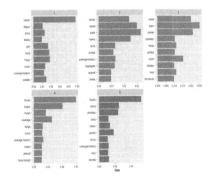

Figure 1.   LDA-based topic modeling results overall rating.

Figure 1 shows the five topics discussed as a whole on the PIXY series Make It Glow review period of September 16, 2018, to September 25, 2019. The topics on Femaledaily were:

1. The results obtained are wet, good products, suitable, medium coverage.
2. Wet results, the color of the product, this product is well liked, but dark.
3. Local products are cheap, oily, suitable.
4. The result of this product is good, suitable, cheap price, light in the face.
5. The results of this product are wet, oxidized, cause acne.

The whole topic discussed is the compatibility of using this product, seen from the words that appear: *good products, medium coverage, much preferred, low prices, suitable, really good.*

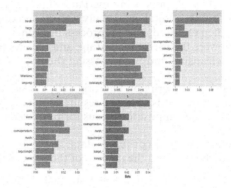

Figure 2. Results of LDA-based topic modeling satisfied rating.

Figure 2 shows the five topics discussed based on the satisfied rating of the PIXY series Make It Glow cushion review period of September 16, 2018, to September 25, 2019. The topics contained on Femaledaily were as follows:

1. Good results, medium coverage, like the product, durable.
2. This product is good, low price, the consumer really likes and is suitable.
3. The results of this product are wet, mild formula, cause acne.
4. The results of this product are very good, coverage medium, low prices, and color of the product.
5. This product is loved by consumers, really good, wet results and medium coverage.

Overall satisfied rating of this product is that consumers feel the results of these products are good, have medium coverage, like the product, durable, cheap prices, lightweight formulations and wet result.

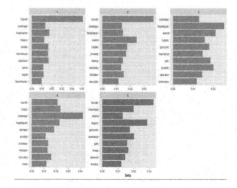

Figure 3. Results of LDA-based topic modeling rating not satisfied.

Figure 3 shows the five topics discussed based on the dissatisfied rating PIXY series Make It Glow cushion review period of September 16, 2018, to September 25, 2019. The topics contained on Femaledaily were as follows:

1. This product is the result of wetness, for the darkness upfront, oily, making a hollow.
2. This product causes acne, looks gray, and oxidizes the skin.
3. The color of darkness on the face, shaking, and gray on the face.
4. Visible putty, oxidation, and darkness upfront.
5. Cause acne and shake it in the face.

The overall dissatisfied rating of this product indicates that consumers feel the product is wet but oxidation is upfront and makes the action cause acne; it is not suitable as it makes the face look ashen or cakey.

5 CONCLUSIONS AND RECOMMENDATIONS

5.1 *Conclusion*

Based on the results of topic modeling conducted by PIXY series Make It Glow users on the Femaledaily website, based on the stages of the research, we can conclude that:

1. The topic discussed as a whole is an expression of praise with the results of this product.
2. Topics discussed based on ratings 5,4, and 3 (satisfied) indicate that consumers feel comfortable with this product, the results are wet, good upfront, medium coverage, lightweight formulas, and low prices.
3. Topics discussed based on ratings 2 and 1 (dissatisfied) indicate that consumers feel this product is not suitable, oxidizing so as to make the face turn gray and dark. The formula also causes zits.

5.2 *Recommendations*

Based on the data tested in this study, in the first stage of product development in management analysis of the perception of success was carried out and can be used as a consideration of the company for its assessment of the manufacture of further products. The author suggests to maintain product formulations that make the product wet, light upfront, and with a medium coverage. The things that need to be improved are formulations so that the product does not undergo oxidation, giving the impression of gray upfront or darkness. We can conclude that this product is good, much preferred by consumers.

REFERENCES

Alamsyah, A., Rizkika, W., Nugroho, D. D., Renaldi, F., & Saadah, S. 2018. Dynamic large scale data on Twitter using sentiment analysis and topic modelling. *International Conference on Information and Communication Technology* (pp. 254–258). IEEE.

Auliya, Z. F., Umam, M. K., & Prastiwi, S. K. 2017. Online customer reviews and rating: Kekuatan Baru pada Pemasaran Online di Indonesia. *EBBANK 8*(1), 89–98.

Onny, I. Y., & Kusumawati, A. 2019. Pengaruh user generated content (UGC) dan brand equity pada green purchase. *Jurnal Administrasi Bisnis 73*(1), 187–195.

Pujianto, A., Mulyati, A., & Novaria, R. 2018. Pemanfaatan big data dan Perlindungan Privasi Konsumen di Era Ekonomi Digital. *Majalah Ilmiah BIJAK 15*(2), 127–137.

Putra, I. K., & Kusumawardani, R. P. 2017. Analisis Topik Infomasi Publik Media Sosial di Surabaya menggunakan Pemodelan Latent Dirichlet Allocation (LDA). *Jurnal Teknik 6*(2), A311–A316.

Rini, E. S. 2013. Peran Pengembangan Produk dalam Meningkatkan Penjualan. *Jurnal Ekonomi 16*(1), 30–38.

Rosi, F., Fauzi, M. A., & Perdana, R. S. 2018. Prediksi rating pada review Produk Kecantikan menggunakan Metode Naïve Bayes dan categorical proportional difference (CPD). *Jurnal Pengembangan Teknologi Informasi dan Ilmu Komputer 2*(5), 1991–1997.

Tong, A., & Chang, H. 2016. A text mining research based on LDA topic modelling. *Computer Science & Information Technology*. 6. 201–210.

Descriptive analysis: Perception index for measuring variables in e-commerce domination

S.H. Komariah & R.Y. Arumsari
Department of Creative Industry, Telkom University, Bandung, Indonesia

ABSTRACT: Visual design is the study of communication applying illustrations, letters, colors, composition, and layout by processing graphic design elements. One's creativity rises when facing a problem in visualizing a design in order to achieve functional, persuasive, artistic, aesthetic, and verbal communication. The public has already been familiar with advertising design. Based on the three variables studied in this research, advertising plays a very important role in business. This study showed that the advertising media variable has a perceptual index value of 72.25, from which we can conclude that the variable of advertising media is in the high category. The results of the study are expected to be the basis of consideration for advertisers in designing an advertisement, i.e. by taking any factors having a dominant impact on the effectiveness of an advertisement into account in reference to the target perception.

1   INTRODUCTION

A growing number of new products and services aims to meet consumers' needs today. Most consumers are satisfied with different products to choose from, yet they are confused as to which products are proper for them. This is why advertisement is important to expand a business.

Development in technology has changed the transaction modes people use to meet their daily needs, i.e. from the conventional system to online systems. This transformation has triggered new business units that use online systems (e-commerce) where sellers and buyers can complete any transaction in virtual shops or markets. This new online business model means that virtual stores are marketed through advertisements broadcast across various media using a model, where the artist or model attempts to ensure the content of an ad is effectively accepted and understood by target communities.

A variety of advertising designs are created and broadcast to make the business unit grow into a big company that is then recognized and accepted by the public as the number of both consumers and producers of the product is increasing (Corvi & Bonera, 2010).

Based on these two conditions, we can observe some interesting points, i.e. creating ads and publishing them. The study reveals the factors that affect making and delivering advertisements and that greatly impact the level of acceptance and public understanding of an ad.

2   METHODOLOGY

Visual design is the study of communication concepts applying illustrations, letters, colors, composition, and layout by processing graphic design elements. One's creativity rises when facing a problem in visualizing design that will achieve functional, persuasive, artistic, aesthetic, and verbal communication.

Advertising is any form of impersonal presentation and promotion of ideas, goods, or services paid for by a single sponsor. In a small company, staff members in a marketing department handle advertising.

The perception of respondents can be measured by using a scoring technique, i.e. by viewing and calculating respondents' answers to items compiled in a questionnaire. The scoring technique performed in this study used a minimum score of 1 and a maximum score of 10 (Amin Bhat & Kensana (2016), Corvi & Bonera (2010), and Shahriari (2015)).

In this study, the researchers used the qualitative method by analyzing the perception of a target society toward an advertisement through a questionnaire. The respondents' answers were processed in order to measure the effectiveness of the perception index of an advertisement. The study was carried out in the following stages: writing a problem statement and performing objective research, conducting a literature review, planning and making a questionnaire, collecting data, carrying out a validity test, processing data, completing an analysis, and generating a conclusion.

Data were collected by giving the questionnaire to 130 respondents. The respondents were students who had ever seen e-commerce ads, especially shopee.com ads, with the advertising tags of shopee and *selalu di hati* (always be in our heart).

The researchers took and observed three ad variables (Corvi & Bonera, 2010):

1. The endorser variable employed four measurement indicators, i.e. the credibility of the endorser (x1), matching the endorser with the audience (x2), matching the endorser with the brand (x3), and endorser appeal (x4).
2. The content variable used five measurement indicators, i.e. attention (x5), interest (x6), desire (x7), decision-making (x8), and cause action (x9).
3. The advertising media variable utilized four measurement indicators, i.e., broadcast quality (x10), media popularity (x11), ad frequency (x12), and media coverage (x13).

Out of 130 questionnaires, 102 questionnaires were considered worthy to process while the study only took 100 students as respondents.

3 RESULTS

3.1 *Descriptive analysis*

This analysis aimed to get the respondents' answers concerning the variables studied using an index analysis technique that described the respondents' perceptions of the items addressed in the questions posed. The scoring technique used in this research used a minimum score of 1 and a maximum score of 10, while a calculation of the respondents' answer index was then conducted with the following formula (Amin Bhat & Kensana, 2016 and Chen & Holsapple, 2013):

$$\text{Index score} = ((\% \, F1x1) + (\% \, F2x2) + (\% \, F3x3) + (\% \, F4x4) + (\% \, F5x5) + (\% \, F6x6) + (\% \, F7x7) + (\% \, F8x8) + (\% \, F9x9) + (\% \, F10x10))/10$$

Where:
F1 = frequency of respondents who answered 1
F2 = frequency of respondents who answered 2
..., F10 = frequency of respondents who answered 10

Using the three-box method, the range of ninety was then divided by three, resulting in a range of thirty, which was then used as the basis in interpreting the index value, as follows:
10.00 – 40.00 = low
40.01 – 70.00 = medium
70.01 – 100.00 = high

3.2 *Endorser variable*

The calculation of the index value for the endorser variable can be seen in Table 1.

Table 1. Perception index of endorser variable.

Variable	Percentage of respondents					Frequency answer					Index
	1	2	3	4	5	6	7	8	9	10	
XI	4	15	19	10	10	13	16	1	1	1	43.8
X2	5	18	21	16	16	11	9	1	1	1	39.9
X3	4	14	20	18	18	20	3	2	2	2	43.7
X4	5	16	17	12	12	20	4	5	1	1	43.1
AV											42.625

From Table 1, we can see that the endorser variable has an index of 42.625, from which we can conclude that the endorser variable belongs in the medium category.

Some respondents' answers indicate that their rating of the endorser of this shopee ad index is in the moderate category.

3.3 *Ad message variable*

The calculation of the index value for the ad message variable can be seen in Table 2.

Table 2. Perception index of ad messages variable.

Variable	Percentage of respondents					Frequency answer					Index
	1	2	3	4	5	6	7	8	9	10	
X5	8	7	11	10	15	18	7	4	10	10	54.9
X6	5	6	9	13	17	17	8	8	7	10	56.6
X7	2	13	17	15	22	6	19	3	2	1	47.0
X8	3	5	5	5	13	19	20	15	12	3	62.5
X9	3	6	10	10	6	15	17	20	8	5	60.6
AV											56.32

Table 2 shows that the advertising message variable has an index value of 56.32, from which we can conclude that the advertising message variable is in the medium category.

The statement of respondents' perceptions supports the quantitative analysis, which states that although the respondents were still considering the advertising message, the advertising message was sufficient so that the respondents decided to try to shop online using shopee.

3.4 *Ad media variable*

Table 3. Perception index of ad media variable.

Variable	Percentage of respondents					Frequency answer					Index
	1	2	3	4	5	6	7	8	9	10	
X10	4	3	3	4	9	13	15	14	20	15	70.5
X11	5	3	2	4	3	11	18	21	17	16	71.7
X12	4	2	2	3	2	14	20	15	18	20	74.2
X13	5	1	1	4	5	16	17	19	12	20	72.6
AVG											72.25

Table 3 shows that the advertising media variable has a perceptual index value of 72.25, from which we can conclude that the variable of advertising media is in the high category.

The description analysis of the statement of respondents' perception is in line with the result of the quantitative analysis, which states that the advertisement media variable is perceived high by the respondents. It is not uncommon for people in different regions of this country to do online shopping using shopee, and it depends on the coverage area of television.

4 DISCUSSION

This study identified three factors that affect the impact of e-commerce advertising. This study then measured the influence of these three factors on the effectiveness of e-commerce advertising. These three factors were ad endorsers, advertising messages, and advertising media. Further discussion is required to examine how the effectiveness of an advertisement may impact the interest and activity of the advertisement viewers in buying advertised products. From other studies, we can see that it is important to also measure the influence of these three factors, including the duration of the ad, the time of the advertisement, and the communication of the advertisement.

5 CONCLUSION

This study, using empirical data, proved that the factors that influence the effectiveness of an advertisement, especially in e-commerce, are the endorser, the advertising message, and the advertising media. The selection of the endorser has a positive impact on the effectiveness of ads in the medium category. This proves that the endorser is one of the factors consumers consider when evaluating the content conveyed in an advertisement. The result shows that the advertising media variable has a perceptual index value of 72.25, from which we can conclude that the advertising media variable is in the high category. Ad messages have a positive effect on the impact of ads in the moderate category, but the attractiveness of advertising is very important as it will improve the communication with the audience. Advertising media have a positive effect on the effectiveness of ads in the high category because the media messenger plays an important role in the communication process. Without media, messages will not reach the audience. Therefore, the selection of appropriate media will greatly determine whether the message to be conveyed can reach the target.

REFERENCES

Amin Bhat, S., & Kensana, K., 2016. A review paper on e-commerce. *Asian Journal of Technology & Management Research 6 (1) pp 16–21.*

Corvi, E., & Bonera, M. 2010. The effectiveness of advertising: A literature review. 10th Global Conference on Business & Economics, Rome, Italy.

Chen, L., & Holsapple, C. W. 2013. e-Business adoption research: State of the art. *Journal of Electronic Commerce Research 14 (3) pp. 261–286.*

Kotler, P. 2000. *Marketing management: Analysis, planning, implementation, and control.* 10th edition. Upper Saddle River, NJ: Prentice Hall.

Shahriari, S., Shahriari, M., & Gheiji, & S. 2015. e-Commerce and its impacts on global trends and markets. *International Journal of Research 3 (4) pp. 49–55.*

Managing Learning Organization in Industry 4.0 – Rachmawati & Hendayani (eds)
© *2020 Taylor & Francis Group, London, ISBN 978-0-367-81920-0*

Measuring the entropy of organizational culture using agent-based simulation

A. Rahman, F. Naufal & S.G. Partiwi
Department of Industrial Engineering, Institut Teknologi Sepuluh Nopember (ITS), Surabaya, Indonesia

ABSTRACT: Organizational culture became one of the main fundamental aspects stated in the strategy of the organization. Unfortunately, most of the culture managers are still fuzzy while measuring the gap of culture implementation and often unable to identify the culture entropy as well. Measuring the organization's culture must be very challenging and close to a subjective tendency of the result. The indicator of each value depends on the intangible responses of the employee; the differences in individual behavior will also drive to the variation of action among the employees. This research applies the human-organization interaction analysis in order to simulate the personal responses while facing some probability actions or decisions related to some specific indicators of integrity. Integrity is the values to be studied by identifying several behavior actions that will present the agent's behavior in simulation. Eight behavior aspects of integrity are built using agent-based modeling; those are honesty, openness and transparency, sense of responsibility for/toward others, abide by rules and regulations, consistency, stability of personal morality, word-action behavior, fairness and justice. Five hundred seventy eight employees of ITS have been involved in culture survey. Based on the year simulation, ITS has a moderate order of entropy for Integrity value implementation. Several scenarios are applied in order to get the pattern of integrity implementation and to propose some opportunities for improvement.

1 INTRODUCTION

Organizational culture has a strong relationship with personal's shared value in the organization. Measuring the culture, we must consider the complexity of people's competencies and behavior, and cope with some extents of uncertainty (Bo & Luoyo, 2008). Organizational culture should be designed as social control by disseminating the norms and values which hold the outcome of the organization (Guiso, et al., 2015). Organizational culture is several patterns of underlying assumption which is formulated and developed as a behavioral patron for all members of the organization (Quinn & Cameron, 2005). Organizational culture might glue the internal member and bring the high spirit of external adaptation (Quinn & Cameron, 2005). Organizational culture must be owned, carried out, maintained, and evaluated systematically.

Organizational values should be defined and designed consciously. Unconscious culture design might bring some dysfunctionalities and disorder of interaction among individuals or groups (Guiso, et al., 2015). However, formulating values which are appropriate with organizational strategy and approaching individual values require a sophisticated approach of assessment. The values, beliefs, and behaviors embedded in each individual are abstract and vague (Soyer, et al., 2007). Even though some previous qualitative and quantitative approaches to evaluate culture have been published, but the opportunity of research to get the fittest and future-oriented formulation is still widely required.

This paper proposes agent-based simulation to measure and predict the entropy phenomena in the organization. Culture evaluation must involve human interaction in the organization. Since the interaction of humans is complex and dynamic, presenting the model of the behavior of humans using agent-based modeling should be more relevant in culture study (Bakhtizin, 2013). Organizational value assessment will be shown more in-depth analysis while the analysis will be run in an extended period. Agent-based modeling provides flexibility in order to analyze the future patterns of culture program implementation.

2 LITERATURE REVIEW

2.1 *Integrity and cultural entropy*

Integrity is the most significant value that makes the organization bigger. Integrity also has a correlation with financial performance and attraction of the organization, but a contradiction with the degree of unionization (Guiso, et al., 2015). Integrity is a complex construction in psychological aspects (Barnard, et al., 2008). Many organizations declare integrity as one of the core values (Barnard, et al., 2008). Integrity has links to various domains of industrial and organizational psychology such as leadership, organizational dynamics, employee welfare, and employee selection (Leroy, et al., 2012). Some weaknesses of integrity study are related to the validity of the integrity construction method. A systematic approach by emphasizing the human behavior and interaction will improve the organization strategic plan (Barnard, et al., 2008).

Cultural entropy has a strong correlation with the disorder condition inside the organization. The higher dysfunctionality an organization has, the higher culture entropy will be. Cultural entropy may reflect the unnecessary energy which must be brought to the organization since non-value-added function or behavior is still prominent (Barrett, 2010). In organizations with high cultural entropy, most of available energy will be absorbed for non-value added operations and behavior, so that the performance becomes low (Barrett, 2010).

Cultural entropy assessment must consider the survival capability of the organization to deal with any changes in the future (Martínez-Berumen, et al., 2014). Cultural entropy is one of risk indicators to ensure long-term sustainability. The high cultural entropy means the low people's knowledge about organizational systems and the high risk of organizational sustainability.

(Martínez-Berumen, et al., 2014) proposed an approach to calculate cultural entropy (S) as in the formula below. Interpretation of cultural entropy level is described in Table 1.

$$S = -\sum\nolimits_{i=1}^{N} p_i \ln(p_i) \tag{1}$$

In which p_i is the cumulative probability of occurrence for each of the scenarios/parameters. The probability density function is set following the normal distribution of data.

In which K is the number of accessible states while ln(K) is the maximum cultural entropy. As proposed by (Martínez-Berumen, et al., 2014), states or scenario of risk can be set from 0 (the lowest risk) up to 10 (the highest risk).

Table 1. Cultural entropy interpretation (Martínez-Berumen, et al., 2014).

S (entropy)	Interpretation
$\frac{3}{4}\ln K < S \leq \ln K$	Highly disordered. The results of the evaluation of organizational sustainability are highly variable.
$\frac{1}{2}\ln K < S \leq \frac{3}{4}\ln K$	Ordered. The results of the evaluation of organizational sustainability are mainly concentrated around a value.
$0 < S \leq \frac{1}{2}\ln K$	Highly ordered. The results of the evaluation of organizational sustainability are grouped around a common point.

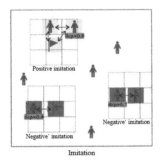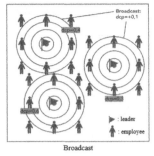

Figure 1. Interaction between employee and the leader.

2.2 *Human-organization interaction*

Broadcasting is one of the interactions between employees in the organization. Performing an intensive broadcasting program may support the organization to cultivate the values and keep the lower entropy of organizational culture. Broadcasting can be arranged with a formal or non-formal activity such as meeting, non-formal discussion, briefing, and interaction using social media. The more frequent broadcasting action is performed, the more effective cultivation of values process will be. Imitation behavior can lead to positive changes or vice versa. A strong influencer employee with positive values must be promoted as an agent of change. But a strong influencer with negative values should be directed and trained so that he or she will not give bad influence to the others. Figure 1 depicts those two interactions behavior in the organization.

3 METHODOLOGY

The study of organizational culture will involve human interaction on a complex scale. Based on previous research, the characteristics and action behavior of integrity have been presented in Table 2. By associating the action behaviors, the questionnaire scorecard is built.

An agent-based model is developed to simulate the various comprehension of the integrity of employees in the organization. Two agents, those are leader and employee, are created in the agent-based simulation. The agent leader performs all integrity characteristics and broadcasting and imitation behavior, while the agent employee performs the same characteristic and behavior, except broadcasting.

This entropy culture evaluation is applied in our university named Institut Teknologi Sepuluh Nopember (ITS) in Surabaya, Indonesia. The integrity survey was conducted using an online questionnaire and accessed by the employee. The questionnaire is designed with 1-5 scales of statements. The total number of respondents who filled the questionnaire is 578 staff. 86% of those respondents have been working for more than five years, 12% have been working between 1-5 years, and the rest have one year work experience. Management of organization is represented by 98 feedbacks and is considered as the leader's values.

4 RESULTS AND DISCUSSION

4.1 *Integrity and interaction characteristic*

Integrity is one of the focuses of corporate culture study since most of the organizations put integrity as their core values. The action behavior for each integrity characteristic is defined as the technical parameter embedded in the simulated agent. Integrity action behavior is also deployed into behavior questions for integrity questionnaire design. Many publications related to integrity have been discussed, and this research has summarized that behavior into eight main characteristics, as presented in Table 2. The interaction behavior, i.e., broadcasting and imitation are also defined in Table 2.

Table 2. Integrity characteristic and action behavior (Sihombing, 2018) (Barnard, et al., 2008) (Martin, et al., 2013).

Integrity characteristic	Integrity action behavior	Integrity characteristic	Integrity action behavior
1. Honesty	Telling the truth and fact Not telling a lie, cheating nor stealing Keeping his word	6. Guided by strong personal moral code/ value	Keeping faith Fulfilling commitments and even exceeding standards of performance Having clear values
2. Openness and transparency	Transparency in dealing with people Not being two-faced Loyalty to all stakeholders	7. Word-action behavior	There is no discrepancy between what you say and actually do Accountable Not doing one thing today and something else the next day
3. Sense of responsibility	Supporting and developing subordinates Accepting a system of rules Putting interests of the organization ahead of personal interests	8. Fair and Just	Equitability and no bias in one's decision making Fair in dealing with others Personal consequence if decisions contradict one's own opinion
4. Abides by rules and regulation	Complying with laws even when no one is looking Living the defined corporate guidelines and laws Attaining what one has set out to do	9. Broadcasting 10. Imitation	Personal motivation spirit Meeting frequency Affected by other's behavior and attitude
5. Value behavior consistency	Consistency of actions, beliefs, moral, and ethical standards Doing the right thing whether someone is looking or not Courage		Duplicate the positive value which has been shown or stated by others Tend to disobey the procedure since the others have been showing the violation

4.2 *Agent-based modeling*

Interaction among employees and the leader in a defined environment has been built. Interaction should be evaluated in order to measure the entropy of integrity. Both the agent leader and agent employee may be evaluated using eight integrity characteristics. The Netlogo, an open source agent-based simulator, has been utilized to develop the integrity value in the organization. Eighty agents are set in to simulate 260 working days interaction among the agents. There are eight steps of evaluation for each agent during the simulation with a random probability of appearances. The evaluation is represented as the 8 integrity characteristics, as depicted in Figure 2. The tree decision diagram describes the step of integrity test.

4.3 *Cultural entropy measurement*

Based on the integrity survey, the initial value of integrity and the probability of P'(xi) as the representation of disordered action is provided in Table 3. The value of integrity tends to be close to the high order of integrity. The initial value shows that integrity characteristic has already met the ideal comprehension in the organization. Furthermore, agent-based simulation is run by applying the probability density function of each integrity characteristic and the behavior of interaction.

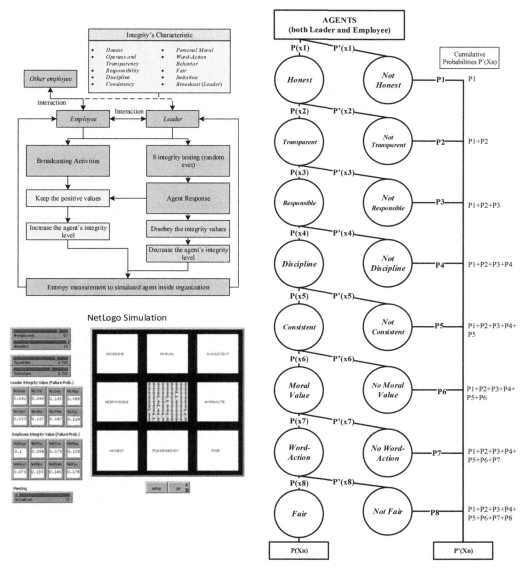

Figure 2. Conceptual Agent Modeling: agent interaction (top left), NetLogo interface (bottom left), and tree of integrity (right).

Table 3. Probability of integrity characteristic proceeded from integrity survey.

Integrity Value	Leader	Employee	Integrity Value	Leader	Employee
Honest	0.032	0.073	Moral Value	0.127	0.155
Transparency	0.046	0.098	Word-act Behavior	0.147	0.186
Responsibility	0.088	0.128	Fair	0.114	0.178
Dicipline	0.061	0.100	Imitation	0.795	0.701
Consistent	0.165	0.179	Broadcasting	0.123	0.181

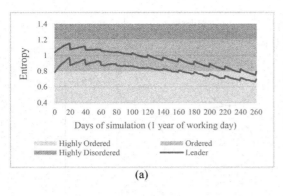
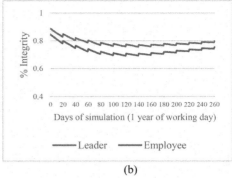

(a)   (b)

Figure 3.   (a) Simulation result for cultural entropy and (b) integrity culture pattern for a year.

Cultural entropy is evaluated by following (Martínez-Berumen, et al., 2014)'s approaches. The interpretation of cultural entropy value may refer to Table 1 in which the number of state (k) is set equal to 5. Range of cultural entropy are divided into 3 condition, those are highly ordered ($0 < S \leq 0.805$), ordered ($0.805 < S \leq 1.207$), and highly disordered ($1.207 < S \leq 1.609$).

One year or 260 working days of agent-based simulation has been run using the Netlogo simulator, as depicted in Figure 3. Integrity culture pattern declines until in the middle of the year; nevertheless, it will rise slowly.

5   DISCUSSION

The integrity characteristics are possible to be evaluated by applying the agent-based simula-tion. The initial integrity performance has been presented on the agent tree of integrity. Agent-based simulation has enabled the prediction of integrity performance. The trend of integrity performance can be analyzed using a specific period of simulation time. Some scen-arios of improvement in culture might be applied by modifying the targeted parameters or behavior of agents. The flexibility of evaluation is offered by this approach by identifying the behavior of the studied agent.

Some scenarios have been applied in order to observe the integrity characteristic in ITS. Leader's negative behavior has significant influence, which makes the entropy of organization increases faster. The positive behavior of leaders may bring a positive impact to reach lower entropy. The role of the leader in spreading positive values is significantly important in order to foster employee integrity. Furthermore, this paper also examines the broadcasting fre-quency by simulating the number of broadcasting. The more frequent the number of broad-casting, the lower cultural entropy will be performed. The every two-week broadcasting is proposed as the optimum broadcasting policy since this policy offers lower entropy taking a faster period of change.

6   CONCLUSIONS

The measurement of organizational culture needs to take into account the complexity of human interaction with various behaviors. Agent-based modeling must involve the inter-actions and behaviors with probabilistic and dynamic parameters. Eight integrity characters have been formulated and simulated to measure the level of integrity of an organization. Cul-tural entropy emphasizes more on the evaluation of organizational strategies and policies. The measurement of cultural entropy provides some indication of functional disorder and organizational risk level. By applying agent-based simulation, various scenarios can be arranged expertly by reducing cultural entropy.

REFERENCES

Bakhtizin, A. R., 2013. Agent Based Modeling of Integration of Organizational Cultures in Mergers and Acquisitions. *Advances in Systems Science and Application*: 333–353.

Barnard, A., Schurink, W. & De Beer, M., 2008. A conceptual framework of integrity. *SA Journal of Psychology*, 34(2): 40–49.

Barrett, R., 2010. *High Performance It's all about entropy*. [Online] Available at: [Accessed 3 February 2019].

Bo, Y. & Luoyo, L., 2008. *Corporate Culture Quantitative Measurement based on Discrete Choquet Fuzzy Integral*. Beijing, IEEE International Conference on Service Operations and Logistics, and Informatics.

Guiso, L., Sapienza, P. & Zingales, L., 2015. The value of corporate culture. *Journal of Financial Economics*, Volume 117: 60–76.

Leroy, H., Palanski, M. & Simons, T., 2012. Authentic Leadership and Behavioral Integrity as Drivers of Follower Commitment and Performance. *Journal of Business Ethics*, 107(3): 255–264.

Martínez-Berumen, H. A., López-Torres, G. C. & Romo-Rojas, L., 2014. *Developing a Method to Evaluate Entropy in Organizational System*. California, Procedia Computer Science.

Martin, G. S. et al., 2013. The meaning of leader integrity: A comparative study across Anglo, Asian, and Germanic cultures. *The Leadership Quarterly*: 445–461.

Quinn, R. E. & Cameron, K. S., 2005. *Diagnosing and Changing Organizational Culture*. Revised ed. San Francisco: Jossey-Bass.

Sihombing, S. O., 2018. Youth perceptions toward corruption and integrity: Indonesian context. *Kasetsart Journal of Social Sciences*: 299–304.

Soyer, A., Kabak, O. & Asan, U., 2007. A fuzzy approach to value and culture assessment and an application. *International Journal of Approximate Reasoning*, Volume 44: 182–196.

Disruption of the workforce in the digital era: A smart store case study

A.I. Munandar & B. Albab
School of Strategic and Global Studies, University of Indonesia, Jakarta, Indonesia

ABSTRACT: Digital development has been destroying the workforce in the retail sector. This research aimed to analyze the causes and effects of labor disruption in the retail sector. This research employed a qualitative method by observing and interviewing employees of retail smart stores. The fishbone analysis technique was conducted so as to describe the digital era's causes and effects on the workforce. The results showed labor disruption in the fields of administration, security, and payment. The labor disruption included a number of disappearing professions, termination of employment, and a necessity for digital skills. Employees need to reskill, upskill, and improve their information technology (IT) skills.

1 INTRODUCTION

Indonesia has entered the era of a digital economy based on computing, information technology (IT), and digital communication. Indonesia is one of the countries with the fastest-growing digital infrastructure in the world; its digital economy is estimated to reach USD 200 billion in 2025 (Azali, 2017). Conventional retail shops in many countries have adopted technology because they have encountered a number of tough challenges: increasingly smart and critical consumers, globalization that attracts more players to the competition arena, and intensifying competition among retail entrepreneurs (PWC, 2017). The tight retail business competition forces each company to adapt and make continuous efficiency improvements. Companies who desire to continue to grow require solid business operations and innovation (Mou, Robb, and DeHoratius, 2018). Retail stores rely on labor-intensive concepts. They can absorb millions of workers, contribute to the advancement of the retail industry, and help grow the national economy (Amin, 2015; Cho, Rutherford, and Park, 2013). One of the developed innovations in the digital era is the smart store. The existence of a smart store in Jakarta is an indication of the effort to implement digital technology in a store's operations. The application of digital technology has an impact on the economy, wages, labor, and tax revenues (King, Hammond, and Harrington, 2017).

Previous researchers have carried out studies on technology disruption (Carvalho et al., 2018; Kostoff, Boylan, and Simons, 2004; Lucas and Goh, 2009; Schuelke-Leech, 2018; Sousa and Rocha, 2019). Another study on disruption was conducted by Kahn and colleagues (2014) in the academic field, Orsi and Santos (2010) on the workforce, and Mazurowski (2019) on radiology labor, and yet studies on retail labor disruption are still quite small, especially related to the industrial revolution 4.0. Other workforce studies focus more on labor perceptions (Brougham and Haar, 2018), organizational performance (Brauer and Laamanen, 2014; Soekiman et al., 2011), wages (Magruder, 2013), and macroeconomics (Cohen and Rettab, 2010; Kis-Katos and Sparrow, 2015; Peluso, 2018). This study concentrated on retail workforce disruption within the current digital era through the case studies of smart stores.

2 LITERATURE REVIEW

Retail stores require workers who prioritize the best service for consumers. Thus, the retail workforce receives training in the art of serving consumers, philosophies, rules, strategies, and promotions, through to company policies (Coulter, 2014). The retail industry is a labor-intensive industry that depends on human labor. Retail will not exist without human labor, and retail needs human labor on an ongoing basis for transactions to continue. But the development of the industrial revolution of human labor began to be corrupted by technology and production methods (Stearns, 2018). Improved technology will affect macroeconomics, labor, the evolution of business models, markets, and industries (Hu et al., 2019; King et al., 2017; Lojeski, 2009; Schuelke-Leech, 2018; Toner, 2011; Valenduc and Vendramin, 2017). In the digital age, retail store operations encounter new challenges and complexities (Mou et al., 2018). Industry 4.0 is a form of manufacturing digitalization that generates a complete change in manufacturing processes, results, business models, services, and work organizations relying on sensor technology, interconnection between telecommunications networks of different telecommunications network providers, and data analysis that enables mass customization, value chain integration, and greater efficiency (Schwab, 2017). Industry 4.0 utilizes smart factories, the Internet of Things (IoT), smart industries, and the internet industry to create sophisticated manufacturing aimed at achieving higher levels of efficiency, operational productivity, and automation.

3 METHODOLOGY

This research employed a qualitative method with a case study of smart stores located in Jakarta. Researchers conducted observations and expert interviews in order to gain primary data. The informants in this study were (1) a smart store manager, (2) a data scientist, (3) technology practitioners, (4) smart store customers, (5) the Indonesian Ministry of Manpower, (6) the Indonesian Digital Entrepreneurs' Association, (7) the Indonesian Employers' Association (APINDO), and (8) academics for a total of twenty-five people. In addition to interviews, observations were also conducted. Observation is a method of collecting data related to a particular event or area through intensive observation of the environment and the people around it (Prunckun, 2014). Observations were carried out by HUMINT (human intelligence) operational techniques for gathering intelligence information that uses human resources, and IMINT (image intelligence) to gather information using cameras to take photographs. Data triangulation was used to strengthen the evidence and to test the validity of the data by combining information from various sources to obtain diverse perspectives. The fishbone analysis technique, or Ishikawa diagram, is a qualitative method that looks at patterns of structured and causal relationships. This analytic technique does not identify symptoms but focuses on the "causes" instead that contribute to the "effects."

4 RESULT AND DISCUSSION

The retail industry has long utilized a variety of digital technologies so as to make it easier for companies and workforces to run day-to-day shop operations, especially providing shopping for their customers. One of them is the JD.ID smart store using the convenience store format with the depth of merchandise concept. The store area is only 270 square meters and is almost the same as the area of the conventional minimarket, which ranges between 150 and 300 square meters. The JD.ID smart store offers multiple categories consisting of various products, ranging from nondurable and consumable products (soft goods) such as snacks, soft drinks, shampoos, liquid soap, baby diapers, and cosmetics, to durable products (durable goods) such as clothes, bags, pillows, bolsters, data cables, and power banks.

The results proved that the application of technologies eliminates a number of professions such as cashiers, security officers, and administrative staff. The technology that takes over the role of cashiers in the payment department is a smart cash register consisting of artificial intelligence, cameras, infrared sensors, computer vision or object recognition, and smartphones. The technology that replaces the role of human security personnel in the security sector is a face recognition system, door scanner, QR code, camera, and smartphone. The technology that eliminates the role of store and warehousing administration staff is big data. These types of technologies collaborate to balance an employee's capacity and replace each of the five human senses.

Based on the fishbone analysis, we divided the causes of the problem into three parts – namely payment, security, and administration – that can be seen in Figure 1. The initial signal of the transformation of digital technology is important to note because it interrupts the workforce in modern retail stores. If human resources are reluctant or late to adapt to change, they are unable to compete. Technology will automatically be estimated to eliminate routine and cognitive manual types of retail jobs (King et al., 2017) that are usually filled by high school/vocational high school graduates such as cashiers, salespeople, shop supervisors, inventory keepers, security officers, customer service providers, general managers, and operational managers, who will slowly disappear.

Automation has cut industry's demand for human labor. These changes will cause workers with low skills who do not get training from companies to be displaced slowly. Workers make a living in the informal sector after losing their jobs. The results of the fishbone analysis explain the causes of corruption in the digital age, as shown in Figure 2. Technological

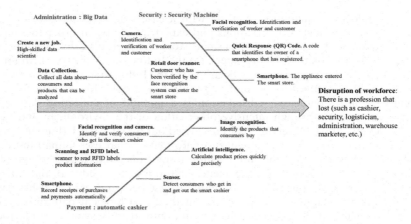

Figure 1.   Fishbone disruption of workforce in smart store.

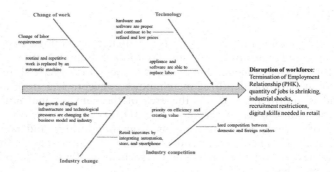

Figure 2.   Fishbone analysis of the disruption of the workforce in the digital age.

118

changes affect business events (King et al., 2017), also bringing significant changes to the way we work, and determining what work will still be available to human employees in the very near future (Brougham and Haar, 2018). An estimated one-third of current jobs will be replaced by smart technology, artificial intelligence, robotics, and algorithms (STARA) by 2025 (Brougham and Haar, 2018).

5 CONCLUSION AND RECOMMENDATIONS

Smart stores require a lot of technology to replace the human workforce. A smart cash register that replaces cashiers at the payment stage consists of technology such as (1) face and camera recognition systems, (2) RFID scanners to read RFID labels, (3) image recognition, (4) infra-red sensors, (5) artificial intelligence, and (6) smartphones and applications. Security machines that replace security personnel at the security stage consist of technology such as (1) face recognition systems, scanner doors, smartphones, and QR codes, and (2) cameras, and big data replaces administrative and warehousing personnel.

The presence of smart stores impacts the labor force in the form of termination of employment, eliminating repetitive routine jobs, reducing permanent jobs, industrial shocks, and reducing the recruitment of new employees. Recommendations can be given in the form of (1) government, employers, associations, and other stakeholders that provide a variety of formal and informal education to enhance new skills (reskilling) and additional skills (upskilling) to the retail workforce; (2) retail technology transfer policies that imitate the working models of the digital era; and (3) an IT-based curriculum.

REFERENCES

Amin, M. 2015. Competition and labor productivity in India's retail stores. *Journal of Asian Economics* 41: 57–68. doi: 10.1016/j.asieco.2015.10.003

Attaran, M., Attaran, S., & Kirkland, D. 2019. The need for digital workplaces: Increasing workforce productivity in the information age. *International Journal of Enterprise Information Systems (IJEIS)* 15(1): 1–23.

Azali, K. 2017. Indonesia's divided digital economy. *Perspective* 70: 1–12. Available at: http://setkab.go.id/inilah-perpres-no–74–.

Brauer, M. & Laamanen, T. 2014. Workforce downsizing and firm performance: An organizational routine perspective. *Journal of Management Studies* 51(8): 1311–1333. doi: 10.1111/joms.12074

Brougham, D. & Haar, J. 2018. Smart technology, artificial intelligence, robotics, and algorithms (STARA): Employees' perceptions of our future workplace. *Journal of Management and Organization* 24(2): 1–19. doi: 10.1017/jmo.2016.55

Carvalho, N. et al. 2018. Manufacturing in the fourth industrial revolution: A positive prospect in sustainable manufacturing. In *Procedia manufacturing*. New York: Elsevier, 671–678. doi: 10.1016/j.promfg.2018.02.170

Cho, Y. N., Rutherford, B. N., & Park, J. K. 2013. Emotional labor's impact in a retail environment. *Journal of Business Research* 66(11): 2338–2345. doi: 10.1016/j.jbusres.2012.04.015

Cohen, S. I. & Rettab, B. 2010. Institutional barriers in labor markets: Examples, impacts, and policies. *Socio-economic Planning Sciences* 44: 93–198. doi: 10.1016/j.seps.2010.07.001

Coulter, K. 2014. *Revolutionizing retail: Workers, political action, and social change*. New York: Springer.

Hammond, R. 2017. *Smart retail: Winning ideas and strategies from the most successful retailers in the world*. London: Pearson.

Hu, X., Zhao, J., Li, H., & Wu, J. 2019. Online footprints of workforce migration and economic implications. Physica A: Statistical Mechanics and its Applications, 528, 121497. doi:10.1016/j.physa.2019.121497

Kahn, M. J., Maurer, R., Wartman, S. A., & Sachs, B. P. 2014. A case for change: disruption in academic medicine. Academic Medicine, 89(9), 1216–1219. doi: 10.1097/ACM.0000000000000418

King, B. A., Hammond, T., & Harrington, J. 2017. Disruptive technology: Economic consequences of artificial intelligence and the robotics revolution. *Journal of Strategic Innovation and Sustainability* 12(2): 53–67. doi: 10.33423/jsis.v12i2.801

Kis-Katos, K. & Sparrow, R. 2015. Poverty, labor markets and trade liberalization in Indonesia. *Journal of Development Economics* 117: 94–106. doi: 10.1016/j.jdeveco.2015.07.005

KOMINFO. 2019. *Laporan Kinerja Kementerian Komunikasi dan Informatika 2018.* Jakarta: KEMENTERIAN KOMUNIKASI DAN INFORMATIKA REPUBLIK INDONESIA.

Kostoff, R. N., Boylan, R., & Simons, G. R. 2004. Disruptive technology roadmaps. *Technological Forecasting and Social Change* 71(1–2): 141–159. doi: 10.1016/S0040-1625(03)00048-9

Levy, M. & Weitz, B. A. 2012. *Retailing management.* 8th edition. New York: McGraw-Hill Irwin.

Lojeski, K. 2009. *Leading the virtual workforce: How great leaders transform organizations in the 21st century [Microsoft executive leadership series].*

Lucas, H. C. & Goh, J. M. 2009. Disruptive technology: How Kodak missed the digital photography revolution. *Journal of Strategic Information System* 18(1): 46–55. doi: 10.1016/j.jsis.2009.01.002

Magruder, J. R. 2013. Can minimum wages cause a big push? Evidence from Indonesia. *Journal of Development Economics* 100: 48–62. doi: 10.1016/j.jdeveco.2012.07.003

Mazurowski, M. A. 2019. Artificial intelligence may cause a significant disruption to the radiology workforce. *Journal of the American College of Radiology* 16(8): 1077–1082. doi: 10.1016/j.jacr.2019.01.026

Mou, S., Robb, D. J., & DeHoratius, N. 2018. Retail store operations: Literature review and research directions. *European Journal of Operational Research*: 399–422. doi: 10.1016/j.ejor.2017.07.003

Orsi, M. J. & Santos, J. R. 2010. Probabilistic modeling of workforce-based disruptions and input-output analysis of interdependent ripple effects. *Economic Systems Research* 22(1): 3–18. doi: 10.1080/09535311003612419

Peluso, N. L. 2018. Entangled territories in small-scale gold mining frontiers: Labor practices, property, and secrets in Indonesian gold country. *World Development* 101: 400–416. doi: 10.1016/j.worlddev.2016.11.003

Piotrowicz, W. & Cuthbertson, R. 2014. Introduction to the special issue information technology in retail: Toward omnichannel retailing. *International Journal of Electronic Commerce*: 5–16. doi: 10.2753/JEC1086-4415180400

Prunckun, H. 2014. *Scientific methods of inquiry for intelligence analysis.* Lanham, MD: Rowman & Littlefield.

PWC. 2017. *10 Retailer investments for an uncertain future.* London. Available at: www.pwc.com/2017totalretail.

Santos, C. et al. 2017. Towards Industry 4.0: Sn overview of European strategic roadmaps. *Procedia Manufacturing* 13: 972–979. doi: 10.1016/j.promfg.2017.09.093

Schuelke-Leech, B. A. 2018. A model for understanding the orders of magnitude of disruptive technologies. *Technological Forecasting and Social Change* 129: 261–274. doi: 10.1016/j.techfore.2017.09.033

Schwab, K. 2017. *The fourth industrial revolution.* Geneva: World Economic Forum.

Soekiman, A. et al. 2011. Factors relating to labor productivity affecting the project schedule performance in Indonesia. *Procedia Engineering* 14: 865–873. doi: 10.1016/j.proeng.2011.07.110

Sousa, M. J. & Rocha, Á. 2019. Skills for disruptive digital business. *Journal of Business Research* 94: 257–263. doi: 10.1016/j.jbusres.2017.12.051.

Stearns, P. N. 2018. *The industrial revolution in world history.* Fourth edition. doi: 10.4324/9780429494475

Toner, P. 2011. Workforce skills and innovation: An overview of the major themes in the literature. *OECD Education Working Papers* 1: 1–74. doi: 10.1787/5kgkdgdkc8tl-en

Valenduc, G. & Vendramin, P. 2017. Digitalisation, between disruption and evolution. *Transfer: European review of labour and research* 23(2): 121–134. doi: 10.1177/1024258917701379

Management of sheep-fighting (*domba adu*) tourism in Rancabango village, Tarogong Kaler, Garut (case study of BUMDES, or village foundations)

D. Qoriah, M.D. Ungkari & H. Muharam
Faculty of Economy, Accounting, Universitas Garut, Garut, Indonesia

ABSTRACT: Tourism potential in the village of Rancabango, Tarogong subdistrict, Garut regency, is very fascinating because it includes the breeding and maintenance of Garut sheep, paraglide hills, the Gunung Putri camping ground, and the sheep-fighting competition. The large tourism potential is accompanied by problems that have not been resolved optimally. These problems include: (1) tourism management that has not been well managed (2) the involvement of the village foundation and local communities has not been organized (3) the youth and housewives' participation in the village is still very minimal. The purpose of this study was to find out how village management is applied to sheep-fighting (*domba adu*) tourism. The research method used was descriptive qualitative research, in order to analyze, explain, and describe it in more depth. Data were obtained through field and library research.

1 INTRODUCTION

Garut's tourism potential is very impressive, consisting of biodiversity, unique and authentic regional culture, *batik* as the beauty of landscapes and historical heritage, and the famous Garut sheep fighting (*domba adu*). The development of tourism has become a reliable sector for promoting economic activities, including other related sector activities. This is intended so that the development and utilization of various national tourism potentials can increase employment opportunities, local revenue, and national income.

In the Regional Regulation No. 29 2011, which concerns Garut regency, from 2011–2031 (1) part of its territory has been planned as a regency tourism strategic zone and as a regional promotion activity center. This anticipates provincial or multiple-district service and acts as a counterweight to the development of the provincial area, in the form of urban Rancabuaya in Caringin district.

The village foundation (BUMDES) is a pillar of social and economic development for responding to the challenges and potential of the current economic system. BUMDES is a foundation owned by the village government and managed by the village and/or inter-village cooperation in order to operate and develop activities in the economic field while at the same time carrying out an orientation to public service and social development so as to improve the welfare of rural communities. BUMDES are formed, managed, and developed jointly by the government and village communities and adapted to the characteristics of local resources and the culture of each village, facilitating, strengthening, maintaining sustainability, and developing productive business and economic activities that can be carried out by village communities. Development policy has also prioritized BUMDES as the manager of village-owned resources for the greatest possible welfare of the community while maintaining the role and existence of villages in regional and national development, including producing abundant rural tourism in the city of Garut.

Management means how a person or an institution manages, starting from planning, organizing, regulating involvement, etc., in order to take advantage of the optimally available potential of the surrounding community. As Caldito, *et al.* (2017) states, "management is a science and art of regulating the process of utilizing human resources and other resources in effective and efficient ways to achieve a certain goal."

The purpose of this study was to determine the impact of management on sheep-fighting tourism potential in the village of Rancabango, Tarogong subdistrict, Garut, especially on:

a. What local policies and strategies village-owned business entities (BUMDES) implement in order to manage tourism potential in their village.
b. How the effectiveness of management has been applied to sheep-fighting tourism.
c. How the local community engages with an implemented strategy.

The urgency of this research was to examine more deeply the management of tourism in the village concerned, so that exploration can be carried out on the factors of management effectiveness, community involvement strategies, and empowerment of local mothers and youth.

The focus of this research was the management of sheep-fighting tourism by village foundations in collaboration with the local community, and the implications of such management.

2 LITERATURE REVIEW

2.1 *Management*

Management is a systematic activity that is mutually proposed to achieve goals. Management of tourist areas is intended to protect the original values when the area is developed. Accommodation, human resources, service products, leadership, products, and packaging facilities should be carefully developed by adopting the original values and involving local residents. This tourism activity will have a positive impact on various aspects of life in the political, economic, social, cultural, and environmental fields. Social, economic, and cultural impacts will be directly felt by the people who reside in a tourist destination area. These social, economic, and cultural impacts include: (1) opening employment opportunities and expanding employment opportunities, (2) growing community economic activities, and (3) increasing the economic income of the community (Hasibuan, 2009).

Peraturan Pemerintah Republik Indonesia (2004) states that adopting a marketing approach in destination management means that specific tourism products are designed to meet the needs of those tourists targeted by the destination. This kind of management of a tourism destination uses SWAT analysis in order to determine the best management system.

Management indicators can be seen in the picture scheme that follows:

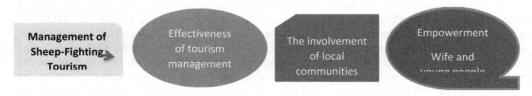

Figure 1. Indicators of tourism management.

While according to Peraturan Daerah Kabupaten Garut, management is a science and an art, management in a tourism village can be done well, which benefits the actors who have common goals, especially the local community. Management collaborates with all arts in order to be in harmony with the desires of human resources as well as the availability of their natural potential.

2.2 Village foundations (BUMDES)

Village foundation entities (BUMDES) are incorporated village businesses managed by the village government. The village government can establish a village foundation in accordance with the needs and potential of the village. The establishment of a village foundation is determined by village regulations. The management of a village foundation consists of the village government and the local village community.

The capital of a village foundation entity can come from the village government, community savings, assistance from the government, provincial government, or regency/city government, loans, or other parties' participation or profit-sharing cooperation on the basis of mutual benefits. Village foundation can make loans, which can be made after obtaining district supervision foundation (BPD) approval.

The establishment of a village-owned business entity is based on Law (6) Concerning Regional Government Article 213, paragraph (1), which states that "villages can establish village foundation entities in accordance with village needs and potential," and this mandate is also listed in Government Regulation (PP) no. 71 of 2005. The establishment of the village business entity is accompanied by efforts to strengthen capacity and be supported by regional policies (district/city) that facilitate and protect the business of rural communities from the threat of competition from large investors. Since village foundation entities are new economic institutions operating in rural areas, they still need a strong foundation to grow and develop. The building foundation for the establishment of BUMDES is the government, both central and regional.

3 METHODOLOGY

3.1 Research methodology

The type of research used in this research was qualitative descriptive research, to analyze, expound on, and describe more deeply the management of village tourism towards sustainable tourism development in village foundation (case studies of sheep-fighting tourism in Rancabango Tarogong Garut). The subjects used in this research were Rancabango Garut sheep-fighting tourism managers both from the government represented by the BUMDES and the local community. The object of this research was to examine the management of village tourism for sustainable tourism development.

In this study, researchers obtained primary and secondary data. Primary data are data in the form of words spoken orally, gestures, or behaviors performed by a subject that can be trusted – namely the subject of research or informants relating to the variables under study or data obtained directly from respondents Siswanto (2015). In this study the researchers obtained the results of observations made by the author and of a literature study. In a way, these secondary data could come from graphic documents such as tables, records, SMS, photos, and other sources Siswanto (2015). Secondary data in this study included documents about the people of Rancabango, documents regarding the sheep fighting, the BUMDES, and management, as well as sustainable tourism development. Data collection techniques in this study comprised interviews and documentation. In this study, researchers used semi-structured interview techniques. According to Sugiyono (2013), a semi-structured interview is an interview that is more free or open in implementation compared to structured interviews. The goal is not to stick to the interview guidelines, thus the problems can be discussed more openly. The documentation in this study was in the form of data collection profiles of Rancabango and of the local community.

4 RESULTS AND DISCUSSIONS

Based on a predetermined research framework, researchers conducted data mining through interviews, observation, documentation, and literature studies; the findings indicated that the management of sheep-fighting (*domba adu*) tourism potential has been well done and organized by involving the local community directly.

a. Effectiveness of tourism management

Based on the results of interviews, 80% of the sample said sheep fighting tourism is managed well. The village community in general, including mothers and young women, feel benefited by the existence of the sheep fighting tour, which is above 65%. So that from good management of sheep-fighting tourism, housewives in this village are not engaged in activities only as housewives. In addition to the sheep fighting management, this village is also supported by other tourism potentials as shown in what follows:

Table 1. Tourism potential of Rancabango Village, Tarogong Garut.

Tourism potential	Total
Hot Spring Tourism	1
Paragliding	1
Camping Ground	1
Ecotourism	1
Agricultural Tourism	1
Hotel	6
Restaurant	6
Villa	4
Swimming Pool	7
Waterpark	1
Sheep-Fighting Ground	1
Domba Iconic Area	3

b. Community involvement strategy

The strategy applied to involve the community in managing the tourism potential of sheep fighting in this village was carried out by forming a local community called RCB/Rancabango, which is intended to manage the sheep fighting. The RCB community itself manages sheep maintenance, sheep training, and sheep-fighting events to the extent of managing parking. Although the involvement of local communities is technically quite dominant, it still receives control and assistance from the village government and other parties, whereas village-owned business entities or BUMDES are not directly involved in managing sheep-fighting tourism in this village. The position of the BUMDES only supports and facilitates things needed at certain times.

c. Empowerment of youth and mothers

At a minimum, the livelihoods of young people and mothers in this village are progressing with the management of sheep-fighting tourism. With the growing recognition of sheep fighting, people from outside the village are increasingly exploring this village to fulfill their curiosity and to enjoy picnics with their families. The following are data on home industries in this village:

Table 2. The number of home industries in Rancabango.

Industri rumahan	Jumlah
Confection	2 families
Tailor	12 families
Workshop	4 families
Bag Craftsman	3 families
Food Craftsman	34 families
Art Craftsman	2 families
Shoes Craftsman	1 family

5   CONCLUSIONS AND RECOMMENDATIONS

Based on the foregoing discussion, we can conclude that management consisting of indicators of management effectiveness, community involvement strategies, and empowerment of youth and mothers in the village of Rancabango Tarogong has been done and implemented quite well thereby enabling them to realize sustainable tourism development, even though several things have not been used properly, especially the wealth of natural and tourism potential that has not been managed and just ignored. The management of sheep-fighting tourism in this village collaborates with various elements in the village to handle quite a lot of tourism potential. The strategy applied in involving the community in managing the tourism potential of sheep fighting in this village is carried out by forming a local community called RCB/Rancabango, which is intended to manage all aspects of the sheep fighting. The mothers and youth in this village are quite involved and empowered, even though not all of them are directly involved in the management of sheep-fighting tourism.

REFERENCES

Arikunto, S. 2016. *Prosedur Penelitian Suatu Pendekatan Praktik*. Jakarta: Rineka Cipta. Depatemen Pendidikan Nasional.
Caldito, L. A., Dimanche, F., Vapnyarskaya, O., and Kharitonova, T.. 2017. Tourism management. *Universidad De Extramadura* 62.
Hasibuan. 2009. *Manajemen Sumber Daya Manusia*. Edisi revisi cetakan ke tiga belas. Jakarta: PT Bumi Aksara.
Peraturan Daerah Kabupaten Garut. Lembar Daerah Kabupaten Garut nomor 74. Seri E.
Peraturan Pemerintah Republik Indonesia. 2004. No 32 tentang Desa.
Suryani, Noak P., Yudhartha, A. 2016. Analisis Manajemen Pengelolaan obyek wisata dalam Mewujudkan Pembangunan Pariwisata Berkelanjutan 1(1).
Siswanto. 2015. *Good corporate governance*. Tata Kelola Perusahaan yang Sehat. Aldrige pustaka.
Sugiyono, 2013. *Metodelogi Penelitian Kuantitatif, Kualitatif Dan R&D*. Bandung: ALFABETA.

Leadership type in youth organizations (case study of the Islamic Association of University Students [HMI])

Kurniana & A.I. Munandar
Department of Leadership Development, School of Strategic and Global Studies, University of Indonesia, Indonesia

ABSTRACT: The right leadership style determines organizational success. The leadership style is applied according to the needs of members and organizational situations. This study aimed to analyze the appropriate leadership style at the Islamic Association of University Students (HMI) in Indonesia. The design of research was simple descriptive. The method used was random sampling. Transformational leadership was dominant compared to transactional leadership. Islamic values support the transformation of cadres in HMI organizations.

1 INTRODUCTION

In an organization, the implementation of work programs is very dependent on the leadership style of the chairperson and his relation to members of the organization. Robbins and Judge (2013) define leadership as the ability to motivate, influence, and enable individuals who are members to contribute to organizational goals. Leadership is what leaders do (Robbins & Coulter, 2010). As with other definitions of leadership, the ability of a leader is to increase the effectiveness of a group (Northouse, 2013). This is in line with the American Management Association survey, which shows that leadership is the main determinant of successful change, followed by corporate values and communication (Gill, 2002). Effectiveness is not possible if leaders and members do not have good solidarity. The current leadership model tends not to be too formal-hierarchical but instead shifts to cooperative patterns (McDowell, Agarwal, Miller, Okamoto, & Page, 2016). Therefore, organizations must identify the right leadership style to increase the productivity of members in implementing work programs.

Many discussions have taken place on transactional and transformational leadership, for example, concerning the relationship between transactional leadership and work safety (Martínez-Córcoles & Stephanou, 2017). The study found that transactional leadership can influence safety performance behavior. Aga (2016) has also conducted research on the influence of transactional leadership as a tool (moderating role) to drive the success of projects in nongovernmental organizations (NGOs) in Ethiopia. Taylor (2016) conducted further research on the role of leadership style in water utility management in water companies in Australia.

Pieterse, van Knippenberg, Schippers, & Stam (2009) and Kidney (2015) have also conducted research on transactional leadership, while Dhaliwal and Hirst (2018) have performed research related to the relationship of transformational leadership and nursing. Alghamdi, Topp, and AlYami (2017) have examined the influence of gender on transformational leadership and job satisfaction among Saudi nurses, and Brewer and colleagues (2016), among others, have studied the effect of transformational leadership on the work of nurses.

However, despite all of these studies, no research has been conducted on Islamic youth organizations and their relation to the realization of work programs because the project and the work programs are not the same. Projects are work systems that may be formed by teams (within or throughout an organization) in order to complete certain tasks under certain time constraints (Manning, 2008). In other words, projects are temporary systems characterized by certain structural properties, specifically task specifications, and are more temporary, whereas work programs are a collection of related and sustainable projects, managed in a coordinated manner so as to obtain benefits that cannot be felt by managing them in part (Wideman, 2014).

2 TRANSFORMATIONAL AND TRANSACTIONAL LEADERSHIP

Cox (2001) lists two basic categories of leadership – namely, transactional and transformational. Downton first made the distinction in 1973, and other researchers began to develop it. Within the HMI organization, details of the dominant leadership style have not yet been verified. However, the HMI youth organization is a nonprofit organization where each member is not given a salary. Transformational leadership tends to prioritize relationship with interactive, visionary, passionate, caring, and empowering traits (Men & Bowen, 2017) not based on giving something (reward). Transformational leadership also focuses on four components: charismatic leadership or ideal influence, inspirational motivation, intellectual stimulation, and individual consideration (Cetin & Kinik, 2015). This is a powerful trick youth organization leaders often use to carry out work programs.

Tyssen, Wald, and Spieth (2014) state that transformational leadership is complementary to transactional leadership. In fact, transactional leadership is considered as a prerequisite needed for transformational leadership to be effective (Aga, 2016). Yukl (1999) has also identified several main weaknesses of transformational leadership. First is the ambiguity that emphasizes the influence and the process. This theory fails to explain the variables related to transformational leadership and positive work outcomes. Second is an excessive emphasis on theories about the leadership process in the realm of communication dyadic (dyadic level), so that the influence of leaders is focused on individuals rather than groups. Third is the lack of clarity of theoretical reasons why it is more rational to distinguish between behaviors that require cognitive stimuli. Another weakness is the insufficient specification of situational variables in transformational leadership, and this theory does not explicitly identify situations where transformational leadership cannot be effectively implemented. Therefore, it is necessary to explore and identify leadership styles among Islamic youth organizations in an effort to realize the goals of the organization.

3 METHODOLOGY

3.1 *Participants*

The sample for this study was taken from active members of the central district Islamic Student Association (HMI) in Jakarta, Indonesia, with ages ranging from sixteen to thirty years old according to the Law of the Republic of Indonesia Youth No. 40 of 2009 in chapter 1, article 1 in the first point. The total number of participants was 100 persons.

3.2 *Research design*

This research was a descriptive study because the research carried out aimed to retrieve direct information that was available in the field about leadership styles within the Islamic student community. Data collection in this study used questionnaires, interviews, and observations on the activities of the organization. Leadership style identification was done by referring to the characteristics of the transformational leadership style and the transactional leadership style base on Cetin and Kinik (2015), which included three of four components: charismatic leadership or ideal influence, inspirational motivation, intellectual stimulation, and individual consideration. The two characteristics examined for transactional leadership were contingent reward and management by exception.

3.3 *Data analysis*

This study employed descriptive analysis in order to analyze the data. Qualitative data on leadership styles in organizations was collected through questionnaire sheets, the Likert scale, interviews, and observations. The leadership-style questionnaire sheet contained nineteen statements about leadership style consisting of six parts, four parts on the characteristics of transformational leadership and two parts on the characteristics of transactional leadership based on the leadership of multifactorial form 6s., while interviews and observations were carried out by giving checklists according to the statements of members and the facts that exist in the field.

4 RESULTS AND DISCUSSION

The results of the questionnaire that was distributed to HMI members indicated that the dominant leadership style applied at HMI is transformational leadership. The answers are illustrated in the following table.

Table 1. The characteristic leadership style that frequently fits in HMI organizations.

Transformational leadership		Transactional leadership	
Idealized influence	3.3	Contingent reward	1.8
Inspirational motivation	3.5	Management by exception	2.8
Intellectual stimulation	3.5		
Individual consideration	3.3		
Average	3.4		2.3

The Likert scale measures both positive and negative responses to a statement. To find out the ideal influence on transformational leadership, several statements were formulated to be assessed based on the Likert scale – whether the leader could hold subordinates' trust, maintain their faith and respect, show dedication to them, appeal to their hopes and dreams, and act as their role model. The members stated that they felt comfortable because there is a spirit of equality that is applied in HMI organizations. The spirit of equality regardless of status is one of the Islamic values that is very visible in HMI organizations. On the other hand, the members are very qualified in science and religion, especially when discussing work agendas. This is what makes the members feel proud to participate in the organization.

Inspirational motivation measures the degree to which leaders provide a vision, use appropriate symbols and images to help others focus on their work, and try to make others feel their work is significant. This motivation derived from the experience of alumni who are always invited in HMI agendas. They can act as speakers or represent professionals. Those who have successfully served in the real world provide motivation for members to emulate what their seniors have done, so this method is effective in encouraging them.

Intellectual stimulation encourages others to be creative in looking at old problems in new ways. This point got the highest score on the questionnaire because the cadres feel they can find a new way and learn new knowledge in HMI organizations. They feel this after attending the training the organizations offer on a regular basis. Besides that, professional institutions can support the development of HMI organization cadres such as legal, educational, tourist, and art institutions, and others.

Individual considerations refer to how an organization pays attention to those who seem less involved in the group. In HMI organizations, instructors are tasked with looking at the development of each individual. The instructors will take notes and give them to leaders at every level of the HMI organization across the district and the province, through to the central management. They will be followed up according to the instructors' notes; for example, if a cadre feels ashamed of not being able to read the Qur'an, then the local leader will help him.

In measuring transactional leadership as according to Bass (2008), transactional leaders use the exchange model of giving rewards for good work or positive outcomes. The HMI organizations rarely give rewards such as a salary, a position, or similar compensation. This is because HMIs are not profit-oriented organizations. The members are taught by doctrine to carry out activities and work programs sincerely in accordance with their constitution, which is abbreviated to "grateful and sincere." This makes the cadres remain loyal to run the program even without material rewards.

5 CONCLUSIONS

The dominant leadership style applied in HMIs is the transformational leadership style. It is influenced by the values of Islam, which are taken from the Qur'an. It is formulated as the basic values of HMI organizations, which is stated in their constitutions. That value influences the cadres to work sincerely as a form of worship to God. This research indicates that the work program can be implemented well even if the cadres do not get enough material rewards.

REFERENCES

Aga, D. A. 2016. Transactional leadership and project success: The moderating role of goal clarity. *Procedia Computer Science* 100, 517–525. doi: 10.1016/j.procs.2016.09.190

Alghamdi, M. G., Topp, R., & AlYami, M. S. 2017. The effect of gender on transformational leadership and job satisfaction among Saudi nurses. *Journal of Advanced Nursing* 74(1), 119–127. doi: 10.1111/jan.13385

Bass, B. 2008. *Bass & Stogdill's handbook of leadership: Theory, research & managerial applications* (4th ed.). New York: Free Press.

Brewer, C. S., Kovner, C. T., Djukic, M., Fatehi, F., Greene, W., Chacko, T. P., & Yang, Y. 2016. Impact of transformational leadership on nurse work outcomes. *Journal of Advanced Nursing* 72(11), 2879–2893. doi: 10.1111/jan.13055

Cox, P. L. 2001. Transformational leadership: a success story at Cornell University. In Proceedings of the ATEM/aappa 2001 conference (Vol. 17, p. 2004).

Dhaliwal, K. K., & Hirst, S. P. 2018. Correctional nursing and transformational leadership. *Nursing Forum*, 1–6. https://doi.org/10.1111/nuf.12314

Dudovskiy, J. 2018. The Ultimate Guide to Writing a Dissertation in Business Studies: A Step-by-Step Assistance.[Research Methodology version].

Gill, R. 2002. Change management – or change leadership? *Journal of Change Management* 3(4), 307–318.

Kidney, R. 2015. *Transformational/transactional leadership. Wiley encyclopedia of management*, 1–4. doi: 10.1002/9781118785317.weom1100

Manning, S. 2008. Embedding projects in multiple contexts: A structuration perspective. *International Journal of Project Management* 26, 30–37.

Martínez-Córcoles, M., & Stephanou, K. 2017. Linking active transactional leadership and safety performance in military operations. *Safety Science* 96, 93–101. doi: 10.1016/j.ssci.2017.03.013

McDowell, T., Agarwal, D., Miller, D., Okamoto, T., & Page, T. 2016, February 29. Organizational design, the rise of teams. Retrieved December 1, 2016, from https://dupress.deloitte.com/dup-us-en/focus/human-capital-trends/2016/organizational-models-network-of-teams.html.

Men, L. R., & Bowen, S. 2017. *Excellence in internal communication management*. New York: Business Expert Press.

Northouse, P. G. 2013. *Leadership: Theory and practice*. Thousand Oaks, CA: Sage.

Pieterse, A. N., van Knippenberg, D., Schippers, M., & Stam, D. 2009. Transformational and transactional leadership and innovative behavior: The moderating role of psychological empowerment. *Journal of Organizational Behavior* 31(4), 609–623. doi: 10.1002/job.650

Robbins, S. P., & Coulter, M. 2010. *Manajemen Edisi Kesepuluh*. Jakarta: Penerbit Erlangga.

Robbins, S. P., & Judge, T. A. 2013. *Organizational behavior* (15th ed.). Upper Saddle River, NJ: Pearson Education.

Taylor, J. 2016. Management of Australian water utilities: The significance of transactional and transformational leadership. *Australian Journal of Public Administration* 76(1), 18–32. doi: 10.1111/1467-8500.12200

Tyssen, A. K., Wald, A., & Spieth, P. 2014. The challenge of transactional and transformational leadership in projects. *International Journal of Project Management*, 32(3), 365–375. doi: 10.1016/j.ijproman.2013.05.010

Wideman, R. M. 2014. A comparison of project, program & portfolio management responsibilities – and who should be responsible for what. Plantation, FL: J. Ross Publishing.

Yukl, G. 1999. An evaluation of the conceptual weaknesses in transformational and charismatic leadership theories. *Leadership Quarterly* 10(2), 285–305.

The strategic roles of Indonesian diaspora scientists for domestic knowledge development

T. Riyani & M. Hanita
School of Strategic and Global Studies, Universitas Indonesia, Jakarta, Indonesia

ABSTRACT: The demands of the industrial revolution, according to Klaus Schwab, have entered stage 4.0 since 2010. The current development of knowledge in Indonesia cannot be separated from the role of existing scientists. The scientists who have the most important roles in helping to develop the industrial revolution 4.0 in Indonesia are Indonesian diaspora scientists. Indonesian diaspora scientists gather annually and produce outcomes and outputs for knowledge development in Indonesia. This research used a qualitative research method and a case study of Indonesian diaspora scientists' meetings for the past four years. The purpose of this research was to analyze the role of Indonesian diaspora scientists in domestic technology development. The results found that Indonesian diaspora scientists play three strategic roles in knowledge development in Indonesia: the field of knowledge development, the area targeted for development, and the output of cooperation.

1 INTRODUCTION

The term *diaspora* is applied to a broad range of migrant populations. A diaspora could include political refugees, labor migrants, expatriates, stable ethnic minorities, and other dispersed groups (Remennick, 2015). Diaspora populations live outside their home country for a long time so that their role in the homeland is very minimal. Nostalgia about the time past in the homeland plays a role in making diaspora members want to contribute to their country (Quayson & Daswani, 2013).

Currently, an estimated 8 million Indonesians live abroad pursuing various professions such as entrepreneurs, researchers, students, professional workers, arts workers, migrant workers, etc. with a per capita income five times that in Indonesia. With a total of around 8 million people, the Indonesian diaspora is in the range of 3% of the total Indonesian population and is spread across all seven continents.

Scholars and researchers have recently begun addressing the development of science and technology in Indonesia. This is in line with the emergence of demands stemming from the industrial revolution, which, according to Klaus Schwab, has entered stage 4.0 since 2010 (Ciffolilli & Muscio, 2018; Sony, 2018; Xu, Xu, & Li, 2018). With the presence of these demands, Indonesia as one part of the world community has contributed to creating new breakthroughs in the fields of science and technology (Aminullah, 2011; Wie, 2006). The proof is that up to now several pioneering companies, or unicorns, have emerged that help the continuity and ease of economic and social transactions of the community, even in remote areas far from a city center.

The development of these technologies is inextricably linked to the role of the Indonesian diaspora of scientists who have carried out renewable initiations by looking at the needs and potential of the community. Some of these initiatives, for example, holding collaborations or cooperation with domestic educational institutions, play an active role in various researches by utilizing local wealth and by cooperating with one or two sons of the nation in diverse modern technology projects. Therefore, the quality of the human resources of the Indonesian people can guarantee the quality of domestic science and technology with reciprocation between the diaspora of Indonesian scientists and local researchers.

2 LITERATURE REVIEW

Diaspora in its development has had a big influence on the development of the home country. Diaspora can introduce the political identity of the home country into the global realm and support domestic goals held by the country of origin (Gamlen, Cummings, & Vaaler, 2019).

Several previous studies have depicted the diaspora as a mouthpiece in the midst of conflicts between nations. Diaspora members are considered competent and can coordinate to face big challenges (Cochrane, Baser, & Swain, 2009; Fair, 2005; Hasić, 2018; Pande, 2017). The diaspora can also develop domestic potential with a variety of capabilities, both scientists in general studies and scientists specifically in the field of technology (Pande, 2014; Tejada, Varzari, & Porcescu, 2013).

Some diasporas fail to build relationships with their home countries (Thandi, 2014). Research on the Punjab paradox has explained that the many generations of Indian Punjab who became a diaspora across the world would be paradoxical because of their failure to establish good engagement with the home country. This makes diaspora even more unique to study because each country has different abilities to empower its human resources.

3 METHODOLOGY

3.1 *Data collection*

The data are the background of researchers present at the World Class Scientist Symposium held by the Ministry of Research, Technology and Higher Education for the past four years.

3.2 *Data analysis*

The data that were obtained are grouped based on their strategic roles. Data grouping was then analyzed according to their respective effects. Data were displayed in tables to explain them.

4 RESULTS AND DISCUSSION

The research found three strategic roles played by Indonesian diaspora scientists in domestic knowledge development. The roles were the field of knowledge development, the area targeted for development, and the output of cooperation.

Table 1.　Fields of knowledge development.

Interests	Scientist
Electrical engineering	1
Energy system	2
Supply chain management	1
Computer science	2
Aerospace	1
Chemical engineering	7
Network and communications	1
Radar and remote sensing	1
Nanotechnology	1
Pharmacy	3
Sociology	1
Architects and urban studies	2
Marine and climate	1

(*Continued*)

Table 1. (*Continued*)

Interests	Scientist
Health and public health	3
Dentistry	1
Physics	3
Metallurgy	1
Molecular biology	2
Food science and technology	2
Biochemistry	1
Applied mathematics	1
Biotechnology	1
Environmental sciences	1
Material engineering	3
Obstetrics and gynecology	1
Robotics and mechatronics	1
Medical	1
Mechanical engineering	4
Sociology	1
Anthropology	1

Based on data gained from the Ministry of Research, Technology and Higher Education as of June 23, 2019, experts in the field of chemical engineering – namely seven scientists – had the highest scientific development compared to other fields. The field of mechanical engineering was represented by four scientists. In addition, the fields of pharmacy, health and public health, physics, and material engineering each had three scientists. The fields of energy systems, computer science, architecture and civil engineering, and molecular biology, as well as technology and food sciences, each had two scientists. The rest – namely electrical engineering, supply chain management, aerospace, network and communications, radar and remote sensing, nanotechnology, sociology, marine and climate, dentistry, metallurgy, biochemistry, applied mathematics, biotechnology, environmental sciences, nursing and midwifery, robotics and mechanical engineering, medicine, and anthropology each had only one scientist.

Table 2. Area targeted for development.

Area	Visit
Bali	2
Bengkulu	1
In Yogyakarta	4
DKI Jakarta	3
West Java	6
Central Java	6
East Java	9
West Kalimantan	1
South Kalimantan	1
Central Kalimantan	1
East Kalimantan	1
Riau Islands	1
Lampung	2
Maluku	1
North Maluku	1
West Nusa Tenggara	1
East Nusa Tenggara	1

(*Continued*)

Table 2. (*Continued*)

Area	Visit
Papua	1
Central Sulawesi	1
South Sulawesi	2
Southeast Sulawesi	1
North Sulawesi	2
West Sumatra	3
South Sumatra	1
North Sumatra	3

Meanwhile, twenty-five provinces became the target of the development of science and technology for the Ministry of Research, Technology and Higher Education. In practice, the ministry visited several areas from the relevant provinces. Taking into account the area and access to the intended province, fourteen of the twenty-five provinces were only be visited once. These included the provinces of Bengkulu, West, South, Central, and East Kalimantan, the Riau Islands, Maluku and North Maluku, East and West Nusa Tenggara, Papua, Central and Southeast Sulawesi, and South Sumatra. The provinces of Bali, Lampung, and South and North Sulawesi were visited twice. The provinces of DKI Jakarta and West and South Sumatra were visited three times. The DI Yogyakarta province was visited four times, West Java and Central Java six times, and East Java nine times.

Table 3. Output of cooperation.

Output	Quantity
Under Review Journal	4
Manuscript Journal	7
Submitted Journal	8
Accepted Journal	3
Published Journal	72

The collaboration between the Ministry of Research, Technology and Higher Education with the Indonesian diaspora scientists in various fields, of course, has had significant results. These results can enrich the treasures of science and technology in Indonesia. So far, many workshops and research funds have been held aiming to support the development and enrichment of science and to promote competition with other countries. In addition, seventy-two journals related to this matter have been published, three received, eight submitted, seven produced as manuscripts, and four placed under review.

5 CONCLUSIONS AND RECOMMENDATIONS

Diaspora scientists make a good contribution for Indonesia. If this contribution is increased, the development of knowledge in the country will improve. The next research that needs to be done is to measure the index of the influence of the existence of the diaspora scientists on state resilience through clearer statistics. The index can be a recommendation for the government to make policies for the advancement of domestic knowledge.

REFERENCES

Aminullah, E. 2011. Long-term forecasting of technology and economic growth in Indonesia. *Asian Journal of Technology Innovation* 15(1): 1–20.

Ciffolilli, A., & Muscio, A. 2018. Industry 4.0: National and regional comparative advantages in key enabling technologies. *European Planning Studies* 26(12): 2323–2343.

Cochrane, F., Baser, B., & Swain, A. 2009. Home thoughts from abroad: Diasporas and peace-building in Northern Ireland and Sri Lanka. *Studies in Conflict & Terrorism* 32(8): 681–704.

Fair, C. C. 2005. Diaspora involvement in insurgencies: Insights from the Khalistan and Tamil Eelam movements. *Nationalism and Ethnic Politics* 11(1): 125–156.

Gamlen, A., Cummings, M. E., & Vaaler, P. M. 2019. Explaining the rise of diaspora institutions. *Journal of Ethnic and Migration Studies* 45(4): 492–516.

Hasić, J. 2018. Post-conflict cooperation in multi-ethnic local communities of Bosnia and Herzegovina: A qualitative comparative analysis of diaspora's role. *Journal of Peacebuilding & Development* 13(2): 31–46.

Pande, A. 2014. The role of Indian diaspora in the development of the Indian IT industry. *Diaspora Studies* 7(2): 121–129.

Pande, A. 2017. Role of diasporas in homeland conflicts, conflict resolution, and post-war reconstruction: The case of Tamil diaspora and Sri Lanka. *South Asian Diaspora* 9(1): 51–66.

Quayson, A., & Daswani, G. 2013. Introduction: Diaspora and transnationalism. In *A companion to diaspora and transnationalism*. Blackwell Publishing. West Sussex: 1–26.

Remennick, L. 2015. Diaspora. In *The Wiley Blackwell encyclopedia of race, ethnicity, and nationalism*. John Wiley & Sons. West Sussex: 1–4.

Sony, M. 2018. Industry 4.0 and lean management: A proposed integration model and research propositions. *Production & Manufacturing Research* 6(1): 416–432.

Tejada, G., Varzari, V., & Porcescu, S. 2013. Scientific diasporas, transnationalism and home-country development: Evidence from a study of skilled Moldovans abroad. *Southeast European and Black Sea Studies* 13(2): 157–173.

Thandi, S. S. 2014. The Punjab paradox: Understanding the failures of diaspora engagement. *Diaspora Studies* 7(1): 42–55.

Wie, T. K. 2006. The major channels of international technology transfer to Indonesia: An assessment. *Journal of the Asia Pacific Economy* 10(2): 214–236.

Xu, L. D., Xu, E. L., & Li, L. 2018. Industry 4.0: State of the art and future trends. *International Journal of Production Research* 56(8): 2941–2962.

Training needs analysis implementation: Dilemmas and paradoxes

B. Fairman
ASEAN Institute of Applied Learning, Jakarta, Indonesia

A. Voak
Deakin University, Melbourne, Australia

U. Sujatmaka
ASEAN Institute of Applied Learning, Jakarta, Indonesia

ABSTRACT: Whilst we acknowledge that many organisations across Asia use Training Needs Assessments (TNAs), also known as Training Needs Analysis, to determine their human resource development requirements in conjunction with training schemes in their workforce, it is observed that this activity occurs with mixed results. It will be argued in this paper that these TNAs, which are conducted in a diverse array of organisational environments in Indonesia, are often employed to further individual rather than institutional goals. In this article, TNAs are examined for their innate potential for making contributions to an organisation in order to provide sound directions for training interventions in Indonesia.

1 INTRODUCTION

This paper focuses on the challenges and benefits of undertaking an applied learning approach to human resource development, with an emphasis placed on the practical issues faced by Australian, as well as local educational developers, when deploying training needs analysis (TNA) tools and methodologies in an Indonesian context. The Australian and Indonesian authors' experiences in analysing workforce development needs for both the Indonesian government and private sectors, will largely shape the discussion. The aim of this analysis is to explore the relevance of detailed training needs assessment in aiding workforce development in Indonesia, and to uncover any unforeseen cultural biases or issues that could potentially emerge. In this latter respect, TNA tools are often developed in an Australian or Western influenced context, and we must ensure that hidden cultural biases do not hinder the application of these otherwise important instruments.

As an entrée to this discussion, the paper will focus upon an examination of an appropriate starting point for conducting and designing a training needs analysis in an Indonesian training context. It will consider the underlying requirements, which determine the most relevant TNA tools and methods which might apply in circumstances appropriate for Government agencies (local and national) and Industry sectors. Whilst these requirements may vary according to the targeted audiences, the similarities and differences in these targeted audiences will help to shape the approach employed, and to avoid any unintended cultural biases which may tend to arise. This paper also reports the results of four TNA pilot investigations which have previously been conducted for Indonesian government instrumentalities. These are TNA approaches for a National Government Body, both District and Provisional governments; a number of local government agencies; and for industry engagements with school curricula. These case studies reveal some of the practical difficulties in articulating client needs across government entities in Indonesia, and they explore, in some depth, the relevance of detailed training needs assessments for the workforce and their subsequent impact upon human resource development needs.

2 TRAINING NEEDS ASSESSMENT

TNA, or Training Needs Analysis as it sometimes called, are deployed to determine if training is the best intervention to a workplace problem or change program (Cekada, 2010). When deployed well, the organisation gathers valuable information and data which assists human resources departments in verifying the most appropriate solution for any the deficiencies in their performance (Cekada, 2010). Leigh et al. (2000) further argued that the proliferation of training needs assessment models, combined with conflicting usage of the terminology, has created the belief, in the popular lexicon, that the term can mean almost anything. Thus, we need to specify, more precisely, what is encapsulated within a TNA. Kaufman (1992) and (1998) describe needs assessment as a formalised process in which recognised gaps between current and desired results are identified, then prioritised as to their importance through a cost benefit analysis, and then ultimately selecting the most important areas for elimination or reduction. Hence, TNAs have a strategic role to play within an organisation, as they can provide clear guidelines as to gaps that need to be remedied (Ferreira and Abbad, 2013), and therefore provision a better understanding of existing and future workforce requirements (Armstrong, 2006). It is a dynamic process, which serves as a valuable enabler for vocational education practitioners when framing discussions and modifications regarding human resources interventions. When deployed effectively, TNA can play a valuable role in directing resources to areas of greatest organisational priority and, ultimately, improving worker productivity and community values (Lawler and Sillitoe, 2013). Further, Salas and Cannon-Bowers (2001) (2001) emphasised while there is wide agreement on the value of TNA within the training process, this phase largely remains as an art rather than as a science.

3 TNA AND ITS IMPACT ON HUMAN RESOURCE DEVELOPMENT

Winfred et al. (2003) contend that needs assessment can more clearly be identified as a three-step process consisting of organisational, task and person analysis. However, the process is often hindered because organisations often have a different human resource development philosophy or agenda to those deploying the TNA. Because organisations are understandably different, no group of organizational needs can be culturally the same, and therefore often deploy significantly different career structures. Indeed, there are different techniques and development processes, together with parallel language used that can potentially have an important impact on data collection within the TNA process (Al-Khayyat, 1998). As a result, organisations need to carefully think about the changes they are both willing and financially able to make, with the needs assessment findings treated as essentially a means to an end rather than an end in themselves (Li, 2000). The needs assessment process is further complicated by the pedagogical approach (Delahaye, 1992) in what is commonly called a 'supply-led approach'. This approach is largely trainer-driven, with their authority to act vesting them with the power of personal interest (Thompson, 1994). Chiu et al. (1999) suggest that because trainers are largely responsible for conducting TNAs and, moreover, determining its potential scope, this can often lead to preferred taxonomy frameworks compiled by the trainer, with participants reduced to merely indicating whether there is a felt need for training for each classified item on the list.

Noe and Schmitt (1986) believe that when deploying such interventions, one must be cognisant of task constraints and also to be particularly attuned to supervisors who provide little or no developmental support. Dierdorff and Surface (2007) therefore contend that task-focused analyses can provide a key role in directly providing the appropriate information to determine the content of a specific training program. Yet, even with perfect analysis, design and enthusiastic trainees, positive change cannot be affected without organisational support Eisenberger et al., (1990). Without such backing, any resultant training will occasion deployment failure because of lack of manifestation of the assumptions, values and beliefs of various key stakeholders within the organisation (Bunch, 2007). Reed and Vocala (2006) articulated this belief when they clearly enunciated that the conduct of the TNA process is comparatively easy when compared to the difficulties faced when dealing with the underpinning issues that require clarity and better understanding of not only the organisational culture but also underlying reasoning behind the change.

Of particular concern is that there are incongruities between the three pillars of human resource development, namely; (i) Industry Needs, (ii) Training and Education programs and (iii) Certification and Accreditation protocols (Figure 1). These are problematic issues because, from an Indonesian perspective, these areas involve a number of unresolved understandings related to the purposes and practice of instituting this sort of analysis. The following questions inevitably arise: what are the issues of implementing TNAs under these Decrees? What do these Decrees actually state, in summary form? How are the TNAs intended to be conducted? What are the current tensions which appear?

This discussion aims to illuminate the uncertainties which might arise in investigating a Human Resource Development initiative based on these three pillars and the potential concerns around developing trustworthy and valuable Indonesian organisations. Importantly, it must be stated that we are asserting that it is only at the intersection of these three pillars (ie. where mutual and effective engagement with each pillar is recognised), where meaningful engagement can occur as represented diagrammatically in Figure 1.

At the centre of the intersection of The Three Pillars, which represents congruity between each developmental activity, we find the desirable position for Human Resource Development outcomes. However it is a common occurrence within training and education institutions in Indonesia, that they develop their curricula without formally consulting industries (Abdullah, 2014). This lack of essential consultation, has an immediate impact upon graduates as they cannot be immediately absorbed into the workforce and cannot work directly in their chosen profession, because they are not adequately prepared for industry requirements and urgent workforce needs (Curtain, 2009). A possible consequence of this lack of industry engagement is that these graduates require further 'vocational skills' development, either through gaining industry experience in a parallel field, or returning to formal vocational skills development or training (Analoui, 1993).

4 CURRENT TNA PRACTICES AND THEIR IMPACTS

The authors of this paper envisage that a successful training and/or education program should be initially evaluated in order to ensure that the material and learning opportunities relate to the intended work environment, and that any change in requirement or miscuing of presentation, can be quickly identified and rectified. Such an interactive program might profitably start with the identification of any gaps in graduate outcomes through a focused Performance/Training Needs Analysis (TNA) (Moore and Morton, 2017). These TNAs will be conducted

Figure 1. Three pillars of human resource development.

by the school or higher education institutions involved, working together with target industries in order to determine what competencies are required by industries in order to make graduates immediately employable. Based on the results of this analysis, which is often referred to as the 'competency gap', the course and its curriculum will consequently be specifically designed to enable students and trainees, together with their teachers, to engage in a learning process in order to gain the appropriate competencies required by the target industries (Paryono, 2017).

We assert here that the quality of TNA implementation, particularly in an overseas context, depends upon three interrelated aspects of practice and context coming together in a formal and integrated way. These interrelated aspects include: (i) respect for the current national and local regulations, (ii) enhancing the quality of existing human resources, (iii) maintaining and increasing the quality of the network between schools and higher education and training institutes (Snepvangers et al., 2018), which sustains the education and training dialogue with the industries that the program is intending to serve. In the current context of increasing global interaction and competition, industries should be regarded as the main clients of the vocational schools and higher education institutes, because, in the wider Indonesian context, most of the graduates will ultimately be seeking employment in area which are experiencing growing production demands. For a number of historical reasons, this 'mind set' of primacy of 'industry engagement' is rare within the Indonesian education sector. It is noted here that there are some strongly competing outcome concerns which make this 'work-ready' mind set difficult to maintain for vocational training providers and higher education institutes (Biech, 2005). Of particular relevance to this discussion is the existence of current Indonesian Government regulations known as Government Decree No 31, (2006) on National Vocational Training System and also the LAN Chairman Decision No 3, (2013)[1]. Both of these crucial documents clearly state the importance of conducting TNA as an integral part of the Human Resource Development system. However, as implied earlier, the realisation and implementation of the requirements of these regulations is still far from the intended outcomes (Sayuti, 2016). To help to contextualise and highlight the extent of this problem, the examination of four case study examples of TNA implementation in an Indonesian context are plat- formed. The selection of material spans the range of training situations from the perspective of industry concerns, local government implications, national government requirements and donor-funded bilateral program intentions.

5 INDUSTRY ENGAGEMENT

5.1 *Vocational school*

A more thorough and detailed examination of the implementation of TNAs is illustrated here through a case study of a regional higher vocational school. This case study illustrates how there was a 'conflict of interest' between industry support for vocational training and the intent of local government's accreditation requirements, which resulted in a dilemma for the management of this vocational school. In this case study, the car producer, which was located near the vocational school, approached the SMK indicating their willingness to contribute physical and human resources to assist the vocational school develop a stronger vocational program. These resources included; foci for content of their learning programs, relevant curriculum and training material, and offers of loans of machinery required for workplace practice. In addition, there was an invitation to use their company's workplace for internship and 'work placements', with the clear intention of making the school students 'work-competent' so that they could be immediately employed in the industry.

---

1 Source, Decree of Lembaga Admistrasi Negara (LAN) No 10, 2010 on Development of Civil Servants Chapter Two-Needs Analysis and Planning Development

This detailed and intimate engagement with an 'industry' sector is a common feature of many developed and developing countries, and is clearly a sensible and generally recommended approach to developing vocationally-ready students. However, in Indonesia, SMKs are required to deliver curricula that have been determined by the local government, since the consequence of not delivering the 'locally endorsed' curricula is that the school would lose their accreditation due to 'non-compliance.' This is part of the structure of the regulatory frameworks that are currently applied in Indonesia (Abdullah, 2014). This example illustrates that schools have to make difficult choices in terms of curriculum implementation, either to support the regulated framework and continue as an accredited school, or engage in a curriculum that meets industry expectations and as a consequence become 'non-accredited', which has significant financial implications.

5.2   *Local government banking sector*

In another illustration, a Local Government Bank decided that, in order to gain information about training needs, their Human Resource Development Department would distribute questionnaires to individual staff asking them to comment on "What they think, and feel they are lacking in terms of competence" and "What training they may require to close this self-perceived gap?" This certainly seemed to be a reasonable proposition to put to their employees, at least on the surface. However, we discovered that this was a standard questionnaire that was previously designed and developed, and is implemented without any apparent change, each year. It is not developed with regard to measuring or linking responses of this TNA to an individual's performance targets or work requirements. As a result of this standardised approach, there is no useful link between the TNA conducted at this worksite and the organisational goals relevant to the organisation's vision and mission. Because there was no evident consequential link between the responses to the TNA conducted with all employees at all levels in the organisation and their identified and real performance indicators, we felt something was missing. Our observation and reflection on this process suggested that it needed a transparent 'link', which would have allowed an evidential determination of what training intervention would be required for various individual's development[2] within the organisation.

5.2.1   *Government agency*

In another case study which was conducted within a government agency, on the completion of a round of training, the resident internal auditor of the agency inquired whether this particular training was based on a 'proper' TNA. The officer responsible for the training stated that training programs were based on the observations and findings during his visits to several local governments, but there was no evidence of these outcomes in the form of a TNA report. As a consequence of the unsatisfactory nature of this situation, a more formal approach to conducting TNAs is now being implemented. In this new approach, the TNAs are developed to cover: the development of a work standard; the development of a training program; and the development of organisational specific TNA questionnaires. These TNA planning and implementation sessions were conducted to identify gaps in competency in each local government, and used these findings as the basis for designing appropriate training programs.

5.3   *Donor funded TNA implementation in a bi-lateral program*

The Indonesia-Australia Specialised Training Project III (IASTPIII), conducted a rapid TNA to determine 'what' the implementing agencies of the Indonesian Government required from the Australian Government's IASTP III in respect of their training needs. The national priorities of the Government of Indonesia were deemed to be the guide for determining the training needs across the provincial focus areas named above. The consequent TNA determined that, throughout the consultation process of local provincial and district agencies, a total of 63 new

---

2 This case illustration was based on the discussions with a highly placed bank official in Jakarta

training requests were identified that met the Government of Indonesia's national priorities. As a result, a number of recommendations were made regarding the training needs and the role of donors, particularly the Australian government, and their engagement with training provisions. The purpose of the TNA was to provide guidance and direction on human resource training and development needs, and describe the importance of taking into account the local needs of provincial and district agencies. However, it was noted that this local input failed considerably in the IASTP context (Scott, 2007), and it was specifically commented that considerably more input from local agencies and 'on the ground' institutes could have created a more targeted and specialised Indonesia training programs.

In attempting to accurately determine what training would be required in a specific area, we suggest that this would ultimately hinge on who was asked for their opinion. It was observed by some respondents in this review, that certain parts of Indonesia seemed to gain greater international attention, and this was particularly noted by AusAID's interest in Nusa Tenggara Tengah. One respondent mentioned the case of Papua, noting that Papua had up to 27 programs with many of them duplicated and repeated by a number of countries (Fairman, 2017). This comment illustrates that expectations of responding to the TNA has little to do with actual evaluation outcomes. In the Indonesian context, there are often different organisational 'arms' conducting the TNA, and the results of a TNA might be influenced by the organisational structure involved.

6 CLOSING REMARKS

It is now recognised that University graduates cannot find relevant work due to their lack of focused skills required by industries. In the workplace, such as a bank, other state-owned or private organisations, or many public institutions, have thus wasted time, money and other resources in conducting training programs which have nothing to do with improving competency in order to bridge the organisation performance gap. There are some important common threads which have emerged from each of the TNA case study examples, and which need to be addressed if TNA practice is to be more carefully shaped to local requirements. These commonalities included: organisational plans were not matched to human resource development needs; here were limited connections between TNA outcomes and future training interventions; and the purpose of carrying out the TNAs can be slightly different for specific circumstances. It is clear from the descriptions above, that the purpose and conduct of the various TNAs within each case study are essentially different. It is important that the conduct of the TNAs is approached from an unbiased perspective and carried out to meet the organisational training requirements rather than meeting an individual's training expectations or requirements. Whilst this comment is not directed particularly at an Indonesian context, we note that similar problems with TNAs conducted internationally can evidence similar outcomes unless care is taken with the analysis process. If this care is not taken, TNAs risk becoming mere formalities in a 'self-justifying' system.

REFERENCES

Abdullah, H. 2014. *VET Training and Industry Partnerships: a Study in East Java, Indonesia*. Victoria University.

Al-khayyat, R. 1998. Training and development needs assessment: a practical model for partner institutes. *Journal of European Industrial Training*, 22, 18–27.

Analoui, F. 1993. *Training and transfer of learning*, Avebury.

Armstrong, M. 2006. *A handbook of human resource management practice*, Kogan Page Publishers.

Biech, E. 2005. *Training for dummies*, John Wiley & Sons.

Bunch, K. J. 2007. Training failure as a consequence of organizational culture. *Human Resource Development Review*, 6, 142–163.

Cekada, T. L. 2010. Training needs assessment: Understanding what employees need to know. *Professional Safety*, 55, 28–33.

Chiu, W., THOMPSON, D., MAK, W.-M. & LO, K. 1999. Re-thinking training needs analysis: A proposed framework for literature review. *Personnel Review*, 28, 77–90.

Curtain, R. 2009. Skills in Demand: Indentifying the Skill Needs of Enterprises in Construction and Hospitality in Timor Leste. Dili: AusAID-Skills Development.

Delahaye, B. 1992. A theoretical context of management development and education. *Smith, B.*

Dierdorff, E. C. & SURFACE, E. A. 2007. Assessing training needs: do work experience and capability matter? *Human performance*, 21, 28–48.

Eisenberger, R., FASOLO, P. & DAVIS-LAMASTRO, V. 1990. Perceived organizational support and employee diligence, commitment, and innovation. *Journal of applied psychology*, 75, 51.

Fairman, B. F. 2017. *Looking for a way out: Skills development and training and its impact on aid practices and their development outcomes, with particular reference to Indonesia and Timor-Leste.* Victoria University.

Ferreira, R. R. & ABBAD, G. 2013. Training needs assessment: where we are and where we should go. *BAR-Brazilian Administration Review*, 10, 77–99.

Foley, G. 2004. *Dimensions of Adult Learning-Adult education and training in a global era.*, Sydney, Allen and Unwin.

Kaufman, R. 1992. *Strategic planning plus: An organizational guide*, Sage publications.

Kaufman, R. 1998. Strategic thinking: A guide to identifying and solving problems (Rev. ed.). *Arlington, VA: American Society for Training & Development.*

Lawler, A. & SILLITOE, J. 2013. Facilitating 'organisational learning'in a 'learning institution'. *Journal of Higher Education Policy and Management*, 35, 495–500.

Leigh, D., WATKINS, R., PLATT, W. A. & KAUFMAN, R. 2000. Alternate models of needs assessment: Selecting the right one for your organization. *Human Resource Development Quarterly*, 11, 87–93.

Li, D. 2000. Needs assessment in translation teaching: Making translator training more responsive to social needs. *Babel*, 46, 289–299.

Moore, T. & MORTON, J. 2017. The myth of job readiness? Written communication, employability, and the 'skills gap'in higher education. *Studies in Higher Education*, 42, 591–609.

Noe, R. A. & SCHMITT, N. 1986. The influence of trainee attitudes on training effectiveness: Test of a model. *Personnel psychology*, 39, 497–523.

Paryono. The importance of TVET and its contribution to sustainable development. AIP Conference Proceedings, 2017. AIP Publishing, 020076.

Reed, J. & VAKOLA, M. 2006. What role can a training needs analysis play in organisational change? *Journal of Organizational Change Management*, 19, 393–407.

Salas, E. & CANNON-BOWERS, J. A. 2001. The science of training: A decade of progress. *Annual review of psychology*, 52, 471–499.

Sayuti, M. 2016. *The Indonesian Competency Standards in Technical and Vocational Education and Training: an Evaluation of Policy Implementation in Yogyakarta Province Indonesia.* The University of Newcastle Australia.

Scott, A. 2007. IASTP Rapid TNAT. Jakarta, Indonesia: Hassells International.

SNEPVANGERS, K., DAVIS, S. & TAYLOR, I. 2018. Transforming Dialogues about Research with Embodied and Walking Pedagogies in Art, Education and the Cultural Sphere. *Embodied and Walking Pedagogies Engaging the Visual Domain*, 1.

Thompson, D. 1994. Developing Managers for the new NHS, Longman.

A literature review of performance appraisal reaction: Predictors and measurement

Y.N. Widiani & N. Dudija
Telkom University, Bandung, Indonesia

ABSTRACT: Performance appraisal is one of the most important aspects of human resource management (HRM). Regardless of the technical soundness of the performance appraisal system, employees' reaction to the performance appraisal itself needs to be a concern. Employee satisfaction with performance appraisal plays a pivotal role in the long-term effectiveness of any performance appraisal implementation, but unfortunately this topic has received little empirical attention. In this paper, a systematic literature review is used to profoundly examine previous research from various available sources. The aim of this study was to deliver a meticulous summary of all available primary research. This research contributes theoretically by highlighting latent factors in employee performance appraisal satisfaction, such as perceived fairness and accuracy. Some suggestions for the design of an appraisal system, appraisal effectiveness, and future research are discussed.

1 INTRODUCTION

An effective performance appraisal system can encourage employees to improve their motivation and work performance (Migiro & Taderera, 2011). However, in order to ensure that performance appraisal can positively affect employee behavior and bring future improvements in organizational performance, employees must first have a satisfied reaction toward the process and output of performance appraisal (Keeping & Levy, 2000; Kuvaas 2006; Saraih et al., 2017). Employees' reactions to performance appraisal are believed to significantly influence the effectiveness and viability of the overall appraisal system (Dobbins, Cardy, & Platz-Vieno, 1990; Murphy & Cleveland, 1995; Roberts & Reed, 1996). Furthermore, Keeping and Levy (2000) argue that it is important to measure employee reactions toward performance appraisal for the following reasons: (a) the notion that reactions represent a criterion of great interest to practitioners, and (b) the fact that reactions have been theoretically linked to determinants of appraisal acceptance and success but have been largely ignored in research. This is supported by the statement of Dobbins and colleagues (1990) that performance appraisal reactions get relatively little empirical attention, even though their role is pivotal for the long-term effectiveness of any performance appraisal system implementation.

Satisfaction has been the most frequently measured appraisal reaction because of its superiority in measuring individual reactions that can assess both fairness cognition and simple affect (Giles & Mossholder, 1990). In addition, this is also based on the outcomes of satisfaction itself that can affect productivity, motivation, and commitment to the organization (Cawley, Keeping, & Levi, 1998). The aim of this study was to deliver a meticulous summary from various available primary research documents regarding satisfaction with performance assessment. This research contributes theoretically by highlighting latent factors in employee performance appraisal satisfaction. In addition, this study also examined satisfaction measurement of performance appraisal based on previous research.

2 LITERATURE REVIEW

2.1 *Performance appraisal satisfaction*

Performance appraisal satisfaction is defined as the degree to which employees perceive that the performance appraisal system represents their contribution to the organization (Giles & Mossholder, 1990). This reaction plays an important role in the appraisal process due to its long-term effect on the acceptance of the appraisal system (Murphy & Cleveland, 1995; Roberts & Reed, 1996), and as a factor that contributes to the validity of the assessment (Lawler, 1967). Research conducted by Blau (1999) and Jawahar (2007) shows that satisfaction with performance appraisal has a significant effect on subsequent overall satisfaction. Similarly, Colquitt, LePine, and Wesson (2015) state that employee overall satisfaction is influenced by psychological conditions where employees have knowledge of their work through a performance evaluation system. Research has shown that satisfaction with performance appraisal positively and significantly influences the work effort of employees and their affective organizational commitment (Jawahar, 2007; Kuvaas, 2006; Naeem, Jamal, & Riaz, 2017; Roberts & Reed, 1996) and can increase employee willingness to improve performance (Sharma & Sharma, 2017). An appraisal process may be designed to motivate employees and inspire their continuous efforts toward reaching goals, but, unless participants are satisfied and support it, the system will ultimately be unsuccessful (Othman, 2014). As Drenth, Thierry, Willems, and de Wolff (1984) suggest, evaluation is a sensitive matter, often eliciting negative psychological responses such as resistance and denial. The literature shows that employees will feel satisfied with performance appraisals only if they meet the criteria of "fairness" (Greenberg, 1986a, 1986b, 1986c; Landy, Barnes, & Murphy, 1978; Sudin, 2011).

2.2 *Predictors of performance appraisal satisfaction*

The perceived fairness and accuracy of the performance appraisal. Many scholars have proven that the perception of justice has a positive effect on employees' satisfaction with performance appraisals (Abdullah, Anamalai, Ismail, & Ling, 2015; Jawahar, 2007; Landy et al., 1978; Roberts & Reed, 1996; Sharma & Sharma, 2017; Sudin, 2011). In a meta-analytical review of justice literature, Colquitt (2001) revealed that fairness perceptions at work are largely affected by justice perceptions where organizational justice can be used as a theoretical support. According to George and Jones (2012: 170), organizational justice theory does not refer to a single theory per se, but rather describes a group of theories that focus on the nature, determinants, and consequences of organizational justice, so there are three forms of organizational justice: distributive justice, procedural justice, and interactional justice. Fairness perceptions influenced by these justice perceptions lead to satisfaction with the performance appraisal system and subsequently to performance appraisal effectiveness (Boswell & Boudreau, 1990; Jawahar, 2007), and justice is also seen as a predictor of the acceptability of the performance appraisal system (Roberts & Reed, 1996). Furthermore, perceived accuracy has also become an important indicator of employee satisfaction with the performance appraisal system (Cawley et al., 1998; Keeping & Levy, 2000; Landy et al., 1978; Taylor et al., 1995). In addition Landy and colleagues (1978) and Greenberg (1986b) identify the ability of the rater to perform accurate evaluations of the ratee's performance as a crucial factor in perceptions of fairness; thus perceived fairness and accuracy are interrelated.

Participation in the performance appraisal. A large number of studies indicate that employee participation in performance appraisal is associated with various outcomes related to the desired appraisal, including appraisal satisfaction and the subsequent acceptance of the appraisal system (Dipboye & de Pontbriand, 1981; Giles & Mossholder, 1990; Greller, 1975, 1978; Greenberg & Folger, 1983; Korsgaard & Roberson, 1995; Landy et al., 1978; Silverman & Wexley, 1984.) As suggested by Roberts (2002) when employees have a meaningful role in the appraisal process, employee acceptance and satisfaction with the appraisal process is strongly enhanced. In his study, Greller (1975) mentions that an invitation by a boss to a subordinate to provide input – which implies a welcome or an acceptance of what the employee has to say – is the largest predictive of satisfaction ratings. Employees involved in the development of performance appraisal processes and systems, as well as having a voice in the implementation of performance appraisal, show more positive reactions to performance appraisal than those who are not involved (Korsgaard & Roberson, 1995; Silverman & Wexley, 1984).

Ratee characteristic. Understanding individual differences in response to performance appraisals can provide an opportunity for organizations to better support the unique developmental needs and career paths of each individual and increase employee retention by increasing employee satisfaction. Studies conducted by Linderbaum and Levy (2010) show that ratee characteristics can influence employee reactions to performance appraisals. Employees with higher feedback orientation tend to take the initiative to seek feedback on performance appraisals and use feedback to achieve their goals. As Cawley and colleagues (1998) mention, individuals who tend to be more defensive when receiving feedback are known to be less satisfied with a performance appraisal session because of their perceptions of the feedback and appraisal as lacking utility and justice.

Performance ratings or feedback. Self-development theory (Shrauger, 1975) shows that individuals will react more positively to higher ranks than to lower ranks. Boswell and Boudreau (2000) argue that performance rating is significantly related to organizational tenure and performance appraisal satisfaction. Similar research conducted by Dusterhoff, Cunningham, and MacGregor (2014) reveals a direct relationship between the favorability of a performance rating and appraisal satisfaction. Positive evaluations are seen as more accurate, are valued more, and are better accepted than negative ratings (McEvoy, Buller, & Roghaar, 1987). Positive ratings elicit positive reactions toward the appraisal (Kacmar, Wayne, & Wright, 1996) and have been related to satisfaction with the appraisal process (Dipboye & de Pontbriand, 1981).

2.3 *Performance appraisal satisfaction measurement*

Appraisal satisfaction has been conceptualized mainly in three ways: (a) satisfaction with the appraisal interview or session, assessing the extent of employees' satisfaction with the performance appraisal discussion, as well as the quality of communication between rater and ratee; (b) satisfaction with the appraisal system, assessing whether the organization should change the appraisal system and whether fewer work problems occur as a result of the system; and (c) satisfaction with performance ratings, which assesses the degree to which employees are satisfied with their latest performance rating and report accurate and fair evaluations (Keeping & Levy, 2000). Several previous studies have separately operationalized these three conceptualizations. Giles and Mossholder (1990) operationalize session and system satisfaction in their research to see predictions of two measures of employees' reaction to performance appraisals. Taylor and colleagues (1995) use system and performance rating satisfaction to measure employee reactions in the form of satisfaction with performance appraisal. A large body of research stretching from the late 1970s to early 2000 has operationalized session satisfaction (Silverman & Wexley, 1984; Korsgaard & Roberson, 1995; Cawley et al., 1998; Jawahar, 2006; Sudin, 2011; Dusterhoff et al., 2014), system satisfaction (Landy et al., 1978; Dipboye & de Pontbriand, 1981; Roberts & Reed, 1996; Blau, 1999; Jawahar, 2007; Sudin, 2011; Othman, 2014; Abdullah et al., 2015; Sharma & Sharma, 2017), and performance rating satisfaction (Ilgen, Barnes-Farrell, & McKeltin, 1979; Kacmar et al., 1996; Kluger and DeNisi, 1996; Tang & Sarsfield-Baldwin, 1996; Jawahar 2006) in studies related to reactions to performance appraisals.

3 CONCLUSION

In conclusion, regardless of the technical readiness of the performance appraisal system, the reaction to the performance appraisal itself needs to be a concern. Studies have shown that employee satisfaction with performance appraisal has a significant effect on overall satisfaction and also affects the work effort of employees and organizational commitment in terms of productivity and motivation. In this paper, the factors influencing satisfaction with performance appraisals are more focused on the individual level such as perceptions of fairness and accuracy, participation in performance appraisals, ratee characteristics, and performance rating or feedback. Many studies have proven that these factors positively and significantly affect satisfaction with performance appraisal. The measurement commonly used to analyze satisfaction with performance appraisal is through three ways that have been conceptualized in several previous studies: (a) satisfaction with appraisal interviews or sessions, (b) satisfaction with the appraisal system, and (c) satisfaction with performance ratings.

As mentioned previously, an effective performance appraisal system can encourage employees to improve their motivation and performance. Some proposals have been made in the literature for improving performance appraisal systems. Javidmehr and Ebrahimpour (2015) suggest that human resources managers and organizational leaders should try to improve the credibility and fairness of the system through addressing bias that exists in their performance appraisal processes. Effective appraisal is underpinned by a relationship of respect where effectiveness is also linked to appraisal processes and information that have clarity, objectivity, and high integrity (Piggot-Irvine, 2003). In the future, further research can focus on the factors influencing satisfaction toward performance appraisal aside on individual level – i.e., appraisal system characteristics, such as types of evaluation measures, rating scales format, appraisal purpose and feedback of appraisal performance, leader-member exchange, etc. Also considering that performance appraisal is implemented in a social context, future research could pay more attention to the social context, e.g., the organization's political atmosphere and cultural setting.

REFERENCES

Abdullah, A. G. K. B., Anamalai, T. A/ L., Ismail, A. B., & Ling, Y. L. 2015. Do perceived organizational justice and trust determine satisfaction of performance appraisal practice? A case of Malaysian secondary school teachers. *Advances in Social Sciences Research Journal* 2(8), 23–37.

Blau, G. 1999. Testing the longitudinal impact of work variables and performance appraisal satisfaction on subsequent overall job satisfaction. *Human Relations* 52(8), 1099–1113.

Boswell, W. R., & Boudreau, J. W. 2000. Employee satisfaction with performance appraisals and appraisers: The role of perceived appraisal use. *Human Resource Development Quarterly* 11(3), 283–299.

Cawley, B. D., Keeping, L. M., & Levy, P. E. 1998. Participation in the performance appraisal process and employee reactions: A meta-analytic review of field investigations. *Journal of Applied Psychology* 83, 615–633.

Colquitt, J. A. 2001. On the dimensionality of organizational justice: A construct validation of a measure. *Journal of Applied Psychology* 86(3), 386–400.

Colquitt, J. A., LePine, J. A., & Wesson, M. J. 2015. *Organizational behavior: Improving performance and commitment in the workplace* (4th ed.). New York: McGraw-Hill.

Dipboye, R. L., & de Pontbriand, R. 1981. Correlates of employee reactions to performance appraisals and appraisal systems. *Journal of Applied Psychology* 66, 248–251.

Dobbins, G. H., Cardy, R. L., & Platz-Vieno, S. J. 1990. A contingency approach to appraisal satisfaction: An initial investigation of the joint effects of organisational variables and appraisal characteristics. *Journal of Management* 16(3), 619–632.

Drenth, P. J. D., Thierry, H., Willems, P. H., & C. J. de Wolff 1984. *Handbook of work and organizational psychology: Vol. 2*. New York: Wiley.

Dusterhoff, C., Cunningham, J. B., & MacGregor, J. N. 2014. The effects of performance rating, leader-member exchange, perceived utility, and organizational justice on performance appraisal satisfaction: Applying a moral judgment perspective. *Journal of Business Ethics* 119(2), 265–273.

George, J. M., & Jones, G. R. 2012. *Understanding and managing organizational behavior* (6th ed.). Upper Saddle River, NJ: Pearson Education.

Giles, W. F., & Mossholder, K. W. 1990. Employee reactions to contextual and session components of performance appraisal. *Journal of Applied Psychology* 75, 371–377.

Greenberg, J. 1986a. Organizational performance appraisal procedures: What makes them fair? In R. J. Lewicki, ed., B. H. Sheppard, ed., & M. H. Bazerman (eds.), *Research on negotiation in organizations, Vol. 1* (pp. 25–41). Greenwich, CT: JAI Press.

Greenberg, J. 1986b. Determinants of perceived fairness of performance evaluations. *Journal of Applied Psychology* 71: 340–342.

Greenberg, J. 1986c. The distributive justice of organizational performance evaluations. In H. W. Bierhoff, et al. (eds.), *Justice in social relations* (pp. 337–351). New York: Plenum Press.

Greenberg, J., & Folger, R. 1983. Procedural justice, participation, and the fair process effect in groups and organizations. In E. B. Paulus (ed.), *Basic group processes* (pp. 235–256). New York: Springer-Verlag.

Greller, M. M. 1975. Subordinate participation and reactions to the appraisal interview. *Journal of Applied Psychology* 60(5), 544–549.

Greller, M. M. 1978. The nature of subordinate participation in the appraisal interview. *Academy of Management Journal* 21, 646–658.

Ilgen, D. R., Barnes-Farrell, J. L., & McKeltin, D. B. 1993. Performance appraisal process research in the 1980s: What has it contributed to appraisals in use? *Organizational Behavior and Human Decision Processes* 54, 321–368.

Javidmehr, M., & Ebrahimpour, M. 2015. Performance appraisal bias and errors: The influences and consequences. I. M. Institute (ed.), *International Journal of Organizational Leadership* 4, 286–301.

Jawahar, I. M. 2006. Correlates of satisfaction with performance appraisal feedback. *Journal of Labor Research* 27(2), 213–236.

Jawahar, I. M. 2007. The influence of perceptions of fairness on performance appraisal reactions. *Journal of Labor Research* 28(4), 735–754.

Kacmar, M. K., Wayne, S. J., & Wright, P. M. 1996. Subordinate reactions to the use of impression management tactics and feedback by the supervisor. *Journal of Managerial Issues* 8, 35–53.

Keeping, L. M., & Levy, P. E. 2000. Performance appraisal reactions: Measurement, modeling, and method bias. *Journal of Applied Psychology* 85, 708–723.

Kluger, A. N., & DeNisi, A. 1996. The effects of feedback interventions on performance: A historical review, a meta-analysis, and a preliminary feedback intervention theory. *Psychological Bulletin* 119(2), 254–284.

Korsgaard, M. A., & Roberson, L. 1995. Procedural justice in performance evaluation: The role of instrumental and non-instrumental voice in performance appraisal discussions. *Journal of Management* 21, 657–669.

Kuvaas, B. 2006. Performance appraisal satisfaction and employee outcomes: Mediating and moderating roles of work motivation. *International Journal of Human Resource Management* 17(3), 504–522.

Landy, F. J., Barnes, J., & Murphy, K. 1978. Correlates of perceived fairness and accuracy of performance appraisals. *Journal of Applied Psychology* 63, 751–754.

Lawler, E. E. 1967. The multirate approach to measuring managerial job performance. *Journal of Applied Psychology* 51, 369–381.

Linderbaum, B. A., & Levy, P. E. 2010. The development and validation of the Feedback Orientation Scale (FOS). *Journal of Management* 36(6), 1372–1405.

McEvoy, G. M., Buller, P. F., & Roghaar, S. R. 1987. User acceptance of peer appraisals in an industrial setting. *Personnel Psychology* 40, 785–797.

Migiro, S. O., & Taderera, M. 2011. Evaluating the performance appraisal system in the bank of Botswana. *African Journal of Business Management* 5(10), 3765–3776.

Murphy, K. R., & Cleveland, J. N. 1995. *Understanding performance appraisal: Social, organizational, and goal-based perspectives*. Thousand Oaks, CA: Sage.

Naeem, M., Jamal, W., & Riaz, M. K. 2017. The relationship of employees' performance appraisal satisfaction with employees' outcomes: Evidence from higher educational institutes. *FWU Journal of Social Sciences* 11(2), 71–81.

Othman, N. 2014. Employee performance appraisal satisfaction: The case evidence from Brunei's civil service. Doctoral Thesis. Faculty of Humanities, University of Manchester.

Piggot-Irvine, E. 2003, June 1. Key features of appraisal effectiveness. *International Journal of Educational Management* 17(6), 254–261.

Prahalad, C. K., & Hamel, G. 1990. The core competence of the corporation. *Harvard Business Review*, 79–91.

Roberts, G. E. 2003. Employee performance appraisal system participation: A technique that works. *Public Personnel Management* 32, 89–98.

Roberts, G. E., & Reed, T. 1996. Performance appraisal participation, goal setting and feedback. *Review of Public Personnel Administration*, Fall, 29–60.

Saraih, U. N., Mohd Karim, K. b., Irza Hanie Abu Samah, C., Amlus, M. H. d., & Aida, N. A. 2017. Relationships between trust, organizational justice and performance appraisal satisfaction: Evidence from public higher educational institutions in Malaysia. ASIA International Multidisciplinary Conference 2017.

Sharma, A., & Sharma, T. 2017. HR analytics and performance appraisal systems: A conceptual framework for employee performance improvement. *Management Research Review* 40(6), 684–697.

Shrauger, J. S. (1975). Responses to evaluation as a function of initial self-perceptions. *Psychological Bulletin*, *82*(4), 581–596.

Silverman, S. B., & Wexley, K. N. 1984. Reaction of employees to performance appraisal interviews as a function of their participation in rating scale development. *Personnel Psychology* 37(4), 703–710.

Sudin, S. 2011. Fairness of and satisfaction with performance appraisal process. *Proceedings of the 2nd International Conference on Business and Economic Research*, 1239–1257.

Tang, T. L., & Sarsfield-Baldwin, L. J. 1996. Distributive and procedural justice as related to satisfaction and commitment. *Advanced Management Journal* 61(3), 3–25.

Taylor, M. S., Tracy, K. B., Renard, M. K., Harrison, J. K., & Carroll, S. J. 1995. Due process in performance appraisal: A quasi-experiment in procedural justice. *Administrative Science Quarterly* 40, 495–523.

Organizational justice: A literature review and managerial implications for future research

A.D. Putri & N. Dudija
Telkom University, Bandung, Indonesia

ABSTRACT: A company plays an important role in managing employees in order to produce the maximum results and encourage a high level of loyalty to the company. One aspect that companies need to pay attention to is managing organizational justice. Organizational justice in principle is balancing the rights and responsibilities of management and employees. Organizational justice has a strong influence on an organization. If an organization cannot treat its employees fairly, then it may negatively impact the organization by reducing employee productivity at work, decreasing the level of employee loyalty, and creating an unhealthy work environment. This study examines the literature on organizational justice, which consists of three main aspects – distributive, procedural, and interactional justice. In addition, in this paper, a systematic literature review is used to analyze previous research. This research contributes theoretically by highlighting latent factors influenced by organizational justice.

1 INTRODUCTION

Human resources is one of the important assets of any organization. Companies need employees who can work more efficiently and effectively in order to achieve company goals. To achieve this, companies need to do things that can increase the employees' level of commitment and loyalty. One of the things that companies can do is to implement organizational justice. Good implementation of organizational justice will make employees satisfied and motivate them to run their duties and responsibilities properly. The sense of fairness that employees experience is based on the satisfaction level within the organization or on the results of their work (Gunawan and Nurmadiansyah, 2017: 53). In addition, organizational justice can affect an employee's attitude and behavior toward an organization. If the attitudes and behavior of an employee are in accordance with the wishes of the organization, it can increase employee job satisfaction so that employees are motivated and also committed to achieving high performance in accordance with organizational expectations (Kristanto, Rahyuda, and Riana, 2014: 310).

2 LITERATURE REVIEW

2.1 *Organizational justice and job satisfaction*

In this section, the author discusses the findings from various research related to distributive, procedural, and interactional justice and its effect on job satisfaction. Research conducted by Iqbal (2013: 54) concerning educational institutions in Pakistan shows that procedural and interactional justice have positive relationships with job satisfaction. However, besides providing satisfaction, distributive justice has a negative relationship with job satisfaction. This shows how employees consider the procedures according to their relationship with the supervisor. This is inversely proportional to the findings Usmani and Jamal (2013: 372) reported in the banking sector in Pakistan, which suggest a significant positive relationship between distributive justice and job satisfaction. However, this does not apply to procedural fairness, which does not have a significant relationship to job satisfaction. Another study founded by Elamin (2012: 83) in Saudi Arabia shows that organizational justice has a related effect to job satisfaction. This study also shows that distributive justice plays a greater role than procedural justice in developing countries like Saudi Arabia. This should be a particular concern for companies to be able to distribute workloads, responsibilities, and fair rewards among employees, to implement processes and procedures based on the norms and culture of Saudi Arabia, and to properly manage relationships between the companies and their employees.

In addition, research that discusses organizational justice toward job satisfaction by Kristanto and colleagues (2014: 326) in the power sector in Bali suggests that distributive justice has a positive but not significant effect on job satisfaction. The insignificance occurred due to several changes in employee income and the implementation of the system in force. Furthermore, this study shows that procedural justice has the most powerful influence on job satisfaction compared to other dimensions. This means that the more employees perceive justice in enforcing the rules, the higher their job satisfaction. Another result from the research states that interactional justice has a positive and significant effect on job satisfaction. This proves that the more justice employees see in interpersonal relationships and access to information, the higher their job satisfaction. Furthermore, other research conducted by Putra and Indrawati (2018: 2034) in Bali in the hospitality sector shows that organizational justice has a positive and significant effect on job satisfaction. This proves that better application of organizational justice by a company will create job satisfaction for its employees. In addition, this study shows that job satisfaction as an intervening variable to observe distribution justice has a relationship with organizational commitment. Research conducted by Khan and Mahboob (2017: 6) in the education sector in Pakistan shows that organizational justice has a significant and positive influence on job satisfaction. In addition, if the organization can implement fair distribution, procedural, and interactional justice, it will lead to increased employee job satisfaction.

Further studies conducted by Bilal, Muqadas, and Khalid (2015: 71) and Zainalipour, Fini, and Mirkamali (2010: 1990) in the education sector in Pakistan and Iran show that all three dimensions of organizational justice have significant positive influence and relationships on job satisfaction. Studies on organizational justice and its relation to job satisfaction have also been conducted by Akram and colleagues (2015: 13) in the banking sector of Pakistan, which shows a positive and significant relationship between distributive justice and job satisfaction. But there is a negative and significant relationship between procedural fairness and job satisfaction. The reason is the employees do not have an empowered voice in decision-making; decisions are made at the top level, but procedural fairness plays a major role in employee perceptions for job satisfaction.

Further research conducted by Beuren and colleagues (2017: 80) in the accounting service provider sector in Brazil shows that organizational justice has a strong positive relationship with job satisfaction. This is because it is related to payments and promotions. In addition, it considers not only the reward but also the procedure for giving benefits and fair treatment, the explanation of the procedure, and the right given to employees to contribute to decisions taken by the organization. Nojani and colleagues (2012: 2904) performed the same research in the education sector in Iran, which showed a significant correlation between organizational justice and job satisfaction. In addition, another study founded by Suifan, Diab, and Abdallah (2017: 1143) in the airline industry in Jordan shows that organizational justice has a significant positive effect on job satisfaction. This proves the importance of using all aspects of organizational justice in the company. This research shows that not only distribution justice has an effect on the company, but fair systems and procedures (procedural justice) and treating employees with dignity and respect (interactional justice) also have an important role to play.

2.2 *Organizational justice and organizational commitment*

Another study conducted by Wiwiek and Sondakh (2015: 75) on traditional family businesses in Indonesia shows that distributive justice has a positive and significant effect on organizational commitment. On the other hand, interesting findings show a positive but insignificant influence between procedural fairness and employee commitment. Subsequent studies conducted by Rai (2013: 276) on the health sector in the United States show that distributive and procedural justice influence organizational commitment. This means that the decisions taken by superiors in a company will influence employee behavior. On the other hand, a study founded by Suifan and colleagues (2017: 1144) in the airline industry in Jordan shows that organizational justice has a significant positive effect on organizational commitment.

Furthermore, research conducted by Jawad and colleagues (2012: 39) in the education sector shows that distribution, procedural, and interactional justice will lead to increased commitment. The study was also supported by research conducted by Rahman and colleagues (2015: 194) on the education sector in Pakistan, which suggested that distribution and procedural justice have a significant positive effect on organizational commitment. The application of distributive and procedural justice can increase the level of commitment of employees. This can alleviate organizational fears of lack of employee loyalty which can later enable success in achieving goals efficiently and effectively. In addition, a study founded by Elamin (2012) in Saudi Arabia shows that procedural justice leads to organizational commitment. But procedural justice has no influence on organizational commitment when the influence of interactional and distributive justice has been controlled.

The research conducted by Bazgir, Vahdati, and Nejad (2018: 11) on the education sector in Iran shows that organizational justice has a positive and significant influence on organizational commitment. This is also supported by research conducted by Turgut, Tokmak, and Gucel (2012: 27) in the education sector. Other findings made by Dorji and Kaur (2019: 439) in the education sector in Bhutan and by Friday and Ugwu (2019: 27) in the Nigerian education sector suggest that organizational justice is significantly and positively related to organizational commitment. Furthermore, a study conducted by Suliman and Kathairi (2013: 98) in the government sector of Dubai shows that procedural and interactional justice has a positive and significant relationship to organizational commitment. In addition, Li (2018: 10) conducted a study in the education sector in China that showed that in order to build employee commitment to the organization, it is very important to apply the principles of justice throughout the organization's management. Another study performed by Luo, Marnburg, and Law (2017: 1179) in the hotel industry in China show that procedural justice has a strong impact on organizational commitment. Companies need to develop procedures that guarantee employee rights because in China, the hotel industry provides the lowest average salary compared to other industries. Therefore, companies need to provide career growth opportunities for employees.

3 CONCLUSION

According to the literature studies presented in this paper, we can conclude that organizational justice has a significant positive effect on job satisfaction and organizational commitment. Besides that, previous studies have revealed negative influences or insignificance between variables. This can occur due to several factors such as differences in the sectors carried out in each study, cultural background, the application of policies in each organization, and the background of each individual. That makes up the difference from one study to another.

In addition, the literature studies that have been presented show that interactional justice has the most influence on job satisfaction compared to other dimensions of justice. This means that companies or organizations need to maintain a high level of interactional justice and make employees feel more comfortable, because employees who have high levels of satisfaction will increase productivity compared to employees who feel less satisfied. In contrast to organizational commitment, procedural justice plays the most important role in increasing commitment. The implementation of procedural justice would have a positive impact on employees by increasing motivation, work engagement, and commitment to work.

4 MANAGERIAL IMPLICATIONS

This study provides a solution for a company's long-term strategy. If a company wants to increase commitment and satisfaction, it is necessary to pay more attention to the development of programs and policies that can encourage justice. Leaders need to apply organizational justice from the perspective of their employees and not just rely on their own observations and assessments. This can be done through surveys, suggestion or complaint boxes, and more transparent policies. This strategy can help to bridge the gap that exists between superiors and subordinates.

In addition, the authors have tried to explain the managerial implications that can be used as reference for an organization in implementing organizational justice in the future in order to increase employee job satisfaction and commitment, which will later have an impact on increasing the full contribution to achieve company goals, including:

1. Distributive justice can be applied in the form of providing incentives, additional benefits, and bonuses to employees as needed, ensuring equitable distribution of workloads and tasks, standardization of salary, and communication systems that can support the distribution. In addition, the distribution system adopted by the company must be based on the contribution that each employee has made to the company.
2. Procedural justice can be carried out by showing consistency, lack of biased behavior, accuracy in decision-making, ethical behavior between management and employees, transparency of reports in each department so as to support the goals and missions of the company. In addition, companies need to involve each member of the organization in the decision-making process. Human resource managers need to ensure that the procedures implemented meet official standards that guarantee that the implementation of decisions is based on objective information, facilitate employees who want to provide advice and responses to feel included in decision-making, provide opportunities for employees to reconsider and change a decision, and determine punishments for unethical acts.
3. Interactional justice can be applied by showing respect and dignity and sharing relevant information between company and employees, maintaining a work environment that reflects mutual respect, treating employees well, improving communication skills through managerial training to enable effective communication, and evaluating and establishing an objective interview system for the promotion of employees.

The higher the perception of fairness, the higher the level of job satisfaction. This finding can help companies in managing their organizations better than before. When employees experience fair treatment from management, it will affect work productivity positively, create attachment to the workplace, and help employees improve their performance. Furthermore, employees who have a high commitment to an organization will improve certain behaviors such as aligning with the goals and advantages of the organization, daring to take risks for the organization, and desiring to remain in the organization. For the company, this will also have an impact by reducing the level of turnover intention and increasing performance and work effectiveness.

5 LIMITATIONS

Several limitations of this study should be addressed. First, this study was limited in its scope by its concentration on the influence of organization justice on job satisfaction and organizational commitment. Second, the dimensions of job satisfaction and organizational commitment are not the same in every study that allows differences in results. Third, the impact of culture was not controlled in this study and may have influenced perceptions of justice.

REFERENCES

Akram, M. U., Hashim, M., Khan, M. K., Zia, A., Akram, Z., Saleem, S., & Bhatti, M. 2015. Impact of organizational justice on job satisfaction of banking employees in Pakistan. *Global Journal of Management and Business Research: Administration and Management* 15(5), 6–16.

Bazgir, A., Vahdati, H., & Nejad, S. H. M. 2018. A study on the effect of organizational justice and commitment on organizational citizenship behavior. *International Journal of Administrative Science & Organization* 25(1), 8–15.

Beuren, I. M., Santos, V. D., Marques, L., & Resendes, M. 2017. Relation between perceived organizational justice and job satisfaction. *Journal of Education and Research in Accounting* 11(sp. art. 4), 67–84.

Bilal, A. R., Muqadas, F., & Khalid, S. 2015. Impact of organizational justice on job satisfaction with mediating role of psychological ownership. *GMAJS* 5(2), 63–74.

Dorji, C., & Kaur, K. 2019. The impact of organizational justice on organizational commitment: A perception study on teachers of Bhutan. *International Journal of Recent Technology and Engineering* 7(6s5), 436–440.

Elamin, A. M. 2012. Perceived organizational justice and work-related attitudes: A study of Saudi employees. *World Journal of Entrepreneurship, Management and Sustainable Development* 8(1), 71–88.

Friday, E. O., & Ugwu, J. N. 2019. Organisational justice and employee commitment of selected private secondary school teachers in Nigeria. International Journal of Management & Entrepreneurship Research 1(1), 18–30.

Gunawan, A., & Nurmadiansyah, M. T. 2017. Pengaruh Keadilan Distributif dan Prosedural Penilaian Kinerja Pada Kepuasan Karyawan BPR Syariah Di Yogyakarta. *Jurnal Manajemen Dakwah* 3(1) 51–64.

Iqbal, K. 2013. Determinants of organizational justice and its impact on job satisfaction, a Pakistan base survey. *International Review of Management and Business Research* 2(1), 48–56.

Jawad, M., Raja, S., Abraiz, A., Tabassum, T. M. 2012. Role of organizational justice in organizational commitment with moderating effect of employee work attitudes. *IOSR Journal of Business and Management* 5(4), 39–45.

Khan, B., & Mahboob, F. 2017. Organizational justice and its impact on job satisfaction in public sector universities of Peshawar. *Arabian Journal of Business and Management Review* 7(5), 1–7.

Kristanto, S., Rahyuda, I. K., & Riana, I. G. 2014. Pengaruh Keadilan Organisasional Terhadap Kepuasan Kerja dan Dampaknya Terhadap Komitmen dan Intensi Keluar Di PT Indonesia Power UBP Bali. *E-Jurnal Ekonomi dan Bisnis Universitas Udayana* 3(6), 308–329.

Li, Y. 2018. Linking organizational justice to affective commitment: The role of perceived supervisor support in Chinese higher education settings. *Asia-Pacific Journal of Teacher Education*, 1–14.

Luo, Z., Marnburg, E., & Law, R. 2017. Linking leadership and justice to organizational commitment. *International Journal of Contemporary Hospitality Management* 29(4), 1167–1184.

Nojani, M. I., Arjmandnia, A. A., Afrooz, G. A., & Rajabi, M. 2012. The study on relationship between organizational justice and job satisfaction in teachers working in general, special and gifted education Systems. *Procedia Social and Behavioral Sciences* 46, 2900–2905.

Putra, I. G. E. S. M., & Indrawati, A. D. 2018. Pengaruh Keadila Organisasi Terhadap Kepuasan Kerja dan Komitmen Organisasional Di Hotel Rama Phala Ubud. *E-Jurnal Manajemen Unud* 7(4) 2010–2040.

Rahman, A., Shahzad, N., Mustafa, K., Khan, M. F., & Qurashi, F. 2016. Effects of organizational justice on organizational commitment. *International Journal of Economics and Financial Issues* 6(S3) 188–196.

Rai, G. S. 2013. Impact of organizational justice on satisfaction, commitment and turnover intention: Can fair treatment by organizations make a difference in their workers' attitudes and behaviors. *International Journal of Human Sciences* 10(2), 260–284.

Suifan, T. S., Diab, H., & Abdallah, A. B. 2017. Does organizational justice effect turnover-intention in a developing country? The mediating role of job satisfaction and organizational commitment. *Journal of Management Development* 36(9), 1137–1148.

Suliman, A., & Kathairi, M. A. 2013. Organizational justice, commitment and performance in developing countries. *Emerald Group Publishing* 35(1), 98–115.

Turgut, H., Tokmak, I., & Gucel, C. 2012. The effect of employees organizational justice perception on their organizational commitment: A university sample. *International Journal of Business and Management Studies* 4(2), 21–30.

Usmani, S., & Jamal, S. 2013. Impact of distributive justice, procedural justice, interactional justice, temporal justice, spatial justice on job satisfaction of banking employees. *Review of Integrative Business & Economics Research* 2(1), 351–383.

Wiwiek & Sondakh, O. 2015. Pengaruh Keadilan Organisasional Pada Motivasi Karyawan dan Komitmen Organisasional. *Jurnal Siasat Bisnis* 19(1), 69–77.

Zainalipour, H., Fini, A. K. S., & Mirkamali, S. M. 2010. A study of relationship organizational justice and job satisfaction among teachers in Bandar Abbas Middle School. *Procedia Social and Behavioral Sciences* 5, 1986–1990.

Influence of work-life balance on organizational commitment and job satisfaction of mothers working as preschool teachers in Cimahi, Indonesia

V.S. Marinda & N. Ramadhan
Widyatama University, Bandung, Indonesia

ABSTRACT: The aim of this study was to investigate the influence of work-life balance on organizational commitment and job satisfaction of mothers working as preschool teachers. A descriptive research method was used to get information about work-life balance, organizational commitment, and job satisfaction. A verificative research method was used to study the influence of work-life balance on organizational commitment and job satisfaction. A survey questionnaire was used to collect the data from mothers working as preschool teachers by taking a sample of 100 preschool teachers who are also mothers in Cimahi, Indonesia. Path analysis revealed that the work-life balance of mothers working as preschool teachers in Cimahi has a significant positive influence on organizational commitment and job satisfaction. The result suggested that work-life balance and job satisfaction cause high organizational commitment.

1 INTRODUCTION

Every employee reaches a level where he identifies himself with his company and decides to continue to contribute actively to the organization. Various conditions that occur in the organization cause each individual to show different levels of desire to contribute to his organization, or organizational commitment. Different factors indicate the level of employees' organizational commitment, such as a long time spent serving or working in the organization.

Individuals in an organization, both men and women, of course, show different organizational commitment. Judging from the current trend, the number of female employees working in an organization continues to grow. In recent years, in Indonesia, the number of workers has increased, especially women, which can be seen through Tingkat Partisipasi Angkatan Kerja (TPAK) data released by the central statistics agency, or Badan Pusat Statistik (BPS), as shown in Table 1.

Based on these data, we can see that the TPAK of women has increased every year. That is, the number of female workers in Indonesia has increased every year. The phenomenon of increasing women's TPAK also occurs among female preschool teachers (PAUD) in Cimahi, Indonesia. Recapitulation data from the Ministry of Education and Culture reveal that in 2018, the number of PAUD was 921, while in 2019, that number had increased by 18.46% to 1,091.

The rising trend is motivated by a variety of factors, especially for PAUD. One of the drivers of this trend is the balance between work and life. When employees perceive that there is no balance in their lives because a lot of time is consumed by work, this will make employees try to consider other work alternatives that allow them to balance roles at work and home (Posig & Kickul, 2004). In addition, the balance between life and work will have an impact on work satisfaction, which ultimately determines employees' level of commitment to the organization. Therefore this study aimed to determine and analyze work-life balance, job satisfaction, and organizational commitment among mothers working as preschool teachers in Cimahi, Indonesia.

Table 1. TPAK 2017–2019.

TPAK	2017	2018	2019
Men	83.05%	83.01%	83.18%
Women	55.04%	55.44%	55.50%

Source: BPS (2019)

2 LITERATURE REVIEW

2.1 Work-life balance

Lockett and Mumford (2009) argue that the purpose of work-life balance is to produce opportunities to gain control over when, where, and how we work so that we can produce the best and also have time to recover and pursue interests outside of our occupation. Akter, Hossen, and Islam (2019) contend that work-life balance is about finding a good match between professional roles and personal activities that are considered important for someone such as recreation, personal activities, family responsibilities, and other social activities. Work-life balance can be measured through several things, according to McDonald and Bradley (2005), namely balance of time given to work and activities outside of work and balance of involvement, related to the level of psychological involvement and commitment inside or outside work.

2.2 Job satisfaction

Veithzal (2008) suggests that job satisfaction is an evaluation that describes someone's attitude of being happy or not at work. Sutrisno (2010) reveals the factors that influence job satisfaction:

a. Psychological factors include an employee's psyche, which includes interests, serenity at work, attitude to work, talents, and skills.
b. Social factors include social interaction among employee and between employees and their superiors.
c. Physical factors include the type of work, work hours and rest time, work equipment, the state of the room, temperature, lighting, air exchange, employee health conditions, age, and so on.

2.3 Organizational commitment

Newstrom and Davis (2015) state that organizational commitment reflects employees' confidence in the mission and goals of the organization, the desire to work hard, and loyalty to the organization. Mayer and Allen (cited in Titisari [2014]) divide organizational commitment into three dimensions:

a. Affective commitment, which refers to emotions that attach to individual employees to identify and involve themselves in the organization.
b. Normative commitment, which refers to employees' feelings about their obligations as employees of the company.
c. Continuous commitment, which refers to employees' awareness related to the consequences that will be obtained when he leaves the organization.

Previous research shows that work-life balance and job satisfaction influence organizational commitment. Akter, Hossen, and Islam (2019) have performed previous research on the effect of work-life balance on organizational commitment with the title "Impact of Work-Life Balance on Organizational Commitment of University Teachers: Evidence from Jashore University of Science and Technology," which states that a positive and significant effect exists between work-life balance and organizational commitment. Dr. Upasna Joshi Sethi (2014) carried out similar research with the title "Influence of Work-Life Balance on Organizational Commitment: A Comparative Study of Women Employees Working in Public and Private Sector Banks," the results of which indicate a positive relationship between work-life balance and organizational commitment. In addition to work-life balance, other research also shows the effect of job satisfaction on organizational commitment. Previous research conducted by Imam, Raza, and Ahmed (2014) with the research title "Impact of Job Satisfaction on Organizational Commitment in Banking Sector Employees of Pakistan" shows that job satisfaction has a significant positive influence on organizational commitment.

Based on the literature review and previous research results that have been submitted in the earlier literature review section, the following framework can be made:

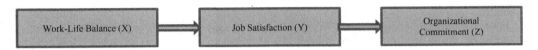

Figure 1. Research paradigm.

3 METHODOLOGY

3.1 *Participants*

Based on the phenomena occurring in the field, the background of the research, and the literature review previously discussed, this study involved mothers working as preschool teachers in Cimahi, Indonesia, as its unit of analysis.

The number of women preschool teachers in Cimahi is 1,091, but no definitive data exist on the number of preschool teachers in Cimahi who are mothers. Because the population of this study is not known with certainty, in determining the number of samples, this study used the Lemeshow formula as follows:

$$n = \frac{z_{1-a/2}^2 P(1-p)}{d^2}$$

n   : number of sample
$Z^2_{1-a/2}$: normal standard (if a: 0.05, Z: 1,960)
P(1−P) : proportion estimation of population (if P:0.5, P(1−p):0.5)
d   : deviation (10%)
number of sample:

$$n = \frac{(1,960)^2(0.25)}{(0,10)^2}$$
$$n = 96.04$$

Therefore this study collected data from a sample with a minimum number of 96 people or rounded up to 100 people and used quota sampling techniques.

3.2 *Measurements*

To ensure that all data used in this study were valid, a validity test with the Pearson Product-Moment Correlation was carried out and all questionnaire items were deemed valid. Besides that, a reliability test was performed with the Cronbach's alpha coefficient (α) and all questionnaire items were found to be reliable.

3.3 Data analysis

This study used descriptive statistical analysis methods to analyze data by describing the data collected from respondents' answers to questionnaire items. Statistical analysis of inference was employed to show the relationship between the work-life balance (X), job satisfaction (Y), and organizational commitment (Z) variables.

Design analysis in this study was divided into two – namely descriptive statistics analysis – by dividing the data into four quality criteria: 0%–<25% (Not Good), 25%–<50% (Moderate), 50%–<75% (Good), and 75%–100% (Very Good). The inferential statistics analysis used path analysis.

Based on the framework and research hypothesis, a simultaneous hypothesis test can be formulated: $Ho:P_{ZX} = P_{ZY} = 0$. Work-life balance has no influence on organizational commitment with the mediation of job satisfaction of mothers working as preschool teachers in Cimahi.

Ha: at least there is a $P_{ZXY} \neq 0$. Work-life balance does influence organizational commitment with the mediation of job satisfaction of mothers working as preschool teachers in Cimahi.

4 RESULTS AND DISCUSSION

The results of the descriptive analysis address the work-life balance (X), job satisfaction (Y), and organizational commitment (Z) variables. Statistical calculations obtained a descriptive analysis, as shown in Table 2.

Table 2. Descriptive analysis of work-life balance (X).

Dimension	Percentage	Criteria
Balance of time	83.5	Good
Balance of involvement	87.1	Very Good
Balance of satisfaction	78.7	Good
Total	83.1	Good

Source: Data processing results (September 2019)

Based on Table 2, the overall quality of work-life balance is good; this is because mothers working as preschool teachers in Cimahi have a good balance in the time spent at work and time spent on social, family, and personal matters. In addition, these mothers' job satisfaction is good. The best quality is involvement; this is due to the fact that mothers working as preschool teachers in Cimahi feel a good balance of life in their involvement in their work environment, and in their social, family, and personal lives.

Table 3. Descriptive analysis of job satisfaction (Y).

Dimension	Percentage	Criteria
Psychological factors	83.7	Good
Physical factors	83.8	Good
Social factors	81.3	Good
Financial factors	67.3	Moderate
Total	79.0	Good

Source: Data processing results (September 2019)

Based on Table 3, the overall quality of job satisfaction is a good; this is because mothers working as preschool teachers in Cimahi are satisfied with psychological factors such as the suitability of work with their interests, skills, and talents. Satisfaction is also felt for physical factors such as comfort in the workplace. Satisfaction was expressed about social factors that include work relationships with colleagues and subordinates. Satisfaction with factors related to finances such as income, bonuses, and working facilities is still moderate.

Table 4. Descriptive analysis of organizational commitment (Y).

Dimension	Percentage	Criteria
Affective commitment	83.0	Good
Normative commitment	74.6	Good
Sustainable commitment	67.6	Moderate
Total	76.1	Good

Source: Data processing results (September 2019)

Based on Table 4, the quality of organizational commitment for mothers working as preschool teachers in Cimahi on the whole has good quality; this is because these mothers have good-quality affective and normative commitments, while their continuing commitments to the workplace were moderate.

This study used path analysis in its statistical analysis, which requires at least interval-scale data, so the ordinal-scale variable data must be changed first to interval data. After the interval-scale research data was obtained, path analysis was used to determine the effect of work-life balance (X/independent variable) on organizational commitment (Z/dependent variable) and job satisfaction (Y/intervening variable).

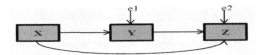

Figure 2. Path coefficient and path diagram model 2.

The model 2 path diagram depicted in Figure 2 shows a direct effect of work-life balance (X) on job satisfaction of 0.240 or 24.0%. The indirect effect of work-life balance (X) on organizational commitment (Z) is 0.578 x 0.463 = 0.268 or 26.8%. The effect of total work-life balance (X) on organizational commitment (Z) and job satisfaction (Y) is 0.162 + 0.268 = 0.43 or 43%.

5 CONCLUSIONS AND RECOMMENDATIONS

Work-life balance of mothers working as preschool teachers in Cimahi, Indonesia, is generally good; this means that these mother experience a good balance between life and work. Job satisfaction of mothers working as preschool teachers in Cimahi is generally good; this means that they are satisfied in carrying out their work. Organizational commitment of mothers working as preschool teachers in Cimahi is generally good; this means that they are comfortable and want to remain and contribute to the school where they teach.

The influence of work-life balance on organizational commitment and job satisfaction of mothers working as preschool teachers in Cimahi is significant and moderate; this means that better work-life balance will increase job satisfaction so these mothers' organizational commitment will be even higher.

Based on these research results and conclusions, the recommendations of this study are that preschools in Cimahi should further increase the level of job satisfaction, especially in financial factors, by reevaluating the compensation and material rewards given to mothers working as preschool teachers in Cimahi so as to make them increasingly committed to the organization. Further research needs to be done related to this so that it can be known what other factors can have a greater influence on organizational commitment than work-life balance and job satisfaction for mothers working as preschool teachers in Cimahi.

ACKNOWLEDGMENT

This study was supported by preschool teachers in Cimahi, Indonesia, in terms of data needs. In addition, this research was also supported by the P2M University of Widyatama and the Ministry of Research and Technology for Higher Education (Kemenristekdikti) research team, so that we got a Penelitian Dosen Pemula Grant in order to progress this research.

REFERENCES

Akter, A., Hossen, M. A., & Islam, Md. N. 2019. Impact of work-life balance on organizational commitment of university teachers: Evidence from Jashore University of Science and Technology. *International Journal of Scientific Research and Management* 7(4), 1073–1079.

Badan Pusat Statistik (BPS). 2019. Keadaan Ketenagakerjaan Indonesia. Accessed August 3, 2019, from https://www.bps.go.id/pressrelease/download.html?nr=MTU2NA%3D%3D&sdfs=ldjfdifsdjkfahi&t woadfnoarfeauf=MjAxOS0wOC0yNCAyMjo1OTo1OA%3D%3D.

Direktorat Jenderal PAUD dan Dikmas. 2019. Jumlah Pendidik PAUD Berdasarkan Jenis Kelamin. Accessed August 3, 2019, from https://manajemen.paud-dikmas.kemdikbud.go.id/Rekap/PAUD-Satuan-Pendidikan.

Imam, A., Raza, A., & Ahmed, M. 2014. Impact of job satisfaction on organizational commitment in banking sector employees of Pakistan. *World Applied Science Journal* 28(2), 271–277.

Lockett, K., & Mumford, J. 2009. *Work/life balance for dummies* (pp. 10–11). Chichester: Wiley.

McDonald, P., & Bradley, L. M. 2005. The case for work/life balance: Closing the gap between policy and practice. 20:20 Series, 15.

Newstrom, J., & Davis, K. 2015. *Human behavior of work: Organizational behavior*. 13th edition. New York: McGraw-Hill College.

Posig, M., & Kickul, J. 2004. Work-role expectations and work-family conflict: Gender differences in emotional exhaustion. *Women in Management Review* 19(7), 373–386.

Sethi, U. J. 2014. Influence of work-life balance on organizational commitment: A comparative study of women employees working in public and private sector banks. *European Journal of Business and Management* 6(34), 215–219.

Sugiyono. 2012. *Metode Penelitian Kuantitatif, Kualitatif dan R&D*. Bandung: Alfabeta.

Sutrisno, E. 2010. *Budaya Organisasi*. Jakarta: Kencana.

Titisari, P. 2014. *Peranan organizational citizenship behavior (OCB)*. Jakarta: Kencana.

Veithzal, R. 2008. *Manajemen Sumber Daya Manusia untuk Perusahaan*. Jakarta: TP. Raja Grafindo Persada.

The strategic roles of overseas Indonesian students' organizations in promoting brain circulation

U.W. Ruhmana & M. Hanita
School of Strategic and Global Studies, Universitas Indonesia, Jakarta, Indonesia

ABSTRACT: Continuing study abroad is becoming increasingly popular among Indonesian students. It is a strong requirement of Indonesia to pursue the international competitiveness related to the Fourth Industrial Revolution (4IR). Overseas Indonesian students' organizations take an important role in supporting the country in achieving this agenda. The purpose of this research was to analyze the role of the Overseas Indonesian Students' Association Alliance (OISAA) in creating and maintaining the brain circulation mindset of the organization members in their homeland. This study adopted a qualitative research method in order to examine the OISAA program by looking at its annual symposium over the past ten years. The results found that the OISAA has contributed to promoting turning brain drain into brain circulation both for Indonesian domestic students and for Indonesian diaspora students. It provides three strategic roles: encouraging students to study abroad then contribute to their homeland, creating community knowledge both in the OISAA and in the homeland, and raising the national resilience indexes in the ideological sector and among human resources.

1 INTRODUCTION

Brain drain is a source of ongoing concern in the Association of Southeast Asian Nations (ASEAN). Brain drain, also known as human capital flight, has been considered a significant inhibiting factor in a country's economic and social development (Batalova, Shymonyak, & Sugiyarto, 2017). This is due to ASEAN nations' large-scale investment in tertiary education, from governments to young people through scholarships. The chance to get an international education is widely opened, and this is in line with the motivation of students in ASEAN countries. They are the human capital who are then faced with numerous opportunities and networks at the international level. Changes in educational levels and the ability of human resources will greatly influence the potential for brain drain.

Indonesia is among the nations that face brain drain. Indonesia should recognize brain drain as a problem that is increasingly developing and relevant due to modern economic development. Large investments in tertiary and international education should not be threats for Indonesia, but an opportunity to increase the number of highly skilled workers who contribute to the country. Therefore, the government should support nationalism among students of the Indonesian diaspora in order to overcome brain drain. The Overseas Indonesian Students' Association Alliance (OISAA) or PPI Dunia has taken this role. As shown in Table 1, college enrollment has increased in large number, even though it still fluctuates. The largest proportional increase was in 2013.

In Indonesia, studies on overseas Indonesian students' associations are uncommon. From 2015 to 2019, the topics of study revolved around students' role in Indonesian politics and on achieving independence, such as the voice of freedom and anticolonialism. We argue that the role of overseas Indonesian students in maintaining the nationalism of diaspora students has been neglected. They still contribute to maintaining Indonesia's independence, especially in the era of international competition. This dearth mainly happened in the failure to protect diaspora students from brain drain or human capital flight. The role of the youth organization

Table 1. Indonesia gross enrollment ratio in tertiary education.

Date	Value	Change (%)
2015	24.3	−22.02
2014	31.1	−0.59
2013	31.3	2.05
2012	30.7	15.67
2011	26.5	9.52
2010	24.2	4.95

Source: World Data Atlas "Indonesian Tertiary Education," accessed November 14, 2019, https://knoema.com/atlas/Indonesia/topics/Education/Tertiary-Education/Gross-enrolment-ratio-in-tertiary-education

is very strategic in helping to maintain nationalism and promoting national identity by transforming brain drain into brain circulation. Our aim in this paper is to begin to fill this gap by examining the role of the Overseas Indonesian Students' Association Alliance (OISAA) in promoting brain circulation among its members.

In conducting this study, this paper gives three contributions to the dynamic of youth organizations, the phenomenon of brain circulation, and the demands of global competition in the era of the Industrial Revolution 4.0. First, this paper highlights how the OISAA represents Indonesia internationally. This paper explores the OISAA's role in encouraging Indonesian diaspora students to maintain their national identity and promote it across the world. Through various organizational programs, the OISAA can unite the visions of Indonesia, both for members of the organization and for citizens of the world. This spirit of nationalism is able to call them to always contribute wherever they are. This discussion also contributes to burgeoning work on the political geographies of children and young people (Skelton, 2013).

Second, this paper shows the importance of creating community knowledge, conducted by the OISAA as the place for sharing and learning culture, which has benefits both for domestic and diaspora students. Third, this paper shows that, through its activity, the OISAA also contributes to Indonesia's national resilience in maintaining the ideology and human resources against the threat of brain drain or human capital flight. Through these three contributions, the OISAA has successfully promoted brain circulation to its members and to students in Indonesia.

2 LITERATURE REVIEW

In 2018–2019, period Indonesia ranked 50th out of 141 countries on the Global Competitiveness Index, down from its previous position at 37th. The government strives to continue to improve the rating with various steps in Industry 4.0. In 2018, the president of Indonesia, Joko Widodo, officially launched "Making Indonesia 4.0" as the strategy to face the Industrial Revolution 4.0. The launch was initiated by the Ministry of Industry of the Republic of Indonesia with the following strategy and road map to improve Indonesia's global competitiveness. Under the "Making Indonesia 4.0," initiative, Indonesia will encourage "10 national priorities" so as to accelerate the manufacturing industry in Indonesia.

One of the national priority points is "quality improvement of human resources." Improving human resources is important to achieve in order to ensure the successful implementation of Making Indonesia 4.0. Indonesia plans to overhaul its educational curriculum with more emphasis on STEAM (science, technology, engineering, the arts, and mathematics), aligning the national education curriculum with the future needs of industry. Indonesia will cooperate with industry actors and foreign governments to improve the quality of vocational schools, while also improving the global workforce mobility program so as to utilize the availability of human resources in accelerating the transfer of capabilities. In this case, the existence of the OISAA is very crucial to help the country implement this agenda successfully.

Youth organizations have a great vision to increase the moral quality of their members, especially their nationalism (Jones, Merriman, & Mills, 2016). As we know, the term *brain drain*, popularized in 1950 (Robertson, 2006), refers to the United States when what was called the "new knowledge economy" played a big role in political and economic issues. Global competition is attracting the best brains from around the world in order to innovate and generate excellent ideas to gain profits and patents.

While discussions of international student mobility often consider responsibility in connection with students' right (Romi, Lewis, and Katz 2009), students studying abroad can contribute many thing to their country. Tseng and Newton (cited in Triana, 2015) list eight characteristics of a successful international student experience: (1) knowing themselves and others, (2) building relationships, (3) expanding their worldview, (4) asking for help, (5) developing cultural and social contacts, (6) establishing relationships with advisors and instructors, (7) English proficiency, and (8) letting go of problems. These characteristics are the competencies and skills of international students that they can share with domestic students as a form of knowledge and culture sharing. Moreover, such educational efforts can be well organized with the country to shape future educational and socioeconomic development as the government's partner due to the needs of Making Indonesia 4.0.

The definition of "brain circulation" from Mahroum (1999) is the two-way flow of experts between the country of origin (COO) and the receiving country. When an expert is in the receiving country, he represents brain gain for that country and brain drain for the COO. But when the expert returns to the COO, he becomes brain gain for the COO and brain drain for the receiving country. Batalova, Shymonyak, and Sugiyarto (2017) mention three important developments in encouraging brain circulation to the country: first, an awareness that the diaspora plays an important role in regional economic growth; second, an educational infrastructure that supports the growth of student mobility in the region; and, third, a recognition of professional qualifications based on the targeted developing field.

Rizvi (2005) contends that beside experts, international students play a significant role in conducting brain circulation as they have a good capability to adapt to international changes and have a chance to bring development to the COO. Gamlen, Cummings, and Vaaler (2019) argue that the diaspora has a significant role in promoting national identity to the world and contributing to achieve national goals. In this case, we claim that the OISAA gives big contributions in promoting brain circulation among its members and domestic students.

3 METHODOLOGY

3.1 *Data collection*

This study adopted a qualitative research method to study the case of the OISAA program by looking at its annual symposium over the past ten years. The result examined all programs the OISAA has done during each period. Drawing on documentary and archival research, the study discussed how the organization has fostered national identities and promoted a brain circulation mindset among its members. The document provided by the OISAA from 2009 to 2019 can also be accessed on its website ppi.id. This research also used participant observers to collect data by becoming a delegate at the eleventh international symposium of the OISAA in Johor Bahru, Malaysia, in 2019.

3.2 *Data analysis*

This study used several data analysis techniques, including (1) data reduction, (2) data display, and (3) drawing conclusions. Data reduction in this study means to classify, sharpen, and direct the data obtained and eliminate data that are deemed unnecessary. Data display came in the form of compilation of data, which can be in the form of narrative text until the process of drawing conclusions as a result of the research.

4 RESULTS AND DISCUSSION

The vision of the OISAA was clearly reflected in its founding background. In September 2007, PPI Australia held an international conference of Indonesian students (KIPI) with the theme "Building the competitiveness of the nation: Returning or serving from afar." The conference was opened by Menpora Adiyaksa Dault and closed by Minister of the Environment Rachmat Witoelar. In this meeting, thirteen student research papers related to the main theme of the conference were presented and discussed to see how and to what extent the phenomenon of brain drain affects the growth and economic development of the country, as well as how to formulate new nationalism that is more relevant and contextual with contemporary conditions. A global network of Indonesian communities abroad, the OISAA, also called Perhimpunan Pelajar Indonesia Se-Dunia (PPI Dunia), was established when the declaration was conducted by the seven PPI Negara members (Australia, India, Malaysia, Italy, Egypt, Netherlands, and Japan). On September 9, 2007, the Indonesian Students' Organization Forum was strengthened into the Overseas Indonesian Students' Association Alliance (OISAA), better known internationally as the Indonesian Students' Association (PPI).

4.1 *Spirit to study abroad and contribute to the homeland*

The OISAA's strategy in increasing the spirit of learning and contribution is implemented in two programs – namely the Festival Luar Negeri/Felari (a foreign festival) and an international symposium. Felari PPI Dunia is a division of the PPI Dunia Daily Board, which has as its main task the implementation of Edufair PPI Dunia on Indonesian campuses. Felari is also responsible for the provision of scholarship information from various countries in the world that have been verified by students or PPI Dunia members.

In carrying out its duties, Felari also has a work program that supports the main task of making infographics, study guides, and scholarships, as well as an open discussion for all Indonesians who plan to study abroad. Other forms besides Edufair – e.g., talk shows/seminars/webinars – are utilized in accordance with the agreement with the organizer. This program has carried out diverse activities in various universities in Indonesia such as Universitas Negeri Solo and Universitas Jenderal Sudirman, and has held many webinars through PPI TV, PPI Radio, and other online groups.

The PPI Dunia International Symposium is the highest decision-making organ of the PPI Dunia and is held annually. To date, the World PPI International Symposium has been held eleven times since 2009: Den-Haag, Netherlands (2009), London, United Kingdom (2010), Kuala Lumpur, Malaysia (2011), New Delhi, India (2012), Bangkok, Thailand (2013), Tokyo, Japan (2014), Singapore (2015), Cairo, Egypt (2016), Warwick, United Kingdom (2017), Moscow, Russia (2018), and Johor Bahru, Malaysia (2019).

This vision continued with the eleventh international symposium of PPI Dunia in Johor Bahru, Malaysia, with the theme of "Sustainable innovation as Indonesia youths' contribution to the independence of the nation." This theme was born out of the results of studies conducted by the PPI Dunia Study and Movement Center under the title "Indonesian Policy Outlook 2019." It simply contained eight points of discussion regarding the challenges of sustainable innovation in Indonesia and was discussed with the Indonesian legislative assembly in the symposium.

This symposium was followed by internal PPI Dunia members, but students in Indonesia also came together so as exchange ideas for the development of Indonesia. This forum then established intensive communication between members of PPI Dunia and Indonesian domestic students. This symposium activity lasted several days with the output contained in several forms of activities in accordance with each commission to be implemented in the next management period and to be recommended to the president of Indonesia.

Both of these programs are very relevant and significant in promoting brain circulation, especially with the current conditions and the need to increase international competitiveness as the challenge for Indonesia. Felari provides great motivation and support for young people to get an international education, and also directs diaspora students to always contribute to the country through symposiums and programs. The young people who take part in this program are inspired to get an international education and begin to think about building the nation after completing their studies.

4.2 *Create community knowledge*

PPI Dunia was established on the basis of the need to develop and promote the unitary state of the Republic of Indonesia, according to the goals and objectives of the nation listed in the opening of the 1945 constitution, "to educate the life of the nation." One of the efforts to advance the national education is to increase the capacity and quality of teachers/educators in Indonesia. Departing from this idea, the PPI Dunia 2018–2019 Education Commission is reconducting its program to help teachers to see the world (Bantu Guru Melihat Dunia [BGMD]). Through this program, teachers have an opportunity to share knowledge and learn culture, which can improve their own knowledge and generate innovative ideas for Indonesia due to the STEAM curriculum.

The role of the teacher is very important to maintain and upgrade the quality of education in Indonesia, especially early childhood education. Helping teachers to see the world is recognized as a strategic way to pursue global competition. The government already has many scholarship programs for teachers to continue study abroad, but not many training programs to improve the quality of teaching like short courses or teachers' exchange programs with other countries.

4.3 *Increase in national resilience indexes*

The OISAA spreads the culture, food, and sports of indigenous Indonesians in all countries around the world through its annual Indonesian Day. In its implementation, the OISAA cooperates with the Indonesian embassies in each country and with local governments in order to support the event's success. Indonesian Day is part of the "PPI Goes to Campus" program held annually to promote many Indonesian cultures internationally, especially among its members and students. This program is a strategic way to cultivate nationalism and introduce Indonesian cultural identity to the people of the world.

Indonesia has many cultures. However, in this era of globalization, not many young people know Indonesian culture well enough to preserve it. Those who continue their studies abroad are dealing with more cultures and must have a reason to still love Indonesia. This is a crucial issue related to Indonesian diaspora students' nationalism. The government still does not consider this a threat to national security, so this raises the potential for brain drain. The government should consider this issue and develop strategic ways to engage students in nationalism.

PPI Dunia made a big contribution to the establishment of Ikatan Ilmuwan Indonesia Internasional (I-4) or the Indonesian International Scientists' Association. PPI Germany initially founded the organization in mid-2007. The purpose of the establishment of I-4 was to accommodate all the potential of Indonesian scientists around the world in terms of science and technology development. Furthermore, I-4 is expected to contribute to solving the nation's problems on a concrete, periodic, and ongoing basis. The establishment of I-4 itself was declared by the Indonesian Students' Association (PPI Dunia) on July 5, 2009, in the International Symposium (SI) PPI Dunia in The Hague, Netherlands, and then formally inaugurated in Jakarta on October 24, 2009.

Currently I-4 has been working with the government of Indonesia through the Directorate General of Science and Technology Resources and the Ministry of Research, Technology and Higher Education by organizing the annual World Class Professor (WCP) program in 2015 and the World Class Symposium of Scholars (Simposium Cendekiawan Kelas Dunia) since 2016. Indonesian diaspora and domestic scientists attended in order to exchange ideas and knowledge as well as to collaborate on various studies. Diaspora scientists also visited campuses in Indonesia for more intensive discussions. Many Indonesian scientists are invited to overseas campuses where the diaspora scientists work so as to collaborate on joint research. It is evident that PPI Dunia has successfully promoted brain circulation in Indonesia.

5   CONCLUSIONS AND RECOMMENDATIONS

The purpose of this study was to investigate the strategic roles of the OISAA in promoting brain circulation. This study filled the gaps in the role of youth organizations and the dynamics of social change. The OISAA has contributed to the country through strategies that enhance the learning spirit and contribute to Indonesia, create community knowledge for sharing knowledge and learning culture, and increase national resilience in the ideological and human resource sectors. Based on the results of these studies, we argue this is the perfect time for governments, academics, and youth in general to realize the importance of overseas youth organizations in helping to defend the diaspora students' nationalism and introducing national identities to the world. Further research can examine the OISAA's empowerment strategy more closely so as to improve Indonesia's ranking on the Global Competitiveness Index through the STEAM curriculum for achieving the Making Indonesia 4.0 agenda.

REFERENCES

Batalova, J., Shymonyak, A., & Sugiyarto, G. 2017. *Firing up regional brain networks: The promise of brain circulation in the ASEAN Economic Community*. Philippines: Asian Development Bank.
Ershov, Y. 2015. National identity in new media. *Precedia* 200(22), 206–209.
Gamlen, A., Cummings, M. E., & Vaaler, P. M. 2019. Explaining the rise of diaspora institutions. *Journal of Ethnic and Migration Studies* 45(4), 492–516.
Gellner, E. 1983. *Nation and nationalism*. Ithaca, NY: Cornell University Press.
Kastoryana, R. 2018. Multiculturalism and inter-culturalism: Redefining nationhood and solidarity. *Comparative Migration Studies* 4(16), 1–5.
Koutit, K. 2013. New entrepreneurship in urban diasporas in our modern world. *Journal of Urban Management* 2(1), 25–27.
Mahroum, Sami. 1999. Highly skilled globetrotters: The international migration of human capital. (DSTI/STP/TIP (99), 2/Final/: 168–185.
Jones, R., Merriman, P., & Mills, S. 2016. Youth organizations and the reproduction of nationalism in Britain: The role of Urdd Gobaith Cymru. *Social and Cultural Geography* 17(5), 714–734.
Rizvi, F. 2005. International education and the production of cosmopolitan identities. *RIHE International Publication series* Presented at Transnational Seminar Series, University of Illinois at Urbana-Campaign, 1–11.
Robertson, S. L. 2006. Brain drain, brain gain and brain circulation. *Globalisation, Societies and Education* 4(1), 1–5.
Romi, S., Lewis, R., & Katz, Y. J. 2009. Student responsibility and classroom discipline in Australia, China, and Israel. *Compare* 39(4), 439–453.
Skelton, T. 2013. Young people, children, politics and space: A decade of youthful political geography scholarship 2003–2013. *Space and Polity* 17, 123–136.
Triana, B. 2015. Cultural demands of the host-nations: International students' experience and the public diplomacy consequences. *Journal of International Students* 5(4), 383–394.

The effect of knowledge sharing on affective commitment: The mediating role of competency development and job satisfaction (A case study of generation Y employees)

A.S. Murat & E.S. Pusparini
Graduate School of Management Science, University of Indonesia, Depok, Indonesia

ABSTRACT: The purpose of this study was to identify the effect of knowledge sharing on affective commitment with the mediating role of competency development and job satisfaction among Generation Y (Gen Y) employees in Indonesia. As a dominant generation in today's workforce, Gen Y shows a tendency to change jobs quickly and displays low affective commitment to the organization. Therefore, this study suggests, there is a positive relation between knowledge sharing and affective commitment to the organization. Three hundred and two respondents from Gen Y were analyzed in this study, four questionnaires were chosen to gather data, and structural equation modeling analysis supported model fitting with the data collected. The findings of this study were that knowledge sharing has a positive impact on effective commitment through the mediating role of competency development and job satisfaction. Knowledge sharing in the workplace can increase employees' competency, and in the same time they can gain satisfaction and finally evoke commitment to the organization.

1 INTRODUCTION

Generation Y (Gen Y), commonly called the Millennial Generation (Murray, Toulson, & Legg, 2011 is the generation of young people born between 1980 and 2000 (Hartman & McCambridge, 2011). It is one of the largest generations in the world today, and it is estimated that in 2025, Gen Y will represent 75% of the global workforce (Kim et al., 2018). In Indonesia, the number of Gen Y members has dominated, reaching 33.75% of the total population (Indonesian Central Statistics Agency, 2018). As they dominate the workforce, Gen Y members born in the early 1980s will hold senior positions and have greater authority in the workplace, and the rest may start their careers (Cattermole, 2018). Gen Y is known to have different work attitudes and behaviors from other generations, so it is very important for organizations to understand how to manage this generation (Kim et al., 2018).

In society itself, stigma states that Gen Y employees have low commitment and often change jobs (Cennamo & Gardner, 2008; Lub et al., 2012, Naim & Lenka, 2018). There is a lack of research, however, on Gen Y's affective commitment (Lub et al., 2012). Affective commitment is seen as important for dedication and loyalty to the organization (Rhoades, Eisenberger, & Armeli, 2001), and is positively related to performance (Gelderen & Bik, 2016). By being committed to the organization, someone will be bound to his work and give maximum performance. This low affective commitment on the part of Gen Y has happened because now workers tend to look for opportunities to learn and develop their competencies (Naim & Lenka, 2017). They are not willing to give more effort if they feel that what they are doing is not meaningful to their career (Jha, Sareen, & Potnuru, 2018), and in the end they show a tendency to leave. So now one of the challenges in human resources management (HRM) is how to increase the commitment of the younger generation (D'amato & Herzfeldt, 2008; Naim & Lenka, 2017).

Apart from the stigma of low affective commitment from Gen Y, it turns out that Gen Y is a generation with a need for self-development and self-actualization (Godshalk & Sosik, 2003; Naim & Lenka, 2017). Gen Y tends to look for an environment that can support them in obtaining and sharing ideas, information, and knowledge (Naim & Lenka, 2017). It is well known that knowledge has become an effective source for organizational performance (Kang, Kim, & Chang, 2008); knowledge sharing in the workplace can facilitate an increase in individual competence and simultaneously create new knowledge (Sveiby, 2001; Trivellas et al., 2015). When organization members create and have the willingness to share their knowledge, it can help individuals in the organization to learn and improve their competence through work experience. Then, the improvement of competency is related to work attitudes (Naim & Lenka, 2017; Zhang et al., 2001), such as affective commitment and job satisfaction.

In addition to knowledge sharing and competency development, one of the main determinants of the presence of affective commitment is job satisfaction (Testa, 2001; Williams & Anderson, 1991; Yew, 2007). As Twenge and colleagues (2010) (cited in Lub et al., 2012) have noted, Gen Y's low level of commitment is also caused by their dissatisfaction with their work. Research on job satisfaction and affective commitment has been widely carried out. Kim and colleagues (2018), who have examined Gen Y employees, state that job satisfaction is a mediator between employees' voice and team-member exchange with affective commitment, thus indicating a positive direct relationship between job satisfaction and affective commitment. This is because when Gen Y employees feel that their desires are heard by the organization, they have a greater sense of ownership in the organization (Brown et al., 2005; Kim et al., 2018).

Based on this discussion, we can assume that the willingness of Gen Y employees to share knowledge can improve their self-competence and job satisfaction, which in turn will encourage their sense of ownership in the organization. For this reason, this study examined the relationship of knowledge sharing with affective commitment mediated by competency development and job satisfaction in Indonesia.

2 LITERATURE REVIEW

2.1 Knowledge sharing

Knowledge sharing is basically a process in which the knowledge and experience of one organizational unit can affect other organizational units (Argote et al., 2000; Wang, Wang, & Liang, 2014). Knowledge sharing is certainly very useful for organizations because in an organization, it is very possible for differences to occur in the level of knowledge between individuals. Knowledge-sharing processes can improve organizational productivity and sustainable excellence (Liao, Fei, & Chen, 2006; Wang et al., 2014). Organizations can successfully carry out knowledge sharing by combining knowledge with their business strategies and by changing the attitudes and behaviors of their members so as to run knowledge sharing consistently (Lin, 2007).

2.2 Affective commitment

Affective commitment is an emotional attachment to the organization, and can be identified through the acceptance of organizational values and the willingness to stay with the organization (Namasivayam & Zhao, 2007). Affective commitment is one of the dimensions of organizational commitment listed by Allen and Mayer (1991), who divide organizational commitment into three dimensions – namely, affective commitment, continuance commitment, and normative commitment (Yucel, 2012). Affective commitment is the most powerful of the three in generating positive employee relations and optimal employee performance (Gao-Urhahn, Biemann, & Jaros, 2016; Meyer &

Herscovitch, 2001; Meyer et al., 2002). With employees affectively committed to the organization, we can assume, they will do their best to achieve organizational goals (Gelderen & Bik, 2016), and will behave according to organizational goals (Mayer et al., 1998; Namasivayam & Zhao, 2007).

Knowledge Sharing and Affective Commitment

Even though much research has explored the influence of affective commitment on knowledge sharing, some research discusses the opposite relationship. Naim and Lenka (2017) state that competency development mediates the relationship between knowledge sharing and affective commitment. From these results, we can assume a positive direct relationship between knowledge sharing and affective commitment.

H1: *Knowledge sharing has a positive significant effect on affective commitment.*

2.3 *Competency development*

Competence is a set of knowledge, skills, abilities, attitudes, and characteristics needed for a job (Naim & Lenka, 2016). Competence leads to an area of expertise. Competence itself has advantages not only for individuals but also for organizations. Competence is one source for organizations to achieve a competitive advantage (Barney, 1991; Heijde & Van der Heijden, 2006; Nordhaug & Gronhaug, 1994; Wright et al., 1994). Competency development is a positive change in a person's level of competence (de Vos et al., 2015; Naim & Lenka, 2017), which is obtained from individual learning processes (Eilstrom & Kock, 2008; Naim & Lenka, 2017). Based on previous research, competency development itself has a contribution in generating positive outcomes such as job satisfaction (Naim & Lenka, 2017; Sumpf, 2010; Zhang et al., 2001), new skills and competencies related to work, and better work performance (Aguinis & Kraiger, 2009; Arthur et al., 2003; Birdi et al., 1997; Kraimer et al., 2011).

Competency Development as Mediating Variable

Oyefolahan and Dominic (2013) note that knowledge management has a positive relationship with competency development, and it is known that knowledge sharing is one component of knowledge management (Reiser & Dempsey, 2012). Similar to this result, another study suggests a strong association between knowledge sharing and competency development (Naim & Lenka, 2016). Hakkarainen (2004) states that knowledge sharing can facilitate employees' competency improvement (Hakkarainen, 2004; Trivellas et al., 2015) and will ultimately result in good work behavior (Sumpf, 2010; Trivellas et al., 2015; Zhang et al., 2001). This behavior can be in the form of job satisfaction and employee loyalty (Naim & Lenka, 2017). Employee loyalty is when an employee can identify commitment to the organization (Somers, 1995; Namasivayam & Zhao, 2007).

In line with the results of the study, Naim and Lenka (2017) state that knowledge sharing has a positive relationship with competency development, which in turn can positively affect affective commitment. From the results of the aforementioned research, we can assume that the sharing of knowledge in a work environment can increase competency development, and ultimately will increase commitment to the organization.

H2: *Competency development mediates the relationship between knowledge sharing and affective commitment.*

2.4 *Job satisfaction*

Job satisfaction represents the worker's view of his work, by showing attitude (Spector, 1997; Yalabik, Rayton, & Raptiet al, 2017), both positive and negative, toward a work role (Greenberg & Baron, 2008). Job satisfaction is an important issue in research on organizational behavior because job satisfaction plays a key role in a person's feelings toward his job (Yucel, 2012). Someone with high job satisfaction will show a positive attitude toward his work and, conversely, someone who is not satisfied with their work will show a negative attitude.

Job Satisfaction as Mediating Variable

Several researchers have discussed that knowledge sharing and job satisfaction have a connection with each other (Jaccob & Roodt, 2007; Rafique & Mahmood, 2018). Most of the studies state that job satisfaction is a strong predictor influencing knowledge sharing (Rafique & Mahmood, 2018; Wu et al., 2003), but in this study we tried to examine the effect of knowledge sharing on job satisfaction. Tong (2014) argues that knowledge sharing mediates the relationship between organizational culture and job satisfaction. That result is supported by Trivellas (2014), who contends that general competency mediates the relationship between knowledge sharing and job satisfaction.

In addition, there have been many studies regarding the relationship of job satisfaction to organizational commitment (Westover et al., 2010; Yunus, et al., 2014), one of which states that job satisfaction is proven to be a proven predictor of organizational commitment (Danovan et al., 2004; Gunlu et al., 2010; Kim et al., 2018). Namasivayam and Zhao (2007) confirm this, stating that affective commitment is positively related to job satisfaction. From the results of this research, we can also assume that the satisfaction felt by someone can produce commitment to the organization. More specifically, one study contends that job satisfaction has a positive relationship with affective commitment (Kwantes, 2009; Noor & Noor, 2007). From the results from these studies, we can assume that job satisfaction can mediate the relationship of knowledge sharing with affective commitment.

H3: *Job satisfaction mediates the relationship between knowledge sharing and affective commitment.*

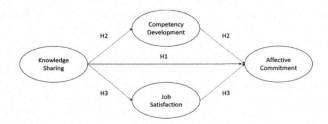

Figure 1.   Conceptual model.

3   METHODOLOGY

The method of this quantitative study was a survey conducted by distributing research questionnaires online through Google. Reliability and validity testing were also carried out so as to test the eligibility of the questionnaire. Reliability analysis can be seen through the Cronbach's alpha coefficient, with provisions above 0.6 (Bagozzy & Yi; Liao et al., 2006). The Cronbach's alpha value of this study was above 0.6, which means that the reliability is high. Validity testing in this study was performed using confirmatory factor analysis, with the provisions of the statement declared valid if the standardized loading factor (SLF) value ≥ 0.50, and the T-value ≥ 1.96. The range of SLF values in the knowledge sharing variable was 0.72–0.95, affective commitment 0.70–0.89, job satisfaction 0.63–0.89, and competency development 0.74–0.85. The overall value was above 0.5, this questionnaire had good validity.

3.1   *Participants*

This study took data from active workers age 19–39 from Gen Y in Indonesia, as many as 302 people.

3.2 *Measurements*

Measurement of the variables in this study was conducted using a questionnaire with a total of thirty-one statement items adopted from several experts in previous studies. The questionnaire of this study used a seven-point Likert scale and assigned rates ranging from 1 (strongly disagree) to 7 (strongly agree).

Knowledge Sharing

The research on knowledge sharing used a questionnaire developed by Hoff and Weenen (2004) (cited in Lin, 2007), which contains seven statement items.

Competency Development

The measurement of competency development in this study adopted a questionnaire developed by Naim and Lenka (2017) and contained fifteen statement items.

Job Satisfaction

The measurement of the job satisfaction variable adopted a questionnaire developed by Brayfield and Rothe (1951) (cited in Yucel, 2012) and contained five statement items.

Affective Commitment

The measurement of the affective commitment adopted a questionnaire developed by Allen and Mayer (1990), which was later adapted and modified by Allen, Mayer, and Smith (1993) (cited in Yucel, 2012). This questionnaire consisted of four statement items representing the affective commitment variable.

3.3 *Data analysis*

Structural equation modeling (SEM) was used in this study because it can simultaneously examine the structure of relationships between constructs in a way similar to multiple linear regression (Hair et al., 2006; Oyefolahan & Dominic, 2013). Then in measuring whether the research model was adequate, confirmatory factor analysis (CFA) was used by employing the Goodness of Fit Index (GoFI).

4 RESULTS AND DISCUSSION

4.1 *Result*

Three hundred and two respondents' data were analyzed in this study. Most of the respondents were female (68%), and 54.5% were 24–28 years old, 12.4% were 19–23 years old, 9.1% were 29–33 years old, and 21.5% were 34–39 years old. In terms of education level the group was dominated by bachelor's degree holders (63%), followed by those holding high school diplomas (22%), master's degrees (10%), diplomas (4%), and doctor's degree (1%). In terms of job rank, most of the respondents were staff (63%), followed by others (15%), then supervisors (9%), managers (7%), assistant managers (4%), general managers (1%), and CEOs (1%). Finally, in the terms of industry, most of the respondents were in the others category (27%), followed by public service (23%), education (12%), finance/banking and tourism/hospitality (9%), consulting (6%), technology/IT (5%), manufacturing (3%), food and beverage and media/entertainment (2%), and, finally, mining, oil and gas, and medical/health (1%).

Before analyzing the relationship between variables, the research model should be analyzed by its Goodness of Fit Index (GoFI) to see if the data matched the model (Wijanto, 2015). The GoFI results can be seen in Table 1.

From Table 1, we can see that most of the criteria are a good fit because they fulfill the provision. In this study, the main criteria that must be met is the value of RMSEA, which, as shown in Table 1, is 0.079, less than 0.08. Therefore, we can conclude that the model is valid and can describe the whole population in this study.

The next step was to analyze the result of the hypothesis test between variables based on the hypothesis developed previously. Figure 2 shows the result of the hypothesis test with coefficient value and T value between variables.

Table 1. Goodness of fit index result.

GoFI	Result	Provision	Conclusion
RMSEA	0.079	RMSEA ≤ 0,08	Good fit
NFI	0.96	NFI ≥ 0,90	Good fit
NNFI	0.97	NNFI ≥ 0,90	Good fit
CFI	0.97	CFI ≥ 0,90	Good fit
IFI	0.97	IFI ≥ 0,90	Good fit
RFI	0.95	RFI ≥ 0,90	Good fit
Standardized RMR	0.09	SRMR ≤ 0,05	Poor fit
GFI	0.79	GFI ≥ 0,90	Poor fit

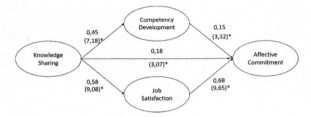

Figure 2. Path diagram.

Hypothesis 1 tested the positive significant relationship between knowledge sharing and affective commitment. The effect of knowledge sharing on affective commitment was positive and significant (β = 0.18; T value = 3.07). Therefore, hypothesis 1 was supported. Next we tested the mediating effect of competency development between knowledge sharing and affective commitment. The effect of knowledge sharing on competency development was positive and significant (β = 0.45; T value = 7.18), as was the effect of competency development on affective commitment (β = 0.15; T value = 3.32). Therefore, hypothesis 2 was supported. Finally, we tested the mediating effect of job satisfaction between knowledge sharing and affective commitment. The effect of knowledge sharing on job satisfaction was positive and significant (β = 0.58; T value = 9.08), as was the effect of job satisfaction on affective commitment (β = 0.68; T value = 9.65). Therefore, hypothesis 3 was supported. These results fulfilled provision by Baron and Kenny (1986), which stated that on testing the mediating effect, the relationship between mediating and dependent variables must also be significant.

4.2 Discussion

This study aimed to investigate the effect of knowledge sharing on affective commitment in Indonesia's Gen Y employees, specifically, the mediating effects of competency development and job satisfaction. The results of this study were as expected. The result proves that there is a positive relationship between knowledge sharing and affective commitment; it indicates that the presence of knowledge sharing in the organization as a daily behavior can boost the affective commitment of Gen Y employees in Indonesia. It is also in agreement with past research that states that knowledge sharing and competency development have a positive impact on affective commitment (Naim & Lenka, 2017). Affective commitment has been known as the key to employee intention to stay, because employees who have higher affective commitment will be emotionally attached to the organization and more likely to stay (Guchait & Cho, 2010), and will give their best performance in order to achieve organizational goals (Mayer et al., 1998; Namasivayam & Zhao, 2007). Knowledge sharing is a process of interaction, where there is exchange of information that becomes knowledge to the person who accepts it, and if it occurs in the workplace, it also can become organizational knowledge. The process of knowledge sharing itself can increase productivity (Liao et al., 2007; Wang et al., 2014), and as Wang and Noe (2010) state, knowledge sharing can provide a sustainable competitive advantage in a highly competitive economy (Rafique & Mahmood, 2018). Therefore, it is important for organizations to create cultures used to knowledge sharing.

The second hypothesis was also supported – that is, competency development has a mediating effect on the relationship between knowledge sharing and affective commitment. This result indicates that employees who have competency development opportunities are highly committed to the organization. This opportunity exists because a process of knowledge sharing occurs in the workplace. This result is consistent with the finding of Naim and Lenka (2017), that competency development mediates the relationship between knowledge sharing and affective commitment. As the key element of knowledge management, knowledge sharing plays a vital role in the learning and development of individuals working in an organization (Bosk & Kim, 2002; Lichtenthaler & Ernst, 2006; Rafique & Mahmood, 2018). Therefore, organizations should develop a strategy that can increase their employees' commitment by focusing on knowledge sharing to increase Gen Y competencies.

Finally, our finding of a mediating effect of job satisfaction on knowledge sharing and affective commitment on Gen Y employee in Indonesia was also supported. It implies that whenever employees feel satisfied with their job, their affective commitment improves. That satisfaction exists because knowledge sharing occurs in the workplace. Gen Y employees have different characteristic and demands; they are educated to collaborate and share their ideas (Cattermole, 2018). As employees seek information and knowledge to finish their tasks, they'll feel satisfied when the task is complete, leading to their emotional attachment to the organization. Looking back to the characteristics of Gen Y employees who actively look for an environment that can support them in obtaining and sharing their ideas, information, and knowledge (Naim & Lenka, 2017), this result confirmed a study by Brown and colleagues (2005) that stated when Gen Y employees feel that their desires are heard by the organization, they feel a greater sense of ownership in the organization (Kim et al., 2018).

5 CONCLUSIONS AND RECOMMENDATIONS

This study aimed to identify the effect of knowledge sharing on affective commitment, with the mediating effect of competency development and job satisfaction. The result pointed out that the outcomes of knowledge sharing, which are competency development and job satisfaction, influence the affective commitment of Gen Y employees in Indonesia. This result indicated that if an organization successfully implements effective knowledge sharing, it will positively increase the competency and job satisfaction of its employees, and finally can make their employees affectively committed to the organization.

As these findings suggest, in order to have a positive impact on employees, the focus of management should be knowledge sharing that stimulates employee to create, capture, and share information, and finally become their new competencies, and also contributing to satisfaction on their job to evoke their commitment to the organization. To do so, organizations should develop a strategy to make employees more involved in knowledge sharing, by creating a culture that stimulates employees to be more involved and willingly to share their skills and ideas with others in the organization.

REFERENCES

Baron, R. M., & Kenny, D. A. 1986. The moderator–mediator variable distinction in social psychological research: Conceptual, strategic, and statistical considerations. *Journal of Personality and Social Psychology* 51. 1173–1182.
Cattermole, G. 2018. Creating an employee engagement strategy for millennials. *Strategic HR Review* 17(6), 290–294.

Gao-Urhahn, X., Biemann, T., & Jaros, S. J. 2016. How affective commitment to the organization changes over time: A longitudinal analysis of the reciprocal relationships between affective organizational commitment and income. *Journal of Organizational Behavior* 37(4), 515–536.

Gelderen, B. R. V., & Bik, L. W. 2016. Affective organizational commitment, work engagement and service performance among police officers. *Policing: An International Journal of Police Strategies & Management* 39(1), 206–221.

Greenberg, J., & Baron, R. 2008. *Behavior in organizations.* 9th edition. Upper Saddle River, NJ: Prentice Hall.

Guchait, P. & Cho, S. 2010. The impact of human resource management practices on intention to leave of employees in the service industry in India: The mediating role of organizational commitment. *The International Journal of Human Resource Management* 21(8), 1228–1247.

Hartman, J., & McCambridge, J. 2011. Optimizing millennials' communication styles. *Business Communication Quarterly* 74(1), 22–44.

Heijde, C. M. V. D., & Van der Heijden, B. I. 2006. A competence-based and multidimensional operationalization and measurement of employability. *Human Resource Management: Published in Cooperation with the School of Business Administration, the University of Michigan, and in alliance with the Society of Human Resources Management* 45(3), 449–476.

Indonesian Central Statistics Agency. 2018. *Gender Statistic: Milenialls Profil Generasi Milenial Indonesia.* Jakarta: Kementerian Pemberdayaan Perempuan dan Perlindungan Anak.

Jha, N., Sareen, P., & Potnuru, R. K. G. 2019. Employee engagement for millennials: Considering technology as an enabler. *Development and Learning in Organizations: An International Journal* 33(1), 9–11.

Kang, Y.-J., Seok-Eun, K., & Gee-Weon, C. 2008 The impact of knowledge sharing on work performance: An empirical analysis of the public employees' perceptions in South Korea. *International Journal of Public Administration* 31(14), 1548–1568.

Kim, M., Choi, L., Borchgrevink, C. P., Knutson, B., & Cha, J. 2018. Effects of Gen Y hotel employees' voice and team-member exchange on satisfaction and affective commitment between the US and China. *International Journal of Contemporary Hospitality Management* 30(5), 2230–2248.

Kraimer, M. L., Seibert, S. E., Wayne, S. J., Liden, R. C., & Bravo, J. 2011. Antecedents and outcomes of organizational support for development: The critical role of career opportunities. *Journal of Applied Psychology* 6(3), 485–500.

Kwantes, C. T. 2009. Culture, job satisfaction and organizational commitment in India and the United States. *Journal of Indian Business Research* 1(4), 196–212.

Liao, S.-h., Fei, W.-c., & Chen, C.-C. 2006. Knowledge sharing, absorptive capacity, and innovation capability: An empirical study of Taiwan's knowledge intensive industries. *Journal of Information Science* 33(3), 340–459.

Lin, H.-F. 2007. Knowledge sharing and firm innovation capability: An empirical study. *International Journal of Manpower* 28(3/4), 315–332.

Lub, X., Nije Bijvank, M., Matthijs Bal, P., Blomme, R., & Schalk, R. 2012. Different or alike? Exploring the psychological contract and commitment of different generations of hospitality workers. *International Journal of Contemporary Hospitality Management* 24(4), 553–573.

Murray, K., Toulson, P., & Legg, S. 2011. Generational cohorts' expectations in the workplace: A study of New Zealanders. *Asia Pacific Journal of Human Resources* 49(4), 476–493.

Naim, M. F., & Lenka, U. 2016. Knowledge sharing as an intervention for Gen Y employees' intention to stay. *Industrial and Commercial Training* 48(3), 142–148.

Naim, M. F., & Lenka, U. 2017. Linking knowledge sharing, competency development, and affective commitment: Evidence from Indian Gen Y employees. *Journal of Knowledge Management* 21(4), 885–906.

Naim, M. F., & Lenka, U. 2018. Organizational learning and Gen Y employees' affective commitment: The mediating role of competency development and moderating role of strategic leadership. *Journal of Management & Organization*, 1–17.

Namasivayam, K., & Zhao, X. 2007. An investigation of the moderating effects of organizational commitment on the relationships between work–family conflict and job satisfaction among hospitality employees in India. *Tourism Management* 28(5), 1212–1223.

Oyefolahan, O. I., & Dominic, P. D. D. 2013. Knowledge management systems use and competency development among knowledge workers: The role of socio-technical antecedents in developing autonomous motivation to use. *VINE: Journal of Information and Knowledge Management Systems* 43(4), 482–500.

Rafique, G. M., & Mahmood, K. 2018. Relationship between knowledge sharing and job satisfaction: A systematic review. *Information and Learning Science* 119(5/6), 295–312.

Reiser, R. A., & Dempsey, J. V. (eds.). 2012. *Trends and issues in instructional design and technology.* Boston, MA: Pearson.

Rhoades, L., Eisenberger, R., & Armeli, S. 2001. Affective commitment to the organization: The contribution of perceived organizational support. *Journal of Applied Psychology 86*(5), 825–836.

Tong, C., Tak, W. I. W., & Wong, A. 2015. The impact of knowledge sharing on the relationship between organizational culture and job satisfaction: The perception of information communication and technology (ICT) practitioners in Hong Kong. *International Journal of Human Resource Studies 5*(1), 19.

Trivellas, P., Akrivouli, Z., Tsifora, E., & Tsoutsa, P. 2015. The impact of knowledge sharing culture on job satisfaction in accounting firms. The mediating effect of general competencies. *Procedia Economics and Finance 19*, 238–247.

Wang, Z., Wang, N., & Liang, H. 2014. Knowledge sharing, intellectual capital and firm performance. *Management Decision 52*(2), 230–258.

Wijanto, S. H. 2015. *Metode Penelitian Menggunkan Structural Equation Modeling Degan LISREL 9.* Jakarta. Lembaga Penerbit Fakultas Ekonomi Universitas Indonesia.

Yalabik, Z. Y., Rayton, B. A., & Rapti, A. 2017. Facets of job satisfaction and work engagement. *Evidence-Based HRM: A Global Forum for Empirical Scholarship 5*(3), 248–265.

Yew, T. 2007. Job satisfaction and affective commitment: A study of employees in the tourism industry in Sarawak, Malaysia. *Sunway Academic Journal 4*, 27–43.

Yucel, I. 2012. Examining the relationships among job satisfaction, organizational commitment, and turnover intention: An Empirical Study. *International Journal of Business and Management 7*(20), 44–58.

Yunus, N. A. M., Rahman, R. A., Aziz, R. A., Noranee, S., & Razak, N. A. 2014. Knowledge sharing on job satisfaction and organizational commitment among customer service representatives. *Proceedings of Knowledge Management International Conference (Kmice)*,1, 58–65.

The effect of employees' psychological and social capital on job satisfaction, organizational commitment, and turnover intention

V.I. Dwanti & D.H. Syahlani
Faculty of Economics and Business, Universitas Indonesia, Indonesia

ABSTRACT: The change of workers in industry has been very fast in recent years. Employees' willingness to work longer for a company, especially in Jakarta, has become a particular challenge. Moreover, individual qualities matter as an employee becomes a concern, such as psychological and social capital that every employee must have to compete. This study aimed to analyze the effects of employees' psychological and social capital on job satisfaction, organizational commitment, and turnover intention. Researchers collected 218 questionnaires and analyzed the data using the path analysis structural equation model. The results showed that psychological capital has a positive effect on job satisfaction and organizational commitment, but psychological and social capital do not affect turnover intention directly. Social capital does not affect job satisfaction. The findings also showed that psychological capital through job satisfaction and organizational commitment can cause turnover intention.

1 INTRODUCTION

Individual employee quality nowadays is not only related to knowledge, skills, abilities, and related experience but is also assessed based on psychological capital such as resilience, self-efficacy, optimism, and expectations of an occupation (Youssef, Luthans, & Aviolo, 2007). Employees must also have social capital. Individual integration of both psychological and social capital is central to the actualization of human potential in the workplace (Youssef et al., 2007). Furthermore, one of the characteristics that workers have today is individualistic, following the shift in the workforce (Adam, 2017). Social capital should be built within an organization through open communication, built trust, authenticity and approval, feedback and recognition, teamwork, and work-life balance (Luthans & Youssef, 2004). For some people, building relationships is a challenge. A previous study found that psychological capital related to job satisfaction and commitment to the organization can cause employees' decisions to stay with the organization or move to another workplace. Dess and Shaw (2001) state that the causes of high turnover consist of the cost of learning, reduced morale, pressure on other current employees, and lack of social capital.

Cost is one of the problems arising from turnover because employees have established social relationships either inside or outside the organization, and there is a price for replacing them (Holtom, Mitchell, & Lee, 2006). Social capital is believed to be important for organizational success, but when someone leaves a company, social relations are disrupted and social capital is lost (Holtom et al., 2006). Thus, this study focused on whether employees' psychological and social capital affect job satisfaction and organizational commitment, which in turn affect their intention to leave or quit the job.

2 LITERATURE REVIEW

2.1 *Psychological capital, job satisfaction, organizational commitment, and turnover intention*

Psychological capital consists of four main factors – namely hope, optimism, self-efficacy, and resilience (Youssef et al., 2007). First, self-efficacy refers to a person's evaluation of a particular task that he or she does and his or her self-confidence in carrying out the task. Second, hope means the energy and plans directed at achieving goals and pathways (Youssef et al., 2007). Third, Youssef and colleagues (2007) define optimism as representing people who expect something positive to occur in the future. Fourth, resilience is the ability to rebound or rise from adversity, conflict, or failure, or even to overcome more positive issues such as progress and responsibility (Youssef et al., 2007).

This study referred to the Minnesota Short Satisfaction Questionnaire (MSQ) short form used in Zopiatis, Constanti, and Theocharous (2014) as a survey instrument consisting of two aspects of job satisfaction – intrinsic and extrinsic satisfaction. Zopiatis and colleagues (2014) define turnover intention as voluntary or involuntary permanent withdrawal from an organization (Robbins & Judge, 2007).

Allen and Meyer (1990) divide organizational commitment into three main components: affective commitment, which is an affective or emotional attachment to an organization such that the strongly committed individual identifies with, is involved in, and enjoys membership in the organization; second, the normative commitment component of organizational commitment will be influenced by an individual's experiences, both prior to familial or cultural socialization, and following entry into the organization or organizational socialization (Wiener, 1982). Third, continuance components of organizational commitment will also develop based on two factors, which are the investments (or side bets) individuals make and a perceived lack of alternatives.

Previous studies have discovered that psychological capital supports organizational commitment and the results of self-efficacy, expectations, and optimism showed a positive commitment. Meanwhile resistance has a negative relationship with commitment (Rego, Lopes, & Nascimento, 2016). Other studies show that psychological capital has significant positive effects on job satisfaction. Employees with high levels of psychological capital can always think positively, expect a brighter future, and remain highly motivated at work so that their cognitive abilities are higher (Jung & Yoon, 2015). Meanwhile, psychological capital has a significant negative effect on turnover intention (Çelik, 2018).

Hypothesis 1: Psychological capital has a significant positive relationship with organizational commitment.

Hypothesis 2: Psychological capital has a significant positive relationship with job satisfaction.

Hypothesis 3: Psychological capital has a significant negative relationship with turnover intention.

Hypothesis 6: Job satisfaction has a significant positive relationship with organizational commitment.

Hypothesis 7: Organizational commitment has a significant negative relationship with turnover intention.

Hypothesis 8: Job satisfaction has a significant negative relationship with turnover intention.

2.2 *Social capital, job satisfaction, and turnover intention*

Luthans and colleagues (2007) explain that important assets in building and managing resilience when faced with difficulties not only prioritize human capital consisting of knowledge, skills, abilities, and experiences but also include vital elements of social capital – namely the ability to build networks or relationships (Youssef et al., 2007). Social capital is defined as the actual amount and potential of resources within, available through, and derived from the networks or relationships owned by an individual or social unit (Nahapiet & Ghoshal, 1998). Social capital, according to Nahapiet and Ghoshal (1998), has three dimensions. The structural dimension is the overall pattern of relationships found in organizations connecting people to each other (Bolino, Turnley, & Bloodgood, 2002). The relation dimension is the type of relationship between individuals in an organization with a focus on the quality or level of connection that is characterized by trust, intimacy, liking, and so on (Bolino et al., 2002). The cognitive dimension is defined as the extent to which employees in social networks share a common perspective or understanding (Bolino et al., 2002).

Requena (2003) divides social capital related to the workplace into five main dimensions – namely trust, social relations, commitment, communication, and influence. In this study, social capital was discussed along with employee job satisfaction, and was expected to predict employee turnover intention at the individual level. Previous research on social capital and job satisfaction has produced a positive reciprocal relationship and has discovered that a high level of social capital significantly increases job satisfaction and professional quality of life.

Social capital is a better indicator of job satisfaction and professional quality of life than the characteristics of workers, organizations, or work environments (Requena, 2003). Professionals prioritize loyalty to their direct workgroups within the organization over loyalty to the organization as a whole, so people who have strong interpersonal relationships with peers have a greater interest in work relationships (Dess & Shaw, 2001). Very high levels of voluntary turnover inhibit an organization's functioning (Dess & Shaw, 2001).

Hypothesis 4: Social capital has a significant positive relationship with job satisfaction.
Hypothesis 5: Social capital has a significant negative relationship with turnover intention.

3 METHODOLOGY

3.1 *Participants*

This study was conducted with employees who work in Jakarta with the professional levels of staff, supervisor, assistant manager, and manager as a sample. Five sets of questionnaires were distributed using an online questionnaire form, and a total of 218 questionnaires was collected. Participants were asked to answer several items that describe the variables of psychological capital, social capital, job satisfaction, organizational commitment, and turnover intention.

3.2 *Measurement and reliability of variables*

Psychological capital was measured using the twenty-four item Psychological Capital Questionnaire (PCQ) by Youssef and colleagues (2007). The measurement used a Likert scale with a range of 1 (strongly disagree) to 7 (strongly agree). The questionnaire consisted of six items about self-esteem, six items about hope, six items about resilience, and six items about optimism. Reliability test results obtained a Cronbach's alpha value of 0.765.

The measurement of the social capital variable used nine items consisting of two items about trust, two about social relations, three about commitment, one about communication, and one about influence, which was developed by Requena (2003). The measurement of job satisfaction used a Likert scale with a range of 1 (strongly disagree) to 7 (strongly agree). The Cronbach's alpha value was 0.789.

Job satisfaction was measured using sixteen items of the MSQ short form developed and used previously by Zopiatis and colleagues (2014). The MSQ short form consists of ten items on internal job satisfaction and six items on external job satisfaction (Zopiatis et al., 2014). The measurement of job satisfaction used a Likert scale with a range of 1 (strongly disagree) to 7 (strongly agree). The Cronbach's alpha value was 0.768.

Organization commitment was measured using the Organizational Commitment Scale (OCS) (Allen & Meyer, 1990). The OCS consists of eight items on affective commitment, eight items on continuance commitment, and eight items on normative commitment. Measurement was made using a Likert scale with a range of 1 (strongly disagree) to 7 (strongly agree). The OCS reliability value was 0.748.

Turnover intention was measured by three items developed and used in previous studies (Zopiatis et al., 2014). A Likert scale was used with a range of 1 (strongly disagree) to 7 (strongly agree). The Cronbach's alpha value on the calculation of turnover intention reliability was 0.865.

4 RESULT AND DISCUSSION

The results of the descriptive statistics of the 218 respondents were categorized by gender, education, professional position, and work experience: 77.5% were female and 22.5% were male; 3.2% held a diploma, 96.3% had a bachelor's degree, and 0.5% had a master's degree; 88.1% 10.1% were supervisors, 0.5% were assistant managers, and 1.4% were managers; 6% had less than a year of work experience, 17% had 1–2 years, 60.1% had 2–5 years, and 17% had 5–10 years.

Analysis of the research model was performed using the structural equation model (SEM) with LISREL 8.80. The results were: RMSEA value = 0.090, GFI = 0.59, NFI = 0.91, NNFI = 0.94, CFI = 0.94, and $\chi2/df$ = 2.73, which then were used to test the hypotheses. The results showed that the research model was in the moderately fit category according to the RMSEA; a reference value of 0.08 or less for the RMSEA indicates a reasonable error of approximation and researchers should not employ a model with an RMSEA value greater than 0.1 (Arbuckle & Wothke, 1999).

The results supported hypotheses 1, 2, 6, and 7. The coefficient value was 0.38 and the t value was 4.24. Hypothesis 1, that psychological capital has a significant positive relationship with organizational commitment, was supported. Hypothesis 2, that psychological capital has a significant positive relationship with job satisfaction, was supported with a coefficient of 0.61 and the t value was 5.48. Hypothesis 6, that job satisfaction has a significant positive relationship with organizational commitment, was supported with a coefficient of 0.42 and the t value was 4.24. Hypothesis 7, organizational commitment has a significant negative relationship with turnover intention, was supported with a coefficient of −0.68 and the t value was −5.32.

Hypotheses 3, 4, 5, 8 were not supported. Hypothesis 3, that psychological capital has a significant negative relationship with turnover intention, was not supported with a coefficient of 0.19 and a t value of 1.5. Hypothesis 4, that social capital has a significant positive relationship with job satisfaction, was not supported with a coefficient of 0.10 and a t value of 1.00. Hypothesis 5, that social capital has a significant negative relationship with turnover intention, was not supported with a coefficient value of −0.08 and a t value of −0.69. Hypothesis 8, that job satisfaction has a significant negative relationship with turnover intention, was not supported with a coefficient value of 0.03 and a t value of 0.26.

These results show that psychological and social capital do not have a direct effect on turnover intention. Besides, psychological and social capital do not have an indirect effect on job satisfaction. The effect of job satisfaction on turnover intention was not significantly negative with a coefficient value of 0.03. Thus, job satisfaction cannot mediate the effect of psychological and social capital on turnover intention. In this case, job satisfaction does not meet the requirements for mediation. Referring to Baron and Kenny (1986), mediation can be fulfilled if the mediator variable and the dependent variable are significant (Shrout & Bolger, 2000).

Also, the results of the analysis show that psychological capital has an indirect effect on turnover intention, through both job satisfaction and organizational commitment as mediators and through organizational commitment. However, job satisfaction and organizational commitment also cannot be calculated as direct mediation effects because psychological capital is not significantly negative with turnover intention or it doesn't meet the fulfillment of mediation requirements according to Baron and Kenny (1986), where independent and dependent variables must be significant.

The results of this study are supported by previous research conducted by Çelik (2018) finding that psychological capital does not affect turnover intention. In addition, this study is supported by Abbas and colleagues (2014), who contend that psychological capital has a significant effect on job satisfaction, but does not have an effect on turnover intention. This study also supports Munyaka's (2012) research, which argues that psychological capital affects organizational commitment, which then significantly predicts turnover intention.

Social capital's impact on turnover intention is not significant, an argument supported by Dess and Shaw's (2001) previous research. Individuals tend to survive within a company by being attached to certain small groups. Whereas social capital has no effect on job satisfaction, in line with the work of Flap and Völker (2001), who state that social capital cannot always increase job satisfaction and only certain types of social capital such as network structure, or the kind of resources relationships provide, will produce satisfaction.

5 CONCLUSION AND RECOMMENDATION

The results of this study indicate that psychological capital positively influences job satisfaction and employee commitment. A high level of job satisfaction is obtained because, with a high level of psychological capital, employees will be able to think positively and remain motivated in work situations (Jung & Yoon, 2015). Psychological capital can significantly contribute to job satisfaction through improvements in performance appraisals, responsibilities, and achievements, while low levels of psychological capital will lead to negative assessments and expectations, reduce intrinsic motivation, and have consequences for extrinsic rewards such as salary, working conditions, and job security (Badran & Youssef-Morgan, 2015). Furthermore, a good level of psychological capital has a positive effect on organizational commitment, which is supported by previous research performed by Luthans and colleagues (2007).

However, psychological capital does not have a negative effect on turnover intention directly. Low psychological capital level does not affect an employee's intention to stop or move from their work. Likewise, a low level of social capital does not affect an employee's job satisfaction or intention to quit work. Employees will try to stay with the company by sticking to certain small groups (Dess & Shaw, 2001). To affect turnover intention, psychological capital needs to be accompanied by other factors – namely organizational commitment.

Limitations of this study include the following concerns. First, further research is expected to be carried out in other geographic areas in Indonesia or applied to other organizations. Second, suggestions for further research are to increase the number of research samples and deep analysis to find other factors that cause turnover intention. Third, further research is expected to explore these issues more deeply through qualitative study or a combination of quantitative and qualitative methods. The next study should define other variables related to turnover intention in the current workforce, remembering the shift in workforce behavior. Fourth, this study tried to capture the causes of turnover intention from the perspective of employees, so future research may determine causes of turnover from an organizational perspective.

REFERENCES

Abbas, M., Raja, U., Darr, W., & Bouckenooghe, D. 2014. Combined effects of perceived politics and psychological capital on job satisfaction, turnover intentions, and performance. *Journal of Management 40*(7), 1813–1830.

Adam, A. 2017. Habis Milenial dan Generasi Z, Terbitlah Generasi Alfa. https://tirto.id/habis-milenial-dan-generasi-z-terbitlah-generasi-alfa-cnEs

Allen, N. J., & Meyer, J. P. 1990. The measurement and antecedents of affective, continuance and normative commitment to the organization. *Journal of Occupational Psychology 63*(1), 1–18.

Arbuckle, J. L., & Wothke, W. 1999. *Amos 4.0 user's guide.* Chicago, IL: SmallWaters Corporation.

Badran, M. A., & Youssef-Morgan, C. M. 2015. Psychological capital and job satisfaction in Egypt. *Journal of Managerial Psychology 30*(3), 354–370.

Baron, R. M., & Kenny, D. A. 1986. The moderator–mediator variable distinction in social psychological research: Conceptual, strategic, and statistical considerations. *Journal of Personality and Social Psychology 51*(6), 1173–1182.

Bolino, M. C., Turnley, W. H., & Bloodgood, J. M. 2002. Citizenship behavior and the creation of social capital in organizations. *Academy of Management Review 27*(4), 505–522.

Çelik, M. 2018. The effect of psychological capital level of employees on workplace stress and employee turnover intention. *Innovar: Revista de ciencias administrativas y sociales 28*(68), 67–75.

Dess, G. G., & Shaw, J. D. 2001. Voluntary turnover, social capital, and organizational performance. *Academy of Management Review 26*(3), 446–456.

Flap, H., & Völker, B. 2001. Goal specific social capital and job satisfaction: Effects of different types of networks on instrumental and social aspects of work. *Social Networks 23*(4), 297–320.

Holtom, B. C., Mitchell, T. R., & Lee, T. W. 2006. Increasing human and social capital by applying job embeddedness theory. *Organizational Dynamics 35*(4), 316–331.

Jung, H. S., & Yoon, H. H. 2015. The impact of employees' positive psychological capital on job satisfaction and organizational citizenship behaviors in the hotel. *International Journal of Contemporary Hospitality Management 27*(6), 1135–1156.

Kompas.com. 2018. Jawab Dengan Jujur Seberapa Puas Anda Dengan Pekerjaan Saat Ini? https://ekonomi.kompas.com/read/2018/12/22/093000526/jawab-dengan-jujur-seberapa-puas-anda-dengan-pekerjaan-saat-ini?page=all

Luthans, F. & Youssef-Morgan, C. (2004). Human, Social, and Now Positive Psychological Capital Management: Investing in People for Competitive Advantage. *Organizational Dynamics.* 33. 143–160.

Luthans, F., Avolio, B. J., Avey, J. B., & Norman, S. M. 2007. Positive psychological capital: Measurement and relationship with performance and satisfaction. *Personnel Psychology 60*(3), 541–572.

Munyaka, S. A. 2012. The relationship between authentic leadership, psychological capital, psychological climate, team commitment and the intention to quit in a South African manufacturing organisation (Doctoral dissertation, Nelson Mandela Metropolitan University).

Nahapiet, J., & Ghoshal, S. 1998. Social capital, intellectual capital, and the organizational advantage. *Academy of Management Review 23*(2), 242–266.

Priherdityo, E. 2016. Milenial Generasi Kutu Loncat Pengubah Gaya Kerja. www.cnnindonesia.com/gaya-hidup/20161215174236-277-179907/milenial-generasi-kutu-loncat-pengubah-gaya-kerja

Rahayu, I. R. S. 2018. Menaker: Generasi Milenial Pilih Berwirausaha Ketimbang Jadi Pekerja. www.inews.id/finance/makro/menaker-generasi-milenial-pilih-berwirausaha-ketimbang-jadi-pekerja/346674

Rahma, A. 2019. Mengapa Milenial Sering Pindah Kerja? www.liputan6.com/bisnis/read/3891097/mengapa-generasi-milenial-sering-pindah-kerja

Rego, P., Lopes, M. P., & Nascimento, J. L. 2016. Authentic leadership and organizational commitment: The mediating role of positive psychological capital. *Journal of Industrial Engineering and Management 9*(1), 129–151.

Requena, F. 2003. Social capital, satisfaction and quality of life in the workplace. *Social Indicators Research 61*(3), 331–360.

Robbins, S. P., & Judge, T. A. (2007). *Organizational behavior (12th ed.).* Upper Saddle River, NJ Pearson Prentice Hall.

Shrout, P. E., & Bolger, N. 2000. Mediation in experimental and nonexperimental studies: New procedures and recommendations. *Psychological Methods 7*(4), 422–445.

Wiener, Y. (1982). Commitment in Organizations: A Normative View. *The Academy of Management Review, 7*(3), 418–428.

Youssef, C. M., Luthans, F., & Avolio, Bruce J. 2007. *Psychological capital: Developing the human competitive edge.* New York: Oxford University Press.

Zopiatis, A., Constanti, P., & Theocharous, A. L. 2014. Job involvement, commitment, satisfaction and turnover: Evidence from hotel employees in Cyprus. *Tourism Management 41*, 129–140.

Managing Learning Organization in Industry 4.0 – Rachmawati & Hendayani (eds)
© 2020 Taylor & Francis Group, London, ISBN 978-0-367-81920-0

The linking of the employee career development program and promotion in PT. XL Axiata Tbk Bandung

A. Silvianita & A.F. Nur
Faculty of Communication and Business, Telkom University, Bandung, Indonesia

ABSTRACT: Human resources is a determining factor for a company, because through quali-fied human resources, the company can compete with others in business. To improve the quality of its workforce, every company must provide strategies to improve its employees' ability in work. Career development has become one of the strategies that a company must consider con-tinuously. The aim of this research was to determine whether career development programs affect the promotion of employees of PT. XL Axiata Tbk Bandung. This study used quantitative methods and descriptive research. To get a complete result, this study analyzed a questionnaire given to thirty employees who acted as respondents and formed a sample. Through using a simple regression analysis as a statistical tool, we know that the career development variable utilized in this research has an effect on the promotion of employees of 50.9% while 49.1% was influenced by another factors not examined in this study. As a suggestion, in the future PT. XL Axiata Tbk Bandung is expected to consider employees' educational background and work experience, which can be appropriate for deciding on promotion. Moreover, the company is also expected to provide equal opportunities for employees to develop their careers.

1 INTRODUCTION

Human resources management (HRM) is a determinant of quality in a company, because without capable human resources, the company cannot face competition. Therefore, every company needs to conduct a career development program continuously. Human resources development through career development programs will improve employee expertise, so there-fore it will be easier for employees to achieve certain goals and professional positions. Through promotion, employees will feel entrusted with a higher position in the office. Promo-tion will also increase their social status, authority, responsibility, and income.

In an effort to maximize its career development program, PT. XL Axiata Bandung holds employee training routinely every year. During 2017 and 2018, thirty employees attended the career development program and training. As a result, most of them reached a "Meet Per-formance" predicate, meaning that he or she demonstrated good performance and is able to meet the work target.

Promotion cannot be separated from the achievement of work targets acquired by employees. If an employee succeeds in achieving the work targets, we can conclude that his or her work performance is in a good position. Employee job performance is one of the main factors to get a promotion. According to the interview results, in 2017, employ-ees' achievement of work targets was 90.3%, and it was 92.60% in 2018. Overall, during two years, the number of employees promoted increased by 2.30%. Based on these phe-nomena, this study focused on the influence of the employees' career development pro-gram on promotion at PT. XL Axiata Tbk Bandung.

2 LITERATURE REVIEW

This research is one of the specific cases of managing human resources in an organization. According to Sedarmayanti (2015: 13), HRM is a policy and action that determines the human aspects of an organization, include selecting, recruiting, or adding employees, providing training, and motivating employees through rewards and assessments.

2.1 Career development

Zailani (2015) mentions that career development can be defined as an employee's personal efforts aimed at carrying out his or her career plan through education, training, job search and acquisition, and work experience. The purpose of career development is to improve and increase employees' professional capabilities so they can give the optimum contributions. According to Siagian (2006: 215), certain factors can influence employees' career development:

1. Job performance: the most important factor to increase and develop employees' career is job performance. Without satisfying work output, they will find it difficult to convince their superintendent to advance them professionally.
2. Loyalty to the organization: an employee's dedication to his or her organization.
3. Mentor and sponsor: a mentor is a person who provides professional advice or suggestions to employees. A sponsor is someone in the company who can create career development opportunities for employees.
4. Support from subordinate: support given by subordinates to their managers.
5. Opportunity to grow: an opportunity for subordinates to increase their ability or continue their education, such as through workshops and courses.

2.2 Promotion

Hasibuan (2014: 108) mentions that promotion is a transfer of jobs to increase an employee's authority and responsibility within an organization, so therefore he or she receives responsibility, rights, status, and more income. Kadarisman (2012: 143) reveals promotion's dimension are:

1. Experience: most employers use job experience as one of the requirements for promotion because employees with a lot of experience are expected to become more innovative in work.
2. Level of education: the higher the level of education, the higher the thinking.
3. Loyalty: measuring loyalty to the organization. Employees with higher loyalty are willing to take a bigger responsibility.
4. Honesty: to get a promotion, honesty is a must.
5. Responsibility: sometimes an organization needs more responsibility, so responsibility becomes the main requirement.
6. Ability to get along: to attain a certain position, employees need the ability to get along with all members of the organization.
7. Job performance: employees can effectively and efficiently reach the work targets for which they are accountable, in terms of both quantity and quality.
8. Initiative and creativity: employees have the initiative and the creativity to reach the organization's goals.

2.3 Research framework

Based on these descriptions, the framework of this research was as follows:

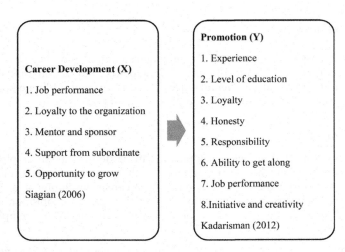

Career Development (X)

1. Job performance

2. Loyalty to the organization

3. Mentor and sponsor

4. Support from subordinate

5. Opportunity to grow

Siagian (2006)

Promotion (Y)

1. Experience

2. Level of education

3. Loyalty

4. Honesty

5. Responsibility

6. Ability to get along

7. Job performance

8. Initiative and creativity

Kadarisman (2012)

Figure 1.   Research Framework.

Based on the literature and relevant previous research, the hypothesis of this study was that a significant relationship exists between career development and promotion at PT. XL Axiata Tbk Bandung.

3   METHODOLOGY

3.1   *Participants*

Since this research used the quantitative method, a questionnaire was used as a tool for collecting data. In general, quantitative research can be defined as business research that discusses research objectives through empirical assessments that involve numerical and analytic approaches (Zikmund et al., 2009). Fitting with the aim of this research, the researcher used two types of questionnaire that represented two different variables that became the focus of this research. Questionnaires were distributed to thirty employees of PT. XL Axiata Bandung. The population of this research was limited; therefore, this research used a census approach for the sample.

3.2   *Data analysis*

To get a comprehensive result, this research used simple linear regression as a statistical tool, with career development as the independent variable (X) and promotion as the dependent variable (Y). Each variable was represented by dimensions and indicators in the questionnaire. Before results were calculated, all the data were checked through statistical steps, such as validity and reliability tests. Based on these calculations, we found that all variables were valid and reliable. The validity results for all questionnaire items were above 0.361. The reliability value for the X variable was 0.773 and reliability for the Y variable was 0.756, which are both higher than Cronbach's alpha (0.60). Table 1 is the result from simple linear regression for this research.

According to Table 1, from the simple linear regression we can conclude that variables X and Y are having a positive relationship. This means that if career development rose by 0.5599, then promotion increased by 1.87%. Furthermore, for the hypothesis testing, from the table we also can conclude that the X variable has significantly influenced the Y variable ($0.00 \leq 0.005$).

Last, from Table 2, we found that the coefficient determination was 50.9%, which means that in this research, career development influenced promotion by 50.9%, while 49.1% was caused by other variables not included in this research.

Table 1. Result of simple linear regression.

Coefficients[a]

Model		Unstandardized Coefficients		Standardized Coefficients		
		B	Std. Error	Beta	t	Sig.
1	(Constant)	1.8728	0.3949		4.7419	0.000
	x	0.5599	0.1038	0.016	5.390	0.000

a. Dependent variable: y

Table 2. Coefficient of determination.

Model summary[b]

Model	R	R Square	Adjusted R Square	Std. Error of the Estimate
1	0.714[a]	0.509	0.492	7.280

a. Predictors: (Constant), X
b. Dependent Variable: Y

4 RESULTS AND DISCUSSIONS

Through simple linear regression, we know that the career development program has significantly influenced promotion for the employees of PT. XL Axiata Tbk. Bandung. These results are in line with the conditions in this firm, which are that from 2017 to 2018, the employees' work target performance increased 2.30%, from 90.30% to 92.60%.

These results are in accordance with previous studies and literature that mention that career development has an influence on employees' promotion in an organization. At the end, it will also increase their working enthusiasm (Rijalulloh, 2017).

5 CONCLUSIONS AND RECOMMENDATIONS

In conclusion, career development has an influence on employees' promotion of 50.9% at PT. XL Axiata Tbk Bandung. To enrich similar research, we recommend further researchers consider other variables that also can influence employees' promotion. We also suggest conducting similar research in other organizations with more respondents, in order to obtain more complete results and flesh out our knowledge of human resources.

REFERENCES

Hasibuan, M. S. P. 2014. *Manajemen Sumber Daya Manusia (Edisi Revisi)*. Jakarta: Bumi Aksara.
Kadarisman. 2012. *Manajemen Pengembangan Sumber Daya Manusia*. Jakarta: Rajawali Pers.
Rijalulloh. 2017. Pengaruh pengembangan karir dan promosi jabatan terhadap semangat kerja. Malang. *Jurnal Administrasi Bisnis* 51(2).
Sedarmayanti. 2015. *Sumber Daya Manusia dan Produktivitas Kerja*. Bandung: CV.Manadar Maju.
Siagian, S., P. 2006. *Sistem Informasi Manajemen*. Jakarta: PT. Bumi Aksara.
Zailani. 2015. Analisis Pengaruh Mutasi dan Promosi Jabatan Pada Peningkatan Karir Pegawai. Surakarta: *Jurnal Ilmu Manajemen dan Akuntansi Terapan (JIMAT)* 6(1).
Zikmund, W. G., Babin, B. J, Carr, J. C., & Griffin, M. 2009. *Business study methods*. 8th ed. USA: South-Western College.

Effect of job satisfaction and training on turnover intention of Starbucks baristas in Bandung City

R.P Ardi & R. Wahyuningtyas
Magister Management Telkom University, Bandung, Indonesia

ABSTRACT: Employee satisfaction and training are one of the most important factors that can determine employee turnover intentions. The data of this research were obtained by distributing questionnaires to 100 respondents who were Starbucks baristas in 17 Bandung City outlets. The data obtained were then analyzed using multiple linear regression methods. The result revealed that job satisfaction and training had a negative and significant influence on employee turnover intentions, both partially and simultaneously. Increased job satisfaction and training was able to reduce employee turnover intentions. Since the job satisfaction has a potential effect on organizational outcomes, features of job enrichment should be considered in designing jobs, so that an effective quality management strategy can be implemented. Several weak positive relationships between high involvement and quality managements with perceived job demands was also investigated in more detailed studies of employee welfare.

1 INTRODUCTION

For Starbucks Indonesia, baristas are a resource that holds the key to implementing the company's business wheels, because they are part of the direct contact with the company's customers. Currently, Starbucks' business in Indonesia is expanding, with more outlets being opened throughout Indonesia. Data obtained from the Starbucks annual report shows an increase in income at other Starbucks outlets in Indonesia.

It can be seen that Starbucks in Indonesia continues to expand its market by adding outlets in various cities in Indonesia. This makes employee turnover in this company very high. Furthermore, looking at the data obtained from the results of interviews with Anastasia Dwiyani, the General Manager of Starbucks Indonesia Human Resources, regarding Starbucks barista turnover rates in Indonesia especially in Bandung City in the period of 2015 to 2017, the turnover rates were high.

Regarding baristas job satisfaction Starbucks, based on the interviews, it was found that most of them were not satisfied with their work. According to Article I Paragraph 9 of Act No. 13 of 2003 Concerning Manpower, job training is an activity to give, obtain, improve, and develop work competencies, productivity, discipline, attitudes, and work ethic. This applies to a certain level of skills and expertise in accordance with the level and job qualifications. Training Starbucks is given to prospective baristas to understand the tasks of their work including understanding each menu.

During the interview, Anastasia Dwiyani said that the training is given to each prospective barista for a month, where prospective baristas will be given direct training from the "star team", the best barista team owned by the company. Moreover, the purpose of this research is to see how job satisfaction and training variables simultaneously or partially affect employee turnover intention variables.

2 LITERATURE REVIEW

2.1 *Job satisfaction*

According to Herzberg (in Broni 2012: 310), job satisfaction is caused by the presence of a set of factors, called motivators, while job dissatisfaction is caused by the absence of a different set of motivators, called the hygiene factor. Motivating factors are related to the aspects contained in the work itself. The factors included in the motivating factor are achievement, recognition, work itself, responsibility, possibility of growth, and advancement. On the other hand, hygiene factors are factors surrounded by the execution of work, which is related to the work context or extrinsic aspects of the workers. Hygiene factors encompass working conditions, interpersonal relations, company policy and administration, job security, and technical supervision. Herzberg also stated that motivating factors can cause someone to shift from dissatisfied to satisfied conditions.

2.2 *Training*

Gomes (2003:197) stated that training is every effort to improve workers' performance on a particular job that becomes their responsibility. In addition, Hollenbeck et al., (2003:251) believe that training is a planned effort to facilitate the learning of work-related knowledge, skills, and behaviors. According to Mangkunegara (2006:46), there are six indicators to measure training, namely instructors, participants, materials, methods, goals, and objectives.

2.3 *Turnover intention*

According to Mathis and Jackson (2004:125), turnover intention is related to job satisfaction and organizational commitment. Turnover intention is the process by which employees leave the organization, and must be replaced immediately. Furthermore, Ronald and Milkha (2014:5) stated that turnover intention is the tendency or intensity of individuals to leave the organization for various reasons and among them is the desire to get a better job. In addition, Mobley (2011:15) stated that turnover intention is the result of an individual evaluation of the continuation of its relationship with the company where it works but has not yet been realized in real action. Mobley (2011:150) also believes that there are three indicators used to measure turnover intention, namely thoughts of quitting, intention to quit, and intention to search for another job.

Therefore, the objective of this research is to discover whether:

H1. Job satisfaction (X1) has a significant effect on turnover intention (Y),

H2. Training (X2) has a significant effect on turnover intention (Y), and

H3. Job satisfaction (X1) and training (X2) has a significant effect on turnover intention (Y).

3 METHODOLOGY

3.1 *Sample and procedure*

The sampling technique used in this research was saturated sampling technique in which all member of the population is used as samples. This technique is frequently used, if the population is relatively small, which is less than 30 people. In addition, saturated samples are also called census terms where all member of the population is sampled (Sugiyono, 2011). In this research, the total population was 100 baristas taken from 17 Starbucks outlets in Bandung City. Therefore, the sample used in this research was the entire population of 100 baristas.

The data collection technique used in this research was a questionnaire. The questionnaire contained a list of written questions that had been formulated beforehand that respondents answered usually in clearly defined alternatives.

3.2 *Method*

The data were analyzed in several stages, starting with conducting validity and reliability tests. After that, a classic assumption test consisting of a normality test was conducted to see whether the residual value is normally distributed or not. The next step was conducting a multicollinearity test to discover whether there are correlations between independent variables in the regression model or not. The heteroscedasticity test was then carried out to see if there were inequalities in residual variance from one observation to another. The data that had been tested were then analyzed using multiple linear regression analysis. With this analysis results, the effect of the independent and dependent variables could be obtained. Then, a partial and simultaneous hypothesis test was carried out to determine the relationship between variables both individually and collectively. Finally, coefficient of determination test was conducted to measure how far the model in this research was able to explain the variation of the dependent variables.

4 RESULTS

4.1 *Multiple linear regression analysis*

After processing the data in multiple linear regression analysis using SPSS, the following results are obtained. The formula for multiple linear regression equations is $Y = a + b_1X_1 + b_2X_2..b_kX_k$. From the table, the results of the coefficient using SPSS above, the following data were obtained:

1. There was a constant number of unstandardized coefficients. From the table above, the number was 52.974, which meant that if there was no job satisfaction (X1) and training (X2) variable, the consistent value of turnover intention (Y) variable was 52.974.
2. (b1) There was a regression coefficient number where the number was -0.586, which meant that every 1% increase in job satisfaction (X1), the turnover intention (Y) would decrease by 0.586.
3. (b2) There was a regression coefficient number where the number was -0.444, which meant that every 1% increase in training (X2), the turnover intention (Y) would decrease by 0.444.
4. Therefore, the regression equation obtained is as follows:

$$Y = 52.974 - 0.586X1 - 0.444X2$$

4.2 *Partial hypothesis test*

1. (t count) -8.735> (t table) 1.988 and significance value (X1) of 0.000 smaller <0.05, which meant that there was an effect of job satisfaction on turnover intention.
2. (t count) -5.015> (t table) 1.988 and significance value (X2) of 0.000 smaller <0.05, which meant that there was an effect of training on turnover intention.

4.3 *Simultaneous hypothesis test*

To see the effect of the independent variables, namely job satisfaction (X1) and training (X2) together on the dependent variable, which was turnover intention (Y), the f test was used The results of the data collection from SPSS are as follows:

1. F count was 767.612, greater > than f table, which was 3.09.
2. The significance value obtained was 0.000, which was smaller < than 0.05.

Therefore, it can be concluded that there was an effect of job satisfaction (X1) and training (X2) variable on turnover intention (Y).

5 CONCLUSION

All in all, the results of this research indicate that there was a significant and negative influence of the satisfaction variable on turnover intention. This result supported the first hypothesis that the job satisfaction variable (X1) significantly affected the turnover intention variable (Y). Moreover, it could be seen that the ranking of factors affecting job satisfaction on motivating indicators factors was as follows: (1) appreciation, (2) opportunity to progress, (3) possibility to develop themselves, (4) responsibility, (5) work itself, (6) success in completing tasks. On the other hand, the hygiene indicator factors showed the following rank: (1) working conditions, (2) feeling safe at work, salary, position, (3) company policy and implementation, (4) supervision techniques, (5) interpersonal relationships. The results of this research were supported by Ladelsky (2013: 1036) who found that job satisfaction was negatively related to turnover intention. The research also showed the fact that job satisfaction indirectly affected voluntary turnover intention.

In addition, this research indicated that there was a significant and negative effect of training on turnover intention. This result supported the second hypothesis, which was the training variable had a significant effect on turnover intention variables. The order of factors affecting the training was (1) training material, (2) training objectives and objectives, (3) training methods, (4) participants, (5) training instructors. This statement was supported by the research conducted by Isabel and Lucy (2013: 46).

Furthermore, this research indicated that there was a significant effect of job satisfaction and training variables on turnover intention, which supported the third hypothesis that there was a significant effect of job satisfaction and training variables on turnover intention. The result was supported by the research of Khawaja et al. (2015: 215) that suggested that job satisfaction mediated the relation between effective training programs, colleague support for training, and turnover intention.

6 DISCUSSION AND IMPLICATIONS

This research discovers that the levels of satisfaction and employee training of Starbucks' baristas in Bandung City were low. On the other hand, the turnover intention of the baristas was high. Furthermore, the results showed that job satisfaction and training variable had a significant and negative effect, both simultaneously and partially, on the turnover intention variable. These results were supported by several previous studies conducted by:

1). Limor (2013), who found that job satisfaction had a negative relation with turnover intention,
2). Sangaran and Jeetesh (2015), who found that job satisfaction had a negative effect on turnover intention.

Moreover, regarding the relation between training and turnover intention, Ali et al. (2017) found that training has a significant and negative effect on employee turnover intention where employee satisfaction of the training is a factor that affects the employees' desire to do the turnover intention.

In addition, research conducted by Cheloti and Jumah (2013) discovered that employee training has a negative and significant effect on employee turnover intention. This is also supported by research conducted by Toh et al. (2017) stated that training can reduce the turnover intention of the lecturers, which means the organization provides training to employees and thus employees feel valued and can reduce employees' intention to leave the current organization.

REFERENCES

Balkin, D., et al. 2001. Managing human resources (4th ed.). Inggris: Pearson.

Bohlander, G, dan snell, S. 2010. Managing human resources. Amerika: Cengage Learning.

Cheng, Y., & Waldenberger, F. (2013). Does Training Affect Employee Turnover Intention? Evidence from China. In Academy of Management Proceedings (Vol. 2013, No. 1, p. 11271). Briarcliff Manor, NY 10510: Academy of Management.

Cherrington, D. 1995. The management of human resources (4th ed). New Jersey: Prentice Hall Inc.

Ghozali, I. (2006). Aplikasi analisis multivariate dengan program SPSS. Badan Penerbit Universitas Diponegoro.

Gomes, F.2003. Manajemen sumber daya manusia. Yogyakarta: Andi Yogyakarta.

Hamidi. 2005. Metode penelitian kualitatif: Aplikasi praktis pembuatan proposal dan laporan penelitian. Malang: UMM Pres.

Harnoto. (2002). Manajemen sumber daya manusia. Jakarta: Prehallindo.

Hasibuan, M. 2008. Manajemen dasar, pengertian, dan masalah. Jakarta: PT Bumi Aksara.

Rotated and mutated comparative employee analysis in Telkomsel Area 2

T.B. Isnandiko & J. Sembiring
Telkom University, Bandung, Indonesia

ABSTRACT: Rotation is the periodic change of employees from one task to another. In this study the author sought to compare the performance of employees who experienced and did not experience rotation in Telkomsel Area 2 Jabodetabek Jabar in 2016. This study was a comparative descriptive study. The results of various test calculations indicated differences between the performance of employees who experienced rotation and mutation and those who did not in Telkomsel Area 2. This study used 194 respondents. This study found an increase in employee performance in terms of quality and quantity, timelines, and interpersonal impact, but it also revealed a decrease in performance after rotating and mutating in terms of cost-effectiveness and monitoring. Suggestion stemming from this research include training and employee gathering and the implementation of reward and punishment.

1   INTRODUCTION

The success of an organization is influenced by employee performance or by the cumulative work an employee achieves in carrying out his tasks in accordance with the responsibilities given to him. For this reason, each company strives to improve employee performance in attaining its stated organizational goals.

Employees are an important resource for organizations because they have the talent, energy, and creativity organizations need to reach their goals. In order to improve the performance of employees, organizations need to pay attention to employees' diverse needs and interests. Motivation is one of the most important aspects in determining a person's behavior, including work behavior. To be able to motivate someone an understanding is needed of how motivation is formed.

Based on these thoughts, this research examined matters relating to work rotation carried out by PT. Telkomsel Area 2's human resource management.

2   THEORETICAL BACKGROUND

Ardana, Mujiati, and Utama (2012) state that human resources management is based on three principles. (1) Human resources are the most valuable and important assets an organization or company owns because the success of an organization is largely determined by the human element. (2) Success is more likely if organizational procedures and regulations relating to human resources are interconnected and benefit all parties. (3) Corporate culture and values, as well as managerial behavior originating from them, have a major influence on achieving the best results. Therefore, the corporate culture and work ethic must be upheld continuously from the top down so that the culture will be accepted and obeyed. According to Sunyoto (2012), placement is "the process or filling of positions or the reassignment of employees to new assignments or positions." Yani (2012) defines placement as the "appointment of employees to occupy or do new work." Robbins and Judge (2008) state that work rotation is a periodic change in workers from one task to another.

The purpose of employee rotation is the following (Hasibuan, 2008):

a. Increase employee work productivity
b. Create a balance between the workforce and the composition of work or positions
c. Expand or increase employee knowledge
d. Eliminate boredom
e. Provide incentives for employees to strive to improve their careers
f. Implement sanctions for violations committed by employees
g. Provide recognition or compensation for achievements
h. Boost work morale
i. Ensure better safety measures
j. Adjust work to the physical condition of employees.

According to Siagian (2010), promotion is when a person is transferred from one job to another whose responsibility is greater, the level of the position hierarchy is higher, and his income is greater. Siswanto (2010) states that transfers involve changing functions, responsibilities, and employment status so that the workforce concerned gets the maximum morale and work performance.

Hasibuan (2008: 94) explains that "performance is the result of work achieved by a person in carrying out tasks assigned to him based on skills, experience, sincerity and time. . . . Performance is a combination of three important factors, namely the ability and interest of a worker, the ability and acceptance of the assignment of delegates and the role and level of motivation of workers." If the performance of each individual or employee is good, it is expected that the company's performance will also be good because the performance of a company is the culmination of the performance of all of its employees.

Performance can be assessed or measured through several indicators. Effectiveness is one important factor related to achievement. Responsibility is an integral part of ownership of authority. Discipline is obedience to the laws and regulations in force. Employee discipline is the obedience of the employee concerned in respecting the employment agreement with the company where he works. Initiative is related to thinking power and creativity in the form of an idea related to the achievement of company goals. The nature of the initiative should receive attention or response from the company, specifically from the direct boss of the employees concerned. In other words, employee initiative is the driving force of progress that will ultimately effect the improvement of employee performance.

Based on these observations, the authors of this study visualize the framework depicted in Figure 1.

3 HYPOTHESIS

Based on this framework, the hypotheses examined in this study were as follows:

H_0: No difference occurs in the performance of PT. Telkomsel Area 2 employees who are rotated and who are not in the rotation at t-cope, 2016.

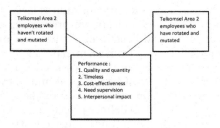

Figure 1. Framework.

H_1: Differences occur in the performance of PT Telkomsel Area 2 employees who are rotated and who are not in the rotation at t-cope, 2016.

4 METHODOLOGY

This research comprised a comparative research method. Descriptive comparative research analyzes the same variables for different samples. Comparison or comparison analysis is a statistical procedure utilized to test differences between two or more groups of data (variables). This test depends on the type of data (nominal, ordinal, interval/ratio) and the sample group.

The population of this study consisted of employees of Telkomsel Area 2, and data were obtained through statistical calculations – namely Slovin's formula. The formula was used to determine the sample size of the known population of 373 employees. The level of precision set in determining the sample was 5%. Slovin's formula was used to evaluate as many as 194 respondents – namely employees who have undergone rotation and mutation and employees who haven't. All questionnaire items in this study were declared valid, according to Sujarweni (2016: 239) If the R-value for n−30 < r count, then the result was valid. All questionnaire items in this study were declared reliable; according to Sujarweni (2016: 239), if the Cronbach's alpha value is > 70, then the result is reliable.

5 DISCUSSION

A descriptive analysis of employees who have not rotated and mutated, and of employees who have, indicated an increase in quality and quantity of 1.65%. The timelines component revealed an increase in value of 1.55%. The interpersonal impact of employees who have rotated and mutated was 0.77%. Employee discipline among those who have rotated and mutated decreased by 1.38%, and the need for supervision was 0.65%, while the results of various test calculations found differences between the performance of employees who have experienced rotation and mutation and the performance of employees who have not.

Table 1. Performance of employees who have not rotated and those who have already rotated.

Performance factors	Performance of employees who have rotated	Performance of employees who have already rotated	Category
Quality and quantity	82.66	82.19	High
Timeliness	84.43	83.85	Very High/High
Cost-effectiveness	90.23	89.53	Very High
Interpersonal impact	87.81	87.25	Very High
Need for supervision	86.50	85.93	Very High
average	86.33	85.93	Very High

Source: Data already processed

The performance of employees who have not rotated was a little higher compared with the performance of employees who have rotated; however, performance was very high in the permanent category. Of the five factors that measure performance, the factor with the largest percentage (more than 89%) was cost-effectiveness. The results of normality testing research show that data normality is not fulfilled because the sig value < 0.5; there are 0.198 before and 0.158 after. A Wilcoxon test was performed because the results of this study were not normally distributed data. Based on the results of the Wilcoxon test and the sig value (according to Sujarweni [2016: 138], if sig < 0.05, then H_0 is rejected), H_1 can be accepted and H_0 can be rejected.

The Z value indicates that H_1 can be accepted and H_0 can be rejected; previous research suggests a Z value of $-11{,}892 < 1.96$ for a total sample of less than 1,000 (Sujarweni, 2016). According to Sujarweni (2016: 138), if the value of Z < Z count, then H_0 is rejected. The results of this study are in line with previous research, which suggests employee job performance has an effect on employee movement.

6 CONCLUSIONS AND SUGGESTIONS

The conclusions obtained from this research are as follows. The performance of employees who have not rotated is very high, the performance of employees who have rotated is a little lower, and the performance of permanent employees is very high (85.75%). Of the five factors that measure performance, cost-effectiveness ranked the highest (located in the very high category), and quantity and quality ranked the lowest. Based on the test results, the obtained counted Z value was larger than the table Z value (H_1 accepted), and this means that the performance of employees who have not rotated differs significantly from the performance of staff who have rotated. The authors also found a decrease in the discipline and supervision needs of employees who have rotated and mutated. Companies should therefore carry out rotation and mutation periodically. Periodic training and company outings, along with rewards and punishments, will allow employees to get refreshment when working, and encourage them to incorporate company values. Future research can test other variables related to the performance of human resources in the company, or examine other work location variables.

REFERENCES

Ardana, I. K., Mujiati, N. W., & Utama, I. W 2012. *Human resource management*. 1st ed. Yogyakarta: Graha Ilmu.
Hasibuan, M. 2008. Human resource management. 10th ed. Jakarta: Bumi Aksara, p.94.
Robbins, S. P., & Judge. 2008. *Organizational behavior*. 13th ed. Jakarta: Salemba Empat, p.649.
Siagian, S. P. 2010. *Role of staff and management*. Jakarta: Bumi Aksara.
Siswanto, B. 2010. *Manajemen tenaga kerja rancangan dalam pendayagunaan dan pengembangan unsur tenaga kerja*. Bandung: Sinar Baru, p.211.
Sujarweni, W. 2016. *Metodologi penelitian bisnis dan ekonomi*. Yogyakarta: Pustaka Baru Pers, pp. 138,239.
Sunyoto, D. 2012. *Human resource management*. 1st ed. Yogyakarta: CAPS.
Yani, M. 2012. *Human resource management*. Jakarta: Mitra Wacana Media, p.74.

The influence of motivation on the creative performance of news television company employees

E. Julianti & M. Mustaqim
Faculty of Economics and Business, University of Indonesia, Depok, Indonesia

ABSTRACT: Performance is one of the most important matters in the creative industry. Despite the pros and cons of rewarding effectiveness, many companies use reward as a tool to enhance their employees' performance. This study focused on the influence of both intrinsic and extrinsic motivation on the creative performance of news television company employees. Creative self-efficacy and reward placed as the moderating variable of extrinsic motivation. Results found that these variables did not have a moderating effect on the influence of reward for creative performance. This analysis enriched previous research concluding that the moderation of creative self-efficacy only occurs in employees who already have high creative self-efficacy (Malik, Choi, & Butt, 2015). The study was conducted on 202 employees of seven news television companies in Indonesia who worked not just as journalists but also in other supporting roles, for instance, business operation and strategy.

1 INTRODUCTION

Employees form an essential component in achieving the mission and vision of a business. They can improve business productivity and profit if they meet companies' requirements. They are assets for a company. We can recognize motivated employees by seeing their faces. When we see smiles on their faces, we can assume that they are happy. Happy employees are full of motivation in doing their tasks. Motivation comes from a Latin word meaning a push or move. Motivation applies to the management of human resources in general and to subordinates in particular. Motivation uses the power and potential of directing subordinates to cooperate productively and successfully achieve the intended purpose.

Motivation is important because it can determine the actions or behavior of a person in conducting a work process (Kreitner & Kinicki, 2010). To increase employees' motivation, companies usually offer several rewards. Research shows that it is important to give rewards to employees because it can stimulate their work motivation (Eisenberger & Shanock, 2003). This strategy is implemented in most Indonesian companies, in particular, the news networks. According to research by Malik and colleagues (2015), several things can be moderators for the effectiveness of the use of these rewards, among them creative self-efficacy (CSE) and the importance of rewards. For this reason, research studies have been conducted to examine how each of these variables affects the creative performance of human resources in the news networks, which have employees of different generations and backgrounds.

2 THEORETICAL REVIEW

Work has been done to understand motivation theory and various research has been published. This paper references the motivation theory of Victor Vroom. He defines motivation as a process governing choice among alternative forms of voluntary activities, a process controlled by the individual. The individual makes choices based on estimates of how well the expected results of a given behavior will match up with or eventually lead to the desired results. Motivation is a product of an individual's expectancy that a certain effort will lead to the intended perform-ance, the instrumentality of this performance to achieving a certain result, and the desirability of this result for the individual, known as *valence*. This theory is often called *expectancy theory*.

Expectancy theory is based on four assumptions (Vroom, 1964). One assumption is that people join organizations with expectations about their needs, motivations, and past experi-ences. These influence how individuals react to the organization. A second assumption is that an individual's behavior is a result of conscious choice – that is, people are free to choose those behaviors suggested by their expectancy calculations. A third assumption is that people want different things from the organization (e.g., good salary, job security, advancement, or chal-lenge). A fourth assumption is that people will choose among alternatives to optimize outcomes for themselves. The expectancy theory, based on these assumptions, has three key elements: expectancy, instrumentality, and valence. A person is motivated to the degree to which he or she believes that (a) effort will lead to acceptable performance (expectancy), (b) performance will be rewarded (instrumentality), and (c) the value of the rewards is highly positive (valence) (Lunen-burg, 2011: 1–2). Expectancy is a person's estimate of the probability that job-related effort will result in a given level of performance. Expectancy is based on probabilities and ranges from 0 to 1. If an employee sees no chance that effort will lead to the desired performance level, the expectancy is 0. On the other hand, if the employee is completely certain that the task will be completed, the expectancy has a value of 1. Generally, employee estimates of expectancy lie somewhere between these two extremes. Meanwhile, instrumentality is an individual's estimate of the probability that a given level of achieved task performance will lead to various work out-comes. As with expectancy, instrumentality ranges from 0 to 1. For example, if an employee sees that a good performance rating will always result in a salary increase, the instrumentality has a value of 1. If there is no perceived relationship between a good performance rating and a salary increase, then the instrumentality is 0. Valence is the strength of an employee's prefer-ence for a particular reward. Thus, salary increases, promotion, peer acceptance, recognition by supervisors, or any other reward might have more or less value to individual employees. Vroom suggests that motivation, expectancy, instrumentality, and valence are related to one another by the equation Motivation = Expectancy x Instrumentality x Valence (Lunenburg, 2011: 2–3).

Motivation is influenced by internal and external factors, or extrinsic and intrinsic motiv-ation. Intrinsic factors are rewards that come from within the individual himself. Intrinsic motivation is influenced by internal factors, while extrinsic motivation is influenced not only by external but also by internal factors, depending on the causal relationship of control of the individual (Deci, Koestner, & Ryan, 1999). Furthermore, in the same article, research has shown the value of being intrinsically motivated in many applied settings such as education, sports, and work environments. Besides, research on intrinsic motivation has focused atten-tion on the more general benefits of support for autonomy and competence for motivated per-sistence, performance, and well-being.

3 RESEARCH QUESTION

Based on the introduction and literature review presented earlier in this paper, the research focused on the following question: how do CSE and the importance of rewards affect the cre-ative performance of human resources in the news networks across different generations and backgrounds?

4 RESEARCH METHOD

In order to answer the research question, this research used both qualitative and quantitative methods. First, the literature about employees' capacity building, motivation, and human resources management theory was analyzed. Subsequently, after analyzing theories, question-naires were arranged and interviewees chosen. Furthermore, the answers of the respondents were analyzed and examined. Respondents consisted of 202 employees of seven news television companies in Indonesia. Finally, the results of this research conclude.

4.1 *Research model*

The researchers used questionnaires, which give indicators that can represent variables for researchers to analyze. The study used six-point Likert scales with five-dimensional sections. Pretests were done on thirty respondents. All the questions obtained valid and reliable results with a Cronbach's alpha value of > 0.8 and a Kaiser-Meyer-Olkin (KMO) value of ≥0.7. Extrinsic motivation questions were based on the research of Yoon. Sung, and Choi (2015), while CSE questions incorporated the modified questions by Malik and colleagues (2015) and Brauckhus and colleagues (2014) based on research developed by Tierney and Farmer (2002). The reward and intrinsic motivation questions were developed by Yoon and colleagues (2015). Subsequently, the creative performance dimension was measured based on items developed by Scott and Bruce (1994). In conducting data collection, the researchers asked the respondents to conduct self-evaluation of creative performance dimensions. Using expectancy theory as a fundamental theory meant integrative frameworks had to be developed in order to see the effects of extrinsic reward on employee's creative performance.

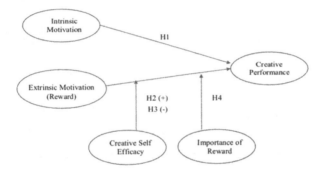

Figure 1. Research model.

4.2 *Hypothesis*

Based on the research model, this study analyzed four hypotheses.

Hypothesis 1: The intrinsic motivation of the employee positively affects the employee's creative performance.

Hypothesis 2: CSE moderates extrinsic motivation in the form of rewards toward employees with a positive impact on creative performance in employees with high CSE.

Hypothesis 3: CSE moderates extrinsic motivation in the form of rewards toward employees with a negative impact on creative performance in employees with low CSE.

Hypothesis 4: Importance of reward moderates extrinsic motivation in the form of rewards in employees who consider a reward important to employee creative performance.

Therefore, the respondents in this study included permanent and contract employees in news television who had worked at least one year. Researchers used purposive or random sampling techniques according to specific criteria. With a total of thirty-three items, the number of samples recommended was at least 165 respondents (Hair et al., 2014), so the total number of 202 in the study is considered adequate.

5 RESEARCH RESULT

This study used simple regression for data processing as well as regression with moderated variables or moderating regression analysis (MRA) using SPSS. In Table 1, we see the effect of intrinsic motivation on creative performance is 0.000 < 0.05 with a calculated t value of 6.873 > T Table 1,973 (obtained through the table T distribution. 025; 199), so we can deduce that intrinsic variable motivation influences creative performance.

Data processing was further done with employees who have a high CSE, which amounted to 192 people, and low CSE, which amounted to 10 people. Employees' CSE was grouped based on the average number of scores, where a score of 4 to 6 was high CSE while underneath was low CSE. Tables 2 and 3 show that after being moderated, the coefficient number in this moderation is 0.651 for high CSE and 0.983 for low CSE. The coefficient number greater than 0.05 is insignificant.

Table 1. Intrinsic motivation relations.

Coefficients Model	Unstandardized Coefficients		Standardized Coefficients		
	B	Std. Error	Beta	t	Sig.
(Constant)	3.111	2.408		1.292	0.198
Intrinsic Motivation	0.522	0.076	0.400	6.873	0.000
Extrinsic Rewards	0.246	0.038	0.372	6.396	0.000

a. Dependent Variable: Creative Performance

Table 2. Regression test with high CSE as moderator.

Coefficients Model	Unstandardized Coefficients		Standardized Coefficients		
	B	Std. Error	Beta	t	Sig.
(Constant)	7.933	12.891		0.615	0.539
ER	0.047	0.366	0.079	0.129	0.897
HCSE	0.271	0.255	0.289	1.062	0.290
ERHCSE	0.003	0.007	0.321	0.454	0.651

a. Dependent Variable: CP

Table 3. Regression test with low CSE as moderator.

Coefficients Model	Unstandardized Coefficients		Standardized Coefficients		
	B	Std. Error	Beta	t	Sig.
(Constant)	0.268	17.243		0.016	0.988
ER	0.413	1.444	0.480	0.286	0.785
LCSE	0.275	0.780	0.299	0.352	0.737
ERLCSE	−0.001	0.034	−0.043	−0.023	0.983

b. Dependent Variable: CP

Different results from previous research can arise due to different research demographics. Previous research was conducted in Pakistan with executive-level network employees. The same steps were also performed on the importance of rewards variable. The significance level is at 0.222, which means there is no moderation of the importance of rewards variable against extrinsic rewards and creative performance relationships.

Table 4. Regression test with intrinsic motivation as moderator.

Coefficients Model	Unstandardized Coefficients		Standardized Coefficients		
	B	Std. Error	Beta	t	Sig.
(Constant)	7.080	5.076		1.395	0.165
ER	0.483	0.177	0.731	2.728	0.007
IR	0.767	0.338	0.390	2.269	0.024
ERIR	−0.014	0.011	−0.440	−1.224	0.222

a. Dependent Variable: CP

6 CONCLUSION

In general, the research results differ from those of previous studies. Based on this study, some conclusions are intrinsic and extrinsic rewards significantly affect creative performance; meanwhile, CSE and importance of rewards do not provide moderation influence over extrinsic rewards and creative performance relationships. A different conclusion is due to several things, including the different demographics of respondents as well as different data collection methodologies through which respondents were asked to measure their performance on the creative variables, resulting in potentially biased responses. Thus, the measurement of better performance was not done by self-evaluation. With the conclusion that extrinsic and intrinsic rewards have a significant influence on improving the creative performance of employees, the form of the benefit provided should be by following the expectations of employees.

7 LIMITATION OF RESEARCH AND ADVICE

This research still has several limitations, including limited research area, limited indicators as research measurements, and a questionnaire technique that potentially produced biased answers. Further research is expected to be done with improvement of this limitation, such as larger research areas, opening up more variables, and using other data collection techniques, like spreading indirect questionnaires to interested parties on research objects, such as corporate or corporate management, and face-to-face interviews. Subsequent research is also expected to be expanded to digital or online media television news companies.

REFERENCES

Brockhus, S., Kolk, T. V., Koeman, B., & Badke-Schaub, P. 2014. The influence of creative self efficacy on creative performance. *The 13th International Design Conference* (pp. 437–444). Dubrovnik, Croatia: DESIGN 2014.
Deci, E. L., Koestner, R., & Ryan, R. M. 1999. A meta-analytic review of experiments examining the effects of extrinsic rewards on intrinsic motivation. *Psychological Bulletin, 125*, 627–668.

Eisenberger, R., & Shanock, L. 2003. Rewards, intrinsic motivation, and creativity: A case study of conceptual and methodological isolation. *Creativity Research Journal, 15*, 121.

Hair, J. F., Black, W. C., Babin, B. J., & Anderson, R. E. 2014. *Multivariate data analysis, Vol. 1*. 7th edition. Harlow: Pearson.

Kreitner, R., & Kinicki, A. 2010. *Organizational behavior*. 9th edition. New York: McGraw Hill.

Lunenburg, F.C. 2011. Expectancy Theory of Motivation: Motivating by Altering Expectations, International Journal of Management, Business, and Administration, *15*(1).

Malik, A. R., Choi, J. N., & Butt, A. N. 2015. Rewards and creative performance: Moderating effects of creative self-efficacy, reward importance, and locus of control. *Journal of Organizational Behavior, 36*, 59–74.

Scott, S. G., & Bruce, R. A. 1994. Determinants of innovative behavior: A path model of individual innovation in the workplace. *Academy of Management Journal, 37*, 580–607.

Tierney, P., & Farmer, S. F. 2002. Creative self-efficacy: Potential antecedents and relationship to creative performance. *Academy of Management Journal, 45*(6), 1137–1148.

Vroom, V. H. 1964. *Work and motivation*. New York: Wiley.

Yoon, H. J., Sung, S. Y., & Choi, J. N. 2015. Mechanisms underlying creative performance: Employee perceptions of intrinsic and extrinsic rewards for creativity. *Social Behavior and Personality: An International Journal, 43*, 1161–1180.

The influence of motivation and competence on employee performance at a Karawang concrete production plant

I. Wigastianto & R. Wahyuningtyas
Telkom University, Bandung, Indonesia

ABSTRACT: PT. Wika Beton PPB Karawang employee performance declined in 2017, as seen in its employee performance evaluation report. The number of employees exceeding performance expectations fell by 12.5%, and the number of employees performing below expectations grew by 1.4%. The goal of this research was to determine motivational, competence, and performance levels, as well as to study the effects of motivation and competence on employee performance. This research was done using the quantitative method. The data analysis was done using the structural equation modeling (SEM) model covariance-based metric (CB-SEM) with AMOS 24. The research results using SEM analysis on the measurement model showed that all indicators had a regression weight of above 70%. Through the structural model based on the connections between its constructs, we can conclude that significant and positive inter-variable connections exist between the three aspect of the variables.

1 INTRODUCTION

Human resources are the main asset of every organization, whether public or private, especially those that are knowledge-based (Sinambela, 2016: 3). Employees, as the gears of an organization, greatly affect the company's future capabilities and competitiveness, as well as its well-being. Thus, the competition between organizations is not in the form of products or within the market, but in their way of thinking (Pella and Inayati in Sule and Wahyuningtyas, 2016: 2).

Through its performance evaluation, PT. Wika Beton management measured the potential competence of each employee. The results of the performance evaluation were then used as consideration for an employees' guidance and improvement program. The results of the 2017 evaluation showed a decline compared to 2016. The number of employees performing above expectations declined by 12.5%, while the number of employees performing below expectations rose by 1.4%.

Aside from performance background, work motivation is also considered important at PT. Wika Beton. The level of work motivation can affect how optimally an employee works. This statement is supported by research done by Zameer and colleagues (2014: 1), who state the importance of motivation for both public and private organizations. Furthermore, they also found that motivation is one of the main factors affecting human resources.

To gauge employees' perception of their work motivation levels, the researchers conducted a survey that resulted in preliminary data with the variable PPB Karawang PT. Wika Beton employees' motivation level. The researchers gave six-question questionnaires to thirty employees. The questions were made based on McClelland's motivation theory. This resulted in a percentage of 74.67%, with the dimension of need for power scoring the lowest.

Besides motivation, another important aspect that affects an employee's performance level, proposed by earlier researchers and experts, is the competence level of the employee. According to Ngo, Jiang, and Loi (2014), competence in human resources management has a positive and significant effect on company performance. This statement is also supported by research

Table 1. Research results recap of PT. Wika Beton Tbk employee performance.

Remarks	Year		
	2015	2016	2017
Below Expectations	23.00	11.10	12.50
Fulfilled Expectations	45.10	28.90	40.00
Above Expectations	31.90	60.00	47.50

done by Mubarok and Putra (2018: 1), which found that competence must be supported by training, development, and motivation for the employees.

Competence in human resources management is an important aspect in the implementation of a human resources management system, as well as its integration in business strategy. PT. Wika Beton performs its competence level evaluation every year. The competence level evaluation is important to assess the need of training for each employee. However, the average competence level of PT. Wika Beton's employees hasn't managed to reach the expected level, which is the above expectation category.

Three theories have been used in the research. The first one was McClelland's theory (in Robbins & Judge, 2015: 131), which was used for the motivation variable. The second was Spencer and Spencer's theory (in Priansa, 2014: 258), which was used for the competence variable. The last one was Bernardin's theory (in Sudarmanto, 2014: 12). Which was used for the performance variable.

2 RESEARCH METHOD

This research was done using the quantitative method. The data collection method consisted of questionnaires and literature studies. The questionnaires were spread to 236 employees of PT. Wika Beton. The sampling method used was saturated sampling. The data analysis was done using the structural equation modeling (SEM) model supported by the covariance-based metric structural equation modeling (CB-SEM) and AMOS 24.

Based on the research concept, the hypotheses were:

H1: Motivation has a significant positive effect on employees' performance.
H2: Competence has a significant positive effect on employees' performance.
H3: Motivation and competence have a significant positive effect on employees' performance.

3 RESEARCH RESULTS

Based on the questionnaire results, the scores for each variable were 82.80% for the motivation variable, 81.42% for the competence variable, and 84.49% for the employee performance variable. These results show that both the motivation and competence variables are classified as high level, while the performance variable is classified as very high level. Based on the descriptive analysis results, the perception of PPB Karawang employees regarding motivation, competence, and performance variables can be classified as high and very high.

In SEM analysis before performing measurement model and structural model tests, it is necessary to calculate the results of the reliability and validity tests. On the validity test, the model is considered according to fitness categories, which are absolute fit, incremental fit, and parsimony fit; the model has a score above the cutoff. On the reliability test, the model can be considered reliable as the construct reliability test of each variable has shown a value above 0.70 (Hair et al., 2010).

According to Santoso (2018: 139), a factor loading value of above 0.70 on standardized regression weight shows that an indicator is part of a construct. The measurement model test proved that the dimensions belonging to the variables are the same dimensions that form those variables. Since the standardized regression weights' estimated values of each dimension–variable relationship is above 0.7, then all dimensions can explain each variable.

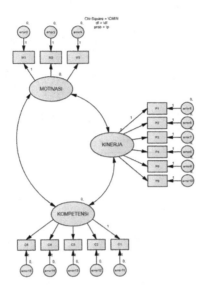

Figure 1. SEM model.

3.1 *Measurement model test*

The next step in SEM analysis after creating a research model form in the AMOS 24 application is to test the validity of the measurement model. The model test equipment is divided into three fit indices – namely absolute fit indices, incremental fit indices, and parsimony fit indices. This is also in line with what Hair (2010: 690) explains: confirmatory factor analysis (CFA) requires at least one of each model's test equipment to be in accordance with its index requirements.

Based on the results of the validity test calculation, it appears that each validity test tool has a fit indicator. Absolute fit indices have a gauge that fits the RMR fit measure. In the incremental fit indices, the fit gauge is found in the NFI. Then the parsimony fit indices have a measuring instrument that is fit on PNFI, PCFI, AIC, ECVI, and HOELTER.

After the validity test, the analysis continued with the reliability test to determine whether the dimensions of each variable are strong to form the variable.

Based on reliability testing, the resulting construct reliability is 0.85 for the motivation variable, 0.90 for the competency variable, and 0.96 for the performance variable. The limit on the reliability test for construct reliability is > 0.70. Therefore, all construct reliability on the variables of motivation, competence, and performance are reliable, or indicate that the variables are consistent and strong in shaping the construct. The variables used in the study are reliable.

After the model is fit, the next step is to measure the relationship between indicators and constructs. This measurement aims to see how strong the relationship is or even if there is no relationship at all. The relationship between indicators and constructs can be measured by construct validity tests (latent variables), and can be performed with a convergent validity test.

In the convergent validity test, the factor loading number can be seen in the standardized regression weights table in AMOS 24 software. In that table, the estimated factor loading figures can be seen from each indicator's relationship with its construct. The results of calculating the relationship of indicators with their constructs can be seen in Table 2.

Table 2. Standardized regression weights.

Indicator with Construct Relationship			Estimate
M1	<—	MOTIVATION	0.84
M2	<—	MOTIVATION	0.70
M3	<—	MOTIVATION	0.88
C1	<—	COMPETENCE	0.79
C2	<—	COMPETENCE	0.78
C3	<—	COMPETENCE	0.77
C4	<—	COMPETENCE	0.86
C5	<—	COMPETENCE	0.82
P1	<—	PERFORMANCE	0.95
P2	<—	PERFORMANCE	0.93
P3	<—	PERFORMANCE	0.91
P4	<—	PERFORMANCE	0.88
P5	<—	PERFORMANCE	0.84
P6	<—	PERFORMANCE	0.79

Based on the data in Table 2, we can see that the average estimated number (factor loading) of each indicator's relationship with its construct is above 0.70. The smallest factor loading figure is 0.70 on the relationship between the M2 indicator and the MOTIVATION variable. The highest factor loading number is 0.95 on the relationship between the P1 indicator and the PERFORMANCE variable.

Based on these results, the factor loading of all relationship indicators and constructs in the research shows that all indicators can explain the existence of each construct. This is due to the factor loading figures in Table 2. Nothing is below 0.70. We can conclude that in this study all indicators can explain each of their constructs.

3.2 *Structural model test*

In this study, the measurement model testing is considered to have passed based on the results obtained from each test. The testing process continued by testing the structural model used in the study, which is testing the relationship between constructs in accordance with the theoretical framework of the research. The process before and when conducting structural test models refers to the points stated by Hair and colleagues (2010: 690), that in SEM analysis it is always recommended to use a two-stage SEM analysis process. The first stage is to test the fit and validity of the constructs used in the measurement model. The second stage is to test the structural theory.

Based on the fit test, we can see that among the absolute fit indices, incremental fit indices, and parsimony fit indices, each of them is categorized as fit. Based on the rule of thumb stated by Hair and colleagues (2010) regarding the validity test, at least one test is categorized as fit from each of the absolute fit indices, incremental fit indices, and parsimony fit indices. Therefore, we can conclude that overall fit testing shows that the structural model used is categorized as fit.

After a structural model can be said to be fit, the next process is to see whether there is a significant and close relationship between the independent variable and the dependent variable.

The process consists of formulating hypotheses, decision-making, and final analysis.

The hypotheses can be formulated based on the number of relationships between the independent and dependent variables in the structural model. Based on the structural model, there are two relations. Therefore, two hypotheses are used:

a. The hypothesis of the relationship of motivation variables with performance variables
   H0: There is no relationship between motivation variables and performance.
   H1: There is a relationship between motivation variables and performance.
b. The hypothesis of the relationship of competence variables with performance variables
   H0: There is no relationship between the competence variable and performance.
   H1: There is a relationship between the competence variable and performance.

In addition to these hypotheses, the relationship between exogenous variables is added. The additional hypothesis is:

c. The hypothesis of the relationship of motivation variables with competence variables
   H0: There is no relationship between motivation variables and competence.
   H1: There is a relationship between motivation variables and competence.

According to Santoso (2018: 169), to make decision-making easier, we use the probability number (P). If $p > 0.05$, H0 is accepted. If $p < 0.05$, then H0 is rejected. Analysis can be done based on the data output estimates for the regression load of each variable.

The structural model test gave a positive and significant result regarding motivation-performance, competence level-performance, and motivation-competence level relationships. The result is shown on the inter-variable relationship regression weight data. The P-value in the regression weight data indicates the presence of a relationship between variables. Each inter-variable relationship shows a P-value close to 0, which proves the presence of a relationship between those variables. The results are shown in Table 3.

Table 3. Regression weight data.

			Estimate	S. E.	C. R.	P
Performance	←	Motivation	1.27	0.16	7.52	***
Performance	←	Competence	0.80	0.06	11.98	***
Competence	←→	Motivation	0.12	8.30	***	0.90

Relationship Motivation → Performance

The p number of the relationships between constructs is ***. This number shows that the number is approaching 0. Therefore, H0 is rejected because $p < 0.05$. This proves the relationship between motivation and performance variables. An employee's work motivation affects his performance; the higher the motivation, the better the performance.

Relationship Competence → Performance

The p number of the relationships between constructs is ***. This number shows that the number is approaching 0. Therefore, H0 is rejected because $p < 0.05$. Then we can conclude that there really isn't a real relationship between the competency and performance variables. Therefore, the competence of an employee does not affect his performance; high competence does not necessarily mean high performance.

Relationship Motivation → Competence

This relationship can be seen in the output estimates in the covariance and correlation sections. Similar to testing the relationship between exogenous and endogenous variables, the base rate for decision-making is based on the p number with a cutoff of 0.05. The resulting covariance number is ***, which means that the number is close to 0. Therefore, H0 is rejected because it is below 0.05. We can conclude that the motivation variable is related to competence. This is also evidenced by the correlation number, amounting to 0.90. The correlation number of the two variables shows that the relationship between the two is very close because it is far above the 0.70 figure.

Meanwhile, the inter-variable correlation of performance with motivation is 0.88 and performance with competence is 0.80. Therefore, we can conclude that based on the structural model test results, relationships exist between motivation and performance as well as motivation and competence variables. Simultaneously, motivation and competence have a positive and significant effect on employees' performance.

4 CONCLUSION

Based on these research results, we can conclude that the level of employee perception of motivation and competence is considered high level, while the perception of performance is considered very high level. Meanwhile, the SEM analysis results proved the existence of a positive and significant relationship between motivation and employees' performance, competence and employees' performance, and both motivation and competence and employees' performance.

We could give two suggestions related to employees' motivation and competence for improving performance. The first one is the theoretical aspect, which is that in further research, two variables (motivation and competence) related to its objects could be improved and made more relevant. Then, for the selection of objects, further research is expected to use the same industry to explore the possibilities of new theories.

The second one is the practical aspect. The organization could do a competency mapping of its employees based on their abilities; this is very useful in that employees can be placed in the right division or section according to their capabilities. In addition, employee competency mapping can optimize training programs because the employees who need to be included in the training are the right people for the type of training.

Based on the results of this research, improving employee performance could be accomplished by paying attention to aspects of motivation and competence that are owned by PPB Karawang employees. In accordance with the recommendations that have been presented, improving employee performance also needs to be considered from theoretical angles. The performance dimensions used in this study were the dimensions of quality, quantity, time, cost, supervision, and relations between employees.

In SEM analysis, seen from the acquisition of numbers based on measurement model testing, the lowest factor loading number is in the dimensions of the relationship between employees (0.79). In a company, the dimensions of relationship between employees in the performance variable can be seen in employee relations or communication between divisions or sections. PPB Karawang requires good coordination from each division so that work activities can run smoothly. For this reason, employees need systematic communication between employees, both horizontally and vertically. Thus, it is expected that employee performance at PPB Karawang can improve.

REFERENCES

Hair Jr., J. F., Black, W. C., Babin, B. J., & Anderson, R. E. 2010. *Multivariate data analysis: A global perspective* (7th ed.). Upper Saddle River, NJ: Pearson Education.

Mubarok, E. S., & Putra, H. 2018. The influence of training, competence, and motivation on employees performance of workers social security agency in Banten Province, Indonesia. *Journal of Economics and Sustainable Development* 9(4), 129–139.

Ngo, H., Jiang, C., & Loi, R. 2014. Linking HRM competency to firm performance: An empirical investigation of Chinese firms. *Personnel Review.* 43(6), 898–914.

Priansa, D. J. 2014. *Perencanaan dan Pengembangan SDM* (1st ed.). Bandung: Alfabeta.

Robbins, S, P., & Judge, T. A. 2015. *Organizational Behavior* (16th ed.). Boston: Pearson.

Santoso, S. 2018. *Konsep Dasar dan Aplikasi SEM dengan AMOS 24.* Jakarta: PT Elex Media Komputindo.

Sinambela, L. P. 2016. *Manajemen Sumber Daya Manusia.* Jakarta: PT. Bumi Aksara.

Sudarmanto. 2014. *Kinerja dan Pengembangan Kompetensi SDM.* Yogyakarta: Pustaka Pelajar.

Sule, E., & Wahyuningtyas, R. 2016. *Manajemen Talenta Terintegrasi.* Yogyakarta: Andi.

Zameer, A., Alis, S., Nisar, W., & Amir, M. 2014. The impact of motivation on employees' performance in the beverage industry of Pakistan. *International Journal of Academic Research in Accounting, Finance and Management Sciences* 4(1), 293–298.

Business process design to improve mean time to recovery radio IP services

P.N. Ayuningtias & P.M. Sitorus
Faculty of Economics and Business, Telkom University, Bandung, Indonesia

Y. Yogaswara
Faculty of Engineering, Pasundan University, Bandung, Indonesia

ABSTRACT: This study aims to analyze the processes on the IP Radio assurance business process that contribute to the value of MTTR Radio IP and evaluate business process that meet the MTTR target of IP Radio with simulation design. This study uses theoretical approach to Business Process Rengineering and Business Process Simulation. This research uses Arena simulators to carry out the design simulation of the IP Radio assurance business process. The results of this study indicate that the dominant processes on the IP Radio assurance business processes are toubleshooting, spare parts request, travel time directly to site, location permission, technical tlosed, and analysis. The assurance business process that can meet the MTTR Radio IP targets is the IP Radio assurance business process that eliminates spare part request process and simplifies the troubleshooting process, spare part request, direct travel time to site, location permission, technical closed, and analysis.

1 INTRODUCTION

1.1 *Background*

IP radio assurance business processes regulate activities in handling IP radio interference and responsible units. IP radio business assurance processes have two vendors to run the business process, namely the fulfillment of spare parts vendors and technician service vendors. Mean time to recovery (MTTR) Radio IP's realization was around 18 hours while the target was 9 hours. MTTR Radio IP's targets have not been achieved due to the long and long coordination paths in fulfilling spare parts and others process.

1.2 *Problem formulation*

Based on previous research by Rinaldi (2015) about improving the efficiency of public administration in Italy through business process reengineering and business process simulation. Business process reengineering (BPR) is fundamental rethinking, radical redesign of a business process to achieve dramatic improvement in performance such as cost, quality, service and speed (Hammer & Champy, 1993). Business process simulation (BPS) is the process of creating and analyzing digital prototypes from physical models to predict their performance in the real world. Simulation modeling is used to help designers understand whether, under what conditions, and in what ways parts can fail and what burdens can be withheld (Abdellatif, 2017). Therefore in this study used a business process reengineering approach and business process simulation as a medium in the design of business assurance processes to improve the Mean Time To Recovery (MTTR) service of IP Radio.

2 LITERATURE REVIEW

2.1 *Business process reengineering*

Business process reengineering (BPR) according to Hammer & Champy (1993), emphasizing the extreme nature of redesign and also identifying desired results. They promote it as "fundamental rethinking and radical redesign of business processes to achieve dramatic improvements in critical performance measures, such as cost, quality, service, and speed. The main emphasis of this approach is the fact that an organization can realize dramatic improvements in performance through a radical redesign of its core business processes.

The main idea of the definition of Hammer & Champy (1993) is breakthrough. Breakthrough is defined as a higher level of organizational performance beyond its current performance in order to achieve its vision. The difference in current performance and breakthrough performance is defined as a performance gap. Watts (2002) explains that performance analysis is another form of process analysis that can be used to define and measure activities. The components of performance analysis, which are cost, time, and quality, can help organizations set priorities for process analysis.

2.2 *Business process simulation*

Business process simulation (BPS) is a tool used to analyze and understand system behavior that helps in decision making. BPS can also help predict system performance under a number of scenarios determined by decision makers (Greasley, 2003). Business process simulation is one of the most widely used operational research applications. This enables understanding the basics of business systems, identifying opportunities for change, and evaluating the impact of proposed changes on key performance indicators (Doomun, 2008). With simulation tools, we can take pictures of dynamic models. Some simulation tools available are Metis, Arena, and SimProcess.

After the data is collected, the process simulator identifies which steps constitute the process bottleneck and other weaknesses in the process design. Based on simulation results, the initial process design can be improved iteratively.

2.3 *Framework*

In this study, the main theories used are BPR and BPS. BPS is a part or tool of BPR quantitatively. The main emphasis of the BPR approach is that an organization can realize dramatic improvements in performance through a radical redesign of its core business processes. In this research, the IP Radio assurance business process has a Mean Time To Recovery (MTTR) that has not reached the target of more than 9 hours. The achievement of MTTR Radio IP depends on the performance of the implementing vendor namely spare part vendor and the technician service vendor. The BPR and BPS theory approaches are used to improve MTTR Radio IP through simulation modeling, simulation scenarios and simulation scenario analysis. The analysis process is done by taking into account managerial and operational impacts and the results of literature studies related to BPS and BPS. The resulting IP Radio business assurance process that can meet MTTR targets is less than 9 hours.

3 METHODOLOGY

3.1 *Research characteristics*

This study uses a mixed methods, namely the quantitative method to understand the processes in the IP Radio assurance business process that contribute to the value of the MTTR Radio IP. And qualitative methods for formulating and evaluating business process recommendations that can meet MTTR Radio IP targets with simulation design.

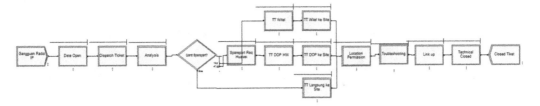

Figure 1. IP radio assurance business process simulation model.

Quantitative methods play a role in obtaining measurable quantitative data that can be descriptive, comparative, and associative in nature. Whereas qualitative methods play a role in proving, deepening, expanding, weakening, and invalidating quantitative data obtained at an early stage.

3.2 Variable operations

In this study a simulation design is carried out involving several variables that will be used to represent the behavior of the model in a real system namely date open, dispatch ticket, analysis, sparepart request, travel time DOP spare part, travel time DOP sparepart to the site, travel time witel, travel time witel to site, travel time directly to site, location permissions, troubleshooting, link up, and tech closed.

3.3 Measurements

In quantitative measurements according to Kelton (2015), verification and validation is the process of ensuring that the simulation model behaves as intended according to the modeling assumptions made. Model validation uses chi square two-sample to ensure that the real system output is the same as the simulation model output.

In qualitative measurements according to Sugiyono (2018), in qualitative research findings or data can be declared valid if there is no difference between what the researcher reports and what happens to the object under study. In this study the validity test was carried out in several ways namely triangulation of sources, triangulation of time, dan Discussion with friends.

3.4 Data analysis

Quantitative data analysis using statistical tests. Kolmogorov Smirnov statistical test is used to identify the probability distribution of each operational variable. Chi square two samples statistical test to test the hypothesis.

According to Sugiyono (2018), data analysis in qualitative research is carried out at the time the data collection takes place and after the data collection is completed within a certain period. Data analysis includes data collection, data reduction, data presentation, and drawing conclusions.

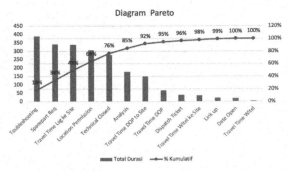

Chart 1.   Pareto diagram.

4   RESULTS AND DISCUSSION

4.1   *Quantitative research results*

4.1.1   *Variable data distribution test*
The data distribution test is the identification of the probability distribution of each operational variable by using the Kolmogorov Smirnov statistical distribution test conducted by the Arena simulator. In this study the data from each variable is entered into the Arena Analyzer Input Analyzer, Kolmogorov Smirnov's D value is represented by the Corresponding p-value (p) in Arena. The limit value of p = 0.05, the probability distribution is more precise if the value of p > 0.05. With a Hypothesis:

H_0: Sample data for processing time duration cannot be approached by a particular distribution
H_1: H_0 isn't right.

Table 1.   Simulation scenarios.

No	Concept	Process	Arena simulation scenario
1	Eliminate	Sparepart request	Remove the spare part request module
2	Simplify	Troubleshooting, sparepart request, travel time directly to the site, location permission, technical closed, dan analysis	Use a constant distribution with minimum values for modules of each dominant process
3	Eliminate & Simplify	Troubleshooting, sparepart request huawei, travel time directly to the site, location permission, technical closed, dan analysis	Remove the Spare Part Request module and use a constant distribution with a minimum value for the module of each dominant process

The criterion used is H0 is rejected if $p > \alpha = 0.05$.

4.1.2   *Modeling simulation models*
The simulation model created duplicates all processes in the IP Radio assurance business process, these processes are translated into modules in the Arena simulator, which are the create module, the process module, the decide module, the dispose module. By using these modules, the simulation model modeling in this study is shown in Figure 1 below

Table 2. Comparative analysis of simulation scenario results.

Competitive advantage	Eliminate	Simplify	Eliminate & Simplify
Time	MTTR 17.28 hours, 4% decrease (MTTR target not reached)	MTTR 7.39 hours, 59.14% decrease (MTTR target reached)	MTTR 5.42 hours, 70% decrease (MTTR target reached)
Quality	MTTR performance 8.16% (quality and performance targets not achieved)	MTTR performance 120.24% (quality and performance targets are achieved	MTTR performance 142.57% (quality and performance targets achieved)
Cost	OPEX costs reduced (merging managed service contracts). Pay penalties to customers	OPEX costs reduced (reduction in transportation costs and technician accommodation). Do not pay penalties to customers	OPEX costs reduced (merging manage service contracts and reducing transportation and technician accommodation costs). Do not pay penalties to customers

4.1.3 *Model verification and validity test/hypothesis testing*
In this research the model verification test is done by ensuring all processes in the Assurance Radio IP business process (which is in a real system) have been modeled in the simulation model. And make sure the simulation model can run without any errors in each module used.

In this research the model validity test is done by comparing the behavior (output) of the Radio IP (real system) assurance business process with the system (output) behavior modeled using the Chi-Square Two Sample statistical test. Testing the validity of the model is also to test the hypothesis in this study.

Output of real system and output of system modeled are then calculated using the Chi Square Two Sample statistical formula as follows:

$$x^2 = \frac{\sum_{i=1}^{r} \sum_{j=1}^{k} \left(o_{ij} - e_{ij}\right)^2}{e_{ij}} \qquad (1)$$

With a Hypothesis:

H_0: There is difference between the behavior (output) of the real system and the behavior of the system being modeled

H_1: There is no difference between the behavior (output) of the real system and the behavior of the system being modeled

Reject H_0 if $X_{2\ count} > X2$ Chi Square table, with a value α: 5%
The calculation results above formula are:
$X_{2\ count}$: 25,4773
$X2_{α/2}$: 23,337 (from the Chi Square table α/2)
So the hypothesis is accepted

4.1.4 *Pareto diagram*
Pareto diagram is a tool to find the dominant cause or factor of a problem. Pareto diagram uses the principle of 80-20. It means that 80% of the accumulated percentage of factors is the dominant factor that must be prioritized while the rest is then.

Based on the total duration of the simulation model, the Pareto diagram is depicted in chart 1. Chart 1 shows that the dominant processes (accumulated percentage up to 80%) in the IP Radio assurance business process are toubleshooting, spare part requests, travel time directly to site, location permission, technical closed, and analysis. Then the processes that need to be improved to improve IP Radio MTTR.

4.2   *Qualitative research results*

4.2.1 *Interview result*
Interviews related to the dominant process to confirm the results of quantitative research include activities in the process, duration of process time, effectiveness of the performance of technicians, effectiveness of coordination, effectiveness of IT Tools, work rules/SOPs, standard processing time and improvement efforts. The results of the interviews reinforce the results of quantitative research

4.2.2 *Simulation scenario*
According to Adriansah (2010) to improve process performance, ESIA theory can be used namely eliminate, simplify, integrate, and automate business processes in organizations. In this study adopted the concepts of eliminate and simplify. Eliminate is eliminating processes that do not provide added value to customers. Simplify is to simplify unnecessary processes. simulation scenarios in the study according to Table 1 below:

4.3   *Analysis scenario simulation result*

All simulation scenarios are run with 365 days length replication and number of replications are 100. According to Heizer (2017), competitive advantage in management operations is defined as a faster response to customers with lower costs and higher quality. In this research aspects of time, quality, and cost, become aspects used to analyze the output of the simulation scenario described in Table 2 below

4.4   *Managerial impact analysis of simulation scenarios*

By combining spare parts supply contracts with technician services, it can reduce the cost of the original contract with two vendors to become just one vendor. In addition, the function of monitoring the vendor's performance is made easier, management focuses on just one vendor.

However, it is necessary to measure the capacity of the technician service vendor in handling the work of fulfilling IP Radio spare parts. If the vendor is able to handle the work, then the amendment to the cooperation agreement with the vendor of the technician service is related to the rights and obligations of Telkom and Mitratel and the making of new work rules in the assurance process of the IP Radio from the amendment. This amendment also needs to pay attention to the sustainability of spare parts availability in all operational areas. Sustainability of spare parts availability and performance of technicians to become vendor KPIs in the amendment.

Setting a standard time on the dominant processes means that the resources of the spare part vendor and the technician service vendor will work at the maximum level so that it will increase the productivity of the two vendor's resources. Increased vendor productivity will also have a positive impact on increasing corporate and regional productivity through the achievement of KPIs.

5   CONCLUSIONS AND RECOMMENDATIONS

The dominant processes in the IP Radio assurance business process are based on the results of quantitative data processing namely toubleshooting, spare part requests, travel time directly to the site, location permission, technical closed, and analysis. The assurance business process that can meet the MTTR Radio IP targets according to the results of the simulation scenario is the eliminate & simplify simulation scenario.

This research can be continued by implementing the simulation scenario that gives positive results on time, quality, and cost in the IP Radio assurance business process. Then do a comparative analysis of the results of the implementation with the simulation results in terms of time, quality, cost and effectiveness of technician resources.

REFERENCES

Abdellatif, M. 2017. Overcoming business process reengineering obstacles using ontology-based knowledge map methodology. *Future Computing and Informatics Journal* 3(1): 7–28. DOI: 10.1016/j.fcij.2017.10.006

Adriansah, D. 2010. *Management Business Process* (Part - 4). Retrieved from: http://doniadriansah.blog spot.com/2010/02/management-business-process-part-4_09.html

Doomun, R. 2002. Business process modelling, simulation and reengineering: call centres. *Business Process Management Journal* 14(6): 838–848. DOI: 10.1108/14637150810916017

Greasley, A. 2003. Using business-process simulation within a business-process reengineering approach. *Business Process Management Journal* 9(4): 408–420. DOI: 10.1108/14637150310484481.

Heizer, J. 2017. *Operations Management Sustainability and Supply Chain Management.* USA: Pearson Education, Inc.

Hammer, C.M.J. 1993. *Reengineering the Corporation: A Manifesto for Business Revolution.* New York, USA: Harper Business Essential.

Kelton, D. 2015. *Simulation with Arena.* Sixth Edition. New York, USA: McGraw-Hill Education. Retrieved from: https://www.academia.edu/35774349/Simulation_with_Arena_6e

Rinaldi, M. 2015. Improving the efficiency of public administrations through business process reengineering and simulation. *Business Process Management Journal* 21(2): 419–462. DOI: 10.1108/BPMJ-06-2014-0054.

Sugiyono. 2018. *Metode Penelitian Kombinasi (Mixed Methods).* Bandung, Indonesia: Alfabeta, CV.

Watts, A.D. 2002. *Business Process Reengineering Fundamentals (with Strategic Planning).* Denver, USA: Mountain Home Training & Consulting, Inc.

Managing Learning Organization in Industry 4.0 – Rachmawati & Hendayani (eds)
© 2020 Taylor & Francis Group, London, ISBN 978-0-367-81920-0

Measurement tool for analyzing the adoption of online tax services in Jayapura, Indonesia

Indrawati, B.P. Tuwankotta & Syarifuddin
Telkom University, Bandung, Indonesia

ABSTRACT: The aim of this study was to provide a measurement tool for analyzing the adoption of online tax services in Jayapura, Indonesia, by exploring and testing the variables that may influence the intention and use behavior of users, especially IHRA employees, in using the online tax service through applying the Unified Theory of Acceptance and Use of Technology 2 (UTAUT2) model, which has been modified. The method used to explore and test the tool were: (1) operationalization of the variable, (2) content validity, (3) face validity, (4) readability, and (5) a pilot test. The data used in this study were collected from 123 respondents. The result showed that the tool for analyzing adoption of online tax services in Jayapura consists of forty-two indicators, of which nine variables are valid and reliable and can be used for further study. The next study that should be done is analyzing the adoption of online tax services in Jayapura by using this founded tool.

1 INTRODUCTION

Since 2011, PT Finnet Indonesia and the government of Jayapura, Indonesia, have collaborated in implementing an online tax service (an electronic payment platform service) in order to collect tax payments from the Indonesian Hotel and Restaurant Association (IHRA) in Jayapura. The implementation of this online tax service is part of the government's effort to make the payment of IHRA taxes transparent. The adoption of the IHRA online tax service is very important because it can increase local tax revenues, which are very important for raising regional income. Understanding the factors that can influence the adoption of the IHRA online tax service is very important. It becomes an input in formulating and developing service features, business processes, corporate strategies, and promotional strategies for PT. Finnet Indonesia as the provider of the IHRA online tax service.

The aim of this study was to provide a measurement tool for analyzing factors behind the adoption of the online tax services in Jayapura by exploring and testing the variables and indicators that can be used to measure the model of this study.

2 LITERATURE REVIEW

Based on previously published literature, this study found that the Unified Theory of Acceptance and Use of Technology 2 (UTAUT2) is the most fit model to employ as the basis for this study. The model is suitable for use in this study since the UTAUT2 model is the latest acceptance theory developed from UTAUT, a synthesis and summary of the previous eight existing theories of technology acceptance (Venkatesh, Morris, Davis, & Davis, 2003). In addition, the UTAUT2 model can explain the acceptance of technology whose context is consumer use (Venkatesh, Thong, & Xu, 2012). In order to fit with the object of this study, the researchers made modifications to the UTAUT2 model by eliminating the Price Value variable and adding a Trust variable to the independent variable where previously the independent variables were Performance Expectancy, Effort Expectancy, Social Influence, Facilitating Conditions, Hedonic Motivation, and

Habit. This research eliminated the independent variable of Price Value from the original UTAUT2 model since in using the IHRA application, the users do not pay any cost, hence Price Value was considered inapplicable. The addition of Trust is based on the consideration that Trust is a factor that has influenced the adoption of mobile banking (Alalwan, Dwivedi, & Rana, 2017), e-learning systems in Qatar and the United States (Masri & Tarhini, 2017), and e-commerce by micro-, small-, and medium-sized enterprises in Indonesia. McKnight, Choudhury, and Kacmar (2002) state that "Trust is orthopedic because it helps consumers overtake perceptions of uncertainty and risk". Trust can be obtained if a person has confidence in the reliability and integrity of a service and feel secure in using it.

The original UTAUT2 model has three moderator factors – namely Age, Gender, and Experience. This research only applied two moderating factors, Age and Gender. The reason for not using Experience as a moderating factor was that such a study must carry out periodic data retrieval methods that the authors did not do in this study.

Figure 1 shows the conceptual modification model of this study.

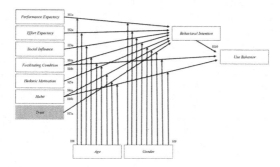

Figure 1.   Modified UTAUT2 model.

3   METHODOLOGY

In order to have a valid and reliable tool with which to measure the research model as presented in Figure 1, this study followed five steps – namely operationalize the variable, content validity, face validity, readability, and convergent validity. First was the operationalize variable, a process of breaking the variables contained in the problem of research into their smallest parts so that it can be known by classification, making it easier to get the data needed for the assessment of research problems (Indrawati, 2015).

Second, fulfilling the content validity test, according to Sekaran and Bougie (2016: 158), "content validity ensures that the measure includes an adequate and representative set of items that tap the concept". Therefore, the authors explored the questionnaire items from previous studies, such as accredited national and international journals. The modified and adjusted variables were also decided through gathering preliminary data. The questionnaire items used in this research have been adapted and modified from previous studies by Venkatesh and colleagues (2003), Venkatesh and colleagues (2012), Indrawati (2015), and Masri and Tarhini (2017).

Third was getting help and suggestions from experts in order to improve the questionnaire items. Fourth was the readability test, which was done to make sure that the items were clear and did not confuse the respondents.

The last was the convergent validity; the criteria used to measure the convergent validity was the Corrected Item – Total Correlation (CITC) score. The CITC was chosen since it meets the convergent validity criteria, which is when the items in the same variable have a high correlation (Indrawati, 2015: 149). Moreover, as cited in Indrawati (2015: 149), Friedenberg and Guilford (1956) suggest that "the minimum correlation coefficient is 0.3 hence it is valid". The sample examined in this convergent validity test comprised 123 users of the IHRA online tax service in Jayapura who have been using the service for at least three months.

4 RESULTS AND DISCUSSION

Based on the first and second processes, which were explained in part 3 of this paper, the result of items on the questionnaire are shown in Table 1.

Table 1. Variables and indicators.

Variables	Indicators	Code
Performance Expectancy (ξ1)	The IHRA online tax service is very helpful for doing IHRA tax payments in my company.	PE1
	Using the IHRA online tax service makes it easier for me to record IHRA tax transactions.	PE2
	Using the IHRA online tax service increases my productivity.	PE3
Effort Expectancy (ξ2)	I can easily learn how to use the IHRA online tax service.	EE1
	The IHRA online tax service is easy to use.	EE2
	It is easy for me to become skilled in using the IHRA online tax service.	EE3
Social Influence (ξ3)	People who are influential to me suggested that I use the IHRA online tax service.	SI1
	People who are close to me think that I ought to use the IHRA online tax service.	SI2
	People around me support me in using the IHRA online tax service.	SI3
Facilitating Conditions (ξ4)	I have an adequate resource to operate the IHRA online tax service.	FC1
	I have an adequate gadget to operate the IHRA online tax service.	FC2
	The customer service center can help me when I have trouble using the IHRA online tax service.	FC3
Hedonic Motivation (ξ5)	I feel comfortable when I use the IHRA online tax service rather than manually inputted tax services.	HM1
	I feel happy when I'm using the transaction monitoring feature on the IHRA online tax service.	HM2
	I feel happy when I'm using the transaction reporting feature on the IHRA online tax service.	HM3
Habit (ξ6)	Using the IHRA online tax service has become a habit for me.	H1
	Using the IHRA online tax service has become a need for me.	H2
	Using the IHRA online tax service has become usual for me.	H3
Habit (ξ6)	It has become an addiction to use the IHRA online tax service for me.	H4
Trust (ξ7)	I believe that the IHRA online tax service is trustworthy for IHRA tax transactions.	T1
	I can rely on the IHRA online tax service for the IHRA online tax.	T2
	I don't have a doubt about the IHRA online tax service compared to other applications.	T3
	The IHRA online tax service has never shown any indication of any fraudulent activity.	T4
	I believe in the security of my company's data that I gave to the IHRA online tax service.	T5
Continuance Intention (η1)	I will always use the IHRA online tax service for monitoring the IHRA tax.	CI1
	I plan to always use the IHRA online tax service in the future.	CI2
	I predict that I will always use the IHRA online tax service for further transactions.	CI3
	I prefer to use the IHRA online tax service rather than other similar applications.	CI4

(Continued)

Table 1. *(Continued)*

Variables	Indicators	Code
Use	I use the IHRA online tax service every time I'm on duty.	UB1
Behavior (η2)	I use the IHRA online tax service every time I do transaction monitoring.	UB2
	I use the IHRA online tax service every time I do transactional reporting.	UB3
	Every time I do tax monitoring, I use the IHRA online tax service.	UB4
	I use the IHRA online tax service every time I do transactional monitoring.	UB5

Table 2. Validity and reliability test results.

Item Code	CITC	Cronbach's Alpha	Item Code	CITC	Cronbach's Alpha
PE1	0.739	0.816	H1	0.891	0.931
PE2	0.534		H2	0.866	
PE3	0.738		H4	0.842	
PE4	0.785		H5	0.753	
PE5	0.320				
EE1	0.547	0.722	T1	0.892	0.923
EE2	0.368		T2	0.680	
EE3	0.536		T3	0.801	
EE4	0.626		T4	0.972	
EE5	0.353		T5	0.746	
SI1	0.816	0.885	BI2	0.609	0.846
SI2	0.695		BI3	0.805	
SI3	0.636		BI4	0.857	
SI4	0.714		BI5	0.505	
SI5	0.781				
FC1	0.616	0,679	UB1	0.868	0.932
FC3	0.463		UB2	0.859	
FC4	0.416		UB3	0.886	
FC5	0.366		UB4	0.746	
HM1	0.427	0,808	UB5	0.755	
HM2	0.492				
HM3	0.554				
HM4	0.727				
HM5	0.822				
H1	0.891	0.931			

*Use IBM SPS Statistic Version for Windows 64-bit

The authors checked the construct validity by using IBM SPSS Statistic 25 as a tool and employed the calculated CITC as a standard for measuring the items. The reliability was checked by the Cronbach's alpha value. Based on the calculation technique the Cronbach's alpha value can be declared reliable if it has a value > 0.70. The study results are presented in Table 2.

Table 2 shows that the validity and reliability test results indicate that forty-two items and nine variables of this measurement model fulfill the requirements of validity and reliability. The H3 (CICT = 0.151), FC (CICT = 0.031), and BI (CITC = 0.131) items were deleted from the questionnaire since the CITC less than the required value.

5 CONCLUSIONS AND RECOMMENDATIONS

This study was conducted with 123 respondents who had used the IHRA online tax service for at least three months in Jayapura, Indonesia. The result shows that the tool for analyzing the adoption of online tax services in Jayapura consists of forty-two indicators, of which nine variables are valid and reliable and can be used for further study. The next study should analyze the adoption of online tax services in Jayapura by using this founded tool.

ACKNOWLEDGMENT

The authors express their gratitude to the Ministry of Research, Technology and Higher Education of Indonesia for financial support.

REFERENCES

Alalwan, A. A., Dwivedi, Y. K., & Rana, N. P. 2017. Factors influencing adoption of mobile banking by Jordanian bank customers: Extending UTAUT2 with trust. *International Journal of Information Management* 37(3): 99–110.

Indrawati. 2015. *Metode Penelitian Manajemen dan Bisnis Konvergensi Teknologi Komunikasi dan Informasi*. Bandung: PT Refika Aditama.

Indrawati and Ariwiati. 2015. Factors affecting e-commerce adoption by micro, small and medium-sized enterprises in Indonesia. Proceedings of the International Conference on E-Commerce and Digital Marketing. July 21–23, 2015, Las Palmas de Gran Canaria, Spain.

Masri, E., & Tarhini, A. 2017. Factors affecting the adoption of e-learning systems in Qatar and USA: Extending the Unified Theory of Acceptance and Use of Technology 2 (UTAUT2). 1–21.

McKnight, D. H., Choudhury, V., & Kacmar, C. 2002. Developing and validating trust measures for e-commerce: An integrative typology. *Information Systems Research* 13(3): 334–359.

Sekaran, U., & Bougie, R. 2016. *Research methods for business: A skill building approach* (7th edn.). London: Wiley.

Venkatesh, V., Morris, M. G., Davis, G. B., & Davis, F. D. 2003. User acceptance of information technology: Toward a unified view. MIS Quarterly, 27(3), 425–478.

Venkatesh, V., Thong, J. Y. L., & Xu, X. 2012. Consumer acceptance and use of 132 information technology: Extending the Unified Theory of Acceptance and Use of Technology. *MIS Quarterly* 36(1), 157–178.

Laboratory assistant assignment problem using Python programming

R. Aurachman
School of Industrial and System Engineering, Telkom University, Bandung, Indonesia

ABSTRACT: An assignment problem is a tool that helps in the decision-making process of resource allocation. Every resource allocated to each job has a certain cost or payoff. Assignment problems help to optimize payoff or minimize costs. This paper presents an example of applying assignment problems using Python programming to solve the assignment of a business simulation laboratory assistant to the class. This paper presents programming examples of assignment problems in Python that will be useful for another implementation.

1 INTRODUCTION

Human resources are different from other resources. Human resources have preferences and desires. This is different from other resources, which are lifeless objects. Humans can be assigned to certain jobs with different levels of performance. The problem of task selection for individuals is a common one in organizational and human resources management.

Mathematical models can support human resources assignment. One such mathematical model is the liner optimization model. The assignment problem optimizes the allocation of resources to a job or a task so that a minimum cost or maximum payoff is obtained. Previous research on the assignment problem has examined multicore assignment (Sudhakar, Adhikari, & Ramesh, 2016), ergonomics (Gebennini, Zeppetella, & Grassi, 2018), energy sharing (Fu, Moran, Guo, Wong, & Zukerman, 2016), product-to-site (Hillebrand, 2019), manufacturing resources (Na, Woo, & Lee, 2016), using a meta-heuristic algorithm (Peters et al., 2019), the traffic problem (Patriksson, 2015), and many more issues. Other research has used a philosophy of assignment rather than algorithms assigning members of the workforce to industry 4.0 jobs (Aurachman, Model Matematika Dampak Industri 4.0 terhadap Ketenagakerjaan Menggunakan Pendekatan Sistem, 2019) (Aurachman, Perancangan Influence Diagram Perhitungan Dampak Dari Revolusi Industri 4.0 Terhadap Pengangguran Kerja, 2018), assigning a server to work at a highway gate (Aurachman & Ridwan, Perancangan Model Optimasi Alokasi Jumlah Server untuk Meminimalkan Total Antrean pada Sistem Antrean Dua Arah pada Gerbang Tol, 2016), assigning resources to a vehicle route (Desiana, Ridwan, & Aurachman, 2016), or assigning a distribution plan (Muttaqin, Martini, & Aurachman, 2017).

This research tried to solve the assignment problem based on the preference of human resources about a job. The case concerned the allocation of laboratory assistant resources in conducting practicum sessions. Practicum assistants were allocated to several class choices. Each assistant had a class preference. The designed mathematical model aimed to optimize the total payoff of these preferences.

2 METHODOLOGY

The research method was carried out through several stages. The first step was to record the preferences of each laboratory assistant in each class. Each assistant was asked to provide an assessment using the Likert scale from numbers 1 to 5 where 1 meant the assistant did not want to assist the class and 5 meant the assistant really wanted to assist the class. Respondents could make a judgment by considering student attitudes and conduciveness in each class.

The next step was mathematical modeling the problem. The mathematical model was generated based on generic form of assignment problem so as to configure the constraint, indexing, and number of variables.

The final step was programming the code in order to optimize the mathematical model. Programming steps were taken using the Python language. Some packages were used to make the coding process simpler. The code was run and then followed by an interpretation of the solution in order to show the optimum solution.

3 RESULT AND DISCUSSION

The programming codes are explained in what follows:

```python
import pulp
model = pulp.LpProblem("Assignment Problem", pulp.LpMaximize)
assistant = [0,1,2,3,4,5,6,7,8,9,10,11,12]
classes = [0,1,2,3,4,5,6,7,8,9,10,11,12]
assign = pulp.LpVariable.dicts("assistant for class,"
                ((i, j) for i in assistant for j in classes),
                lowBound=0,
                cat='Biner')
c = ([2,4,4,4,4,4,4,4,3,4,4,4,4],
    [2,4,4,4,4,4,4,4,3,4,4,3,4],
    [2,3,3,3,3,3,3,3,3,3,3,3,3],
    [2,4,4,4,4,5,4,5,4,4,4,4,4],
    [2,4,4,4,4,4,4,4,4,4,4,4,4],
    [3,4,4,4,4,4,4,4,4,4,4,4,4],
    [2,4,4,4,5,5,5,5,4,4,4,4,4],
    [2,3,3,3,3,3,3,3,3,3,3,3,3],
    [3,4,4,4,4,4,4,4,4,4,4,4,4],
    [3,4,3,4,4,4,4,5,4,4,3,3,3],
    [3,4,3,4,4,4,4,5,4,4,3,3,3],
    [3,4,4,4,4,4,4,4,4,4,4,4,4],
    [2,3,3,3,3,3,3,3,3,3,3,3,3])
model += (
    pulp.lpSum([
        c[i][j]*assign[(i,j)]
        for i in assistant for j in classes])
)
for i in assistant:
    model += pulp.lpSum([assign[i,j] for j in classes]) == 1

for j in classes:
    model += pulp.lpSum([assign[i,j] for i in assistant]) == 1
model.solve()
pulp.LpStatus[model.status]
for var in assign:
    var_value = assign[var].varValue
    if var_value==1:
        print("assistant," var[0], "for class," var[1], "are," var_value)
print("payoff optimal are,"pulp.value(model.objective))
```

The results of the optimum solution of that optimization code follow:
assistant 0 for class 10 are 1.0
assistant 1 for class 12 are 1.0
assistant 2 for class 0 are 1.0
assistant 3 for class 5 are 1.0
assistant 4 for class 2 are 1.0
assistant 5 for class 3 are 1.0
assistant 6 for class 4 are 1.0
assistant 7 for class 6 are 1.0
assistant 8 for class 1 are 1.0
assistant 9 for class 7 are 1.0
assistant 10 for class 8 are 1.0
assistant 11 for class 11 are 1.0
assistant 12 for class 9 are 1.0

Solutions resulting from Python gave a label of 1 for each assistant assigned to the class. For example, assistant 0 for class 10 is 1.0 meant that assistant 0 is assigned to assist practicum in class 10. Or we can draw the assignment table in Figure 1.

With that solution, the optimum payoff was 51. Not every assistant got the optimum payoff preferences, like assistant G, who was assigned to his most favorable class, class C-05. Some assistants got the second-best preferences, like assistant K, who was assigned to class C-08, even though his most favorable class was C-07. Even assistant C got his worst choice, which was C-00. But overall the whole system got the optimum accumulated payoff from all the assistants.

4 CONCLUSIONS AND RECOMMENDATIONS

Using the assignment problem method, an optimal solution was obtained. In this research, the laboratory assistants' assignment to the practice class was illustrated using the assignment problem. Payoff was based on an assessment of the subjectivity of each assistant to the class.

The authors have several suggestions for continuation of this research. We can create a model that includes the preferences of each student in a class. The model can be improved in order not only to optimize the payoff but also to maximize the minimum payoff of all assistants. We can use a ranking scale rather than the Likert scale for payoff, or maybe we can consider using the ratio scale supported by the analytical hierarchy process.

	C-00	C-01	C-02	C-03	C-04	C-05	C-06	C-07	C-08	C-09	C-10	C-11	C-12
A	2	4	4	4	4	4	4	3	4	4	4	4	4
B	2	4	4	4	4	4	4	4	3	4	4	3	4
C	2	3	3	3	3	3	3	3	3	3	3	3	3
D	2	4	4	4	4	5	4	5	4	4	4	4	4
E	2	4	4	4	4	4	4	4	4	4	4	4	4
F	3	4	4	4	4	4	4	4	4	4	4	4	4
G	2	4	4	4	5	5	5	5	4	4	4	4	4
H	2	3	3	3	3	3	3	3	3	3	3	3	3
I	3	4	4	4	4	4	4	4	4	4	4	4	4
J	3	4	3	4	4	4	4	5	4	4	3	3	3
K	3	4	3	4	4	4	4	5	4	4	3	3	3
L	3	4	4	4	4	4	4	4	4	4	4	4	4
M	2	3	3	3	3	3	3	3	3	3	3	3	3

Figure 1. Assignment table.

REFERENCES

Aurachman, R. 2018. Perancangan Influence Diagram Perhitungan Dampak Dari Revolusi Industri 4.0 Terhadap Pengangguran Kerja. *Jurnal Teknologi dan Manajemen Industri 4*(2): 7–12.

Aurachman, R. 2019. Model Matematika Dampak Industri 4.0 terhadap Ketenagakerjaan Menggunakan Pendekatan Sistem. *Jurnal Optimasi Sistem Industri*: 14–24.

Aurachman, R., & Ridwan, A. Y. 2016. Perancangan Model Optimasi Alokasi Jumlah Server untuk Meminimalkan Total Antrean pada Sistem Antrean Dua Arah pada Gerbang Tol. *JRSI (Jurnal Rekayasa Sistem dan Industri) 3*(2): 25–30.

Desiana, A., Ridwan, A. Y., & Aurachman, R. 2016. *Penyelesaian Vehicle Routing Problem (vrp) Untuk Minimasi Total Biaya Transportasi Pada Pt Xyz Dengan Metode Algoritma Genetika. eProceedings of Engineering.* Bandung: Telkom University.

Fu, J., Moran, B., Guo, J., Wong, E. W., & Zukerman, M. 2016. Asymptotically optimal job assignment for energy-efficient processor-sharing server farms. *EEE Journal on Selected Areas in Communications, 34*(12): 4008–4023.

Gebennini, E., Zeppetella, L., & Grassi, A. 2018. Optimal job assignment considering operators' walking costs and ergonomic aspects. *International Journal of Production Research 56*(3): 1249–1268.

Hillebrand, B. 2019. A Multi-site Facility Layout and Product-to-Site Assignment Problem: 427–433. Cham: Springer.

Muttaqin, B. M., Martini, S., & Aurachman, R. 2017. Perancangan Dan Penjadwalan Aktivitas Distribusi Household Product Menggunakan Metode Distribusi Requirement Planning (DRP) Di PT. XYZ Untuk Menyelaraskan Pengiriman Produk Ke Ritel. *JRSI (Jurnal Rekayasa Sistem dan Industri)*: 56–61.

Na, B., Woo, J.-E., & Lee, J. 2016. Lifter assignment problem for inter-line transfers in semiconductor manufacturing facilities. *International Journal of Advanced Manufacturing Technology*: 1615–1626.

Patriksson, M. 2015. *The traffic assignment problem: Models and methods.* Mineola, NY: Courier Dover Publications.

Peters, J., Stephan, D., Amon, I., Gawendowicz, H., Lischeid, J., & Salabar, L. 2019. Mixed integer programming versus evolutionary computation for optimizing a hard real-world staff assignment problem. *Proceedings of the International Conference on Automated Planning and Scheduling 29*(1): 541–554.

Sudhakar, C., Adhikari, P., & Ramesh, T. 2016. Process assignment in multi-core clusters using job assignment algorithm. Second International Conference on Computational Intelligence & Communication Technology (CICT). IEEE.

Analysis of bank stability, competition, family ownership, and multiple large shareholders in Indonesia

R.K. Koeswiyono & C.A. Utama
University of Indonesia, Jakarta, Indonesia

ABSTRACT: This thesis aims to analyze the relationship of organizational structure and competition to banking stability in Indonesia in the period 2007 to 2017. The bank samples are obtained from banks listed on the stock exchange. The findings support the competition–fragility theory and reveal that family ownership has a positive relation with bank stability. As for multiple large shareholders (MLSs), the results show that MLSs strengthen the relation between family ownership and banking stability.

1 INTRODUCTION

Many researchers in the world have already studied the relation between competition and bank stability. The results of this research vary. Some of these results state that with higher competition, bank stability will also increase (Beck et al., 2006). On the other hand, some researchers have also found that with higher competition levels, bank stability will decrease, or become more fragile (Jiménez et al., 2013). This contradiction has not become a topic to research until now.

Besides competition, other variables such as ownership influence bank stability (Lee and Hsieh, 2014). The structure of family ownership is also characterized by the presence of family members in the ranks of top management, so there is no conflict between company owners and management (Demsetz and Villalonga, 2001) and agency problems are reduced. Management by family members provides a stronger motivation regarding the sustainability of the company than if it is managed by professionals (Gomej-Mejia et al., 2007). Other studies suggest that family businesses achieve higher profits when the founders are actively involved in management (Miller and Le Breton-Miller, 2007).

Another characteristic of the family company is the shareholders, who are the long-term investors, because most families who own a company consider the company an asset that must be passed on to the next generation, so that investment is oriented to the long term (Casson, 1999). Other researchers such as Claessens and colleagues (2002) have found that the performance of family companies tends to be poor. This is caused by less professional management by family members (Bloom and Van Reenen, 2006).

Not all banks in Indonesia have a single shareholder but instead have several large shareholders that can be referred to as multiple large shareholders (MLSs). The MLS mentioned by the authors is the second largest shareholder. Related to agency problems, Attig, Guedhami, and Mishra (2008) examine where the existence of MLSs can influence the decisions of the first largest shareholders. Assuming majority family ownership can improve banking stability in Indonesia, then MLSs can weaken that relationship. This will happen if the second largest shareholder has a different opinion from the first largest shareholder, so a conflict of interest occurs. These concerns are the reasons the writers have conducted this research.

2 LITERATURE

The relationship of competition to banking stability has been a complex and broad debate for both academics and policy makers in the past two decades, especially since the global financial crisis in 2007–2008 (Beck, 2008; Beck et al., 2010; Carletti, 2008). Yeyati and Micco (2007) used a sample of commercial banks from eight countries in Latin America during the period 1993–2002. The study found a positive relationship between bank risk and competition, or, in other words, that competition had a positive impact on banking stability.

A study conducted by Fu, Lin, and Molyneux (2014) collected data from fourteen Asia Pacific countries during the period 2003–2010 in order to investigate the effect of competition on bank fragility as measured by the default probability and Z-score of banks. The results show that competition encourages financial fragility.

Lee and Hsieh (2014) conducted a study to find out the relationship of ownership and banking stability by exploring data collected on Asian banks. The results included the discovery of an inverse U-shaped relationship between foreign ownership and stability.

Family-owned companies are known to have inefficient performance. The founders did not diversify their portfolios, and they tended to avoid risks so that the profits derived were less than optimal. Family firms tend to invest in long-term projects with the aim of increasing efficiency in their investment policies (Stein, 1989).

In addition, the management of a family company is carried out by family members or parties who have connections with the owner so that the family company lacks expertise and is less competitive than nonfamily companies (Morck et al., 2000). With these weaknesses, family companies tend to have higher risk. Laeven and Ross (2009) found that family banks tend to have poor governance and high risk.

As mentioned, MLSs can influence the decisions of the first shareholder (Attig et al., 2008). According to Afriani, Utama, and Amarullah (2016), the second largest shareholder has the highest potential to influence the first largest shareholder.

Referring to the aforementioned research, we knew that a decision maker is related to bank performance. When examining ownership, we tried to find the relation of multiple decision makers to bank stability in Indonesia.

3 DATA AND VARIABLES

The data used in this study were banking data taken from the official websites of each bank listed on the Indonesian stock exchange. The data period used was from 2007 to 2017. It is expected that the eleven-year data range can adequately represent and reflect the relationship between financial stability and banking competition.

This study used a fixed effect model. For the first step, we estimated using Competition and Family as independent variables, Bank Stability as a dependent variable, and Size and Noninterest Income as controls.

In general, the first empirical model in this study was as follows:

$$Bank\ Stability_{i,t} = \alpha + \beta_1 Lerner\ Index_{i,t} + \beta_2 OS_{i,t} + \beta_3 Size_{i,t} + \beta_4 NIE_{i,t}$$

The first model aimed to estimate the relation of competition (Lerner Index) and family ownership (OS) to bank stability.

For the second step, we estimated using Competition and OS*MLS as independent variables, Bank Stability as a dependent variable, and Size and Noninterest Income as controls.

The second empirical model in this study was as follows:

$$Bank\ Stability_{i,t} = \alpha + \beta_1 Lerner\ Index_{i,t} + \beta_2 OS_{i,t} + \beta_3 OS_{i,t} * MLS_{i,t} + \beta_4 Size_{i,t} + \beta_5 NIE_{i,t}$$

The second model was intended to estimate the relation of competition (Lerner Index) and family (OS) * (MLS) to bank stability. This model's purpose was also to estimate the impact of MLSs on the first largest shareholder, which is in this research was the family.

Table 1. Dependent variable, independent variables, and controls.

Variable	Definition	Previous research
Bank Stability (Z-score)	A larger value indicates the risk of a small bank as a whole and has a higher bank stability.	Bharath and Shumway (2008), Soedar-mono (2013)
Lerner Index (Lerner)	The bank's nonstructural indicators of bank competition, measured by the Lerner index, use the fixed effect method, with higher values indicating less competition in the banking sector.	Claessens and Laeven (2004), Maudos and Fernández de Guevara (2004), Fernández de Guevara et al. (2005), Berger et al. (2009) and Maudos and Solís (2009)
Family Ownership (OS)	Family ownership is the largest shareholder (percentage).	Wang (2006), Maury and Pajuste (2005), Laeven and Levine (2008); Attig et al. (2008)
Multiple Large Shareholder (MLS)	Second largest shareholder (percentage)	Maury and Pajuste (2005); Laeven and Levine (2008); Attig et al. (2008); Afriani et al. (2016)
SIZE	Size of bank Normal log of total assets in millions of IDR	Flannery (2008); Wibowo (2016)
NIE	Noninterest Income	Wibowo (2016)

4 RESULT AND ANALYSIS

Table 2. Statistic descriptive.

Variable	Observation	Minimum	Maximum	Mean	Std. Deviation
Z-score	360	0.04	1.22	0.66	0.22
Lerner Index	360	0.00	0.66	0.21	0.14
OS	360	0.00	0.89	0.26	0.28
MLS	360	0.00	0.49	0.04	0.10
Size	360	13.56	20.84	17.10	1.76
NIE	360	0.00	0.41	0.09	0.06

The minimum value of the Z-score of 0.04 from Table 2 indicates the value of issuers that suffered losses due to allowance for possible losses of earning assets. Banks with this value are the Pearl Bank and the Banten Regional Development Bank, which have performed poorly for a few years.

The high Z-score of banking issuers shown in Table 2 are owned by major banks such as BCA, BRI, BNI, and Mandiri in the range 0.8–1.22. These banks tend to have a stable Z-score that reflects the level of profits with low volatility, good leverage, and high profitability.

Table 2 shows that an average of 26% of banking issuers are owned by family shareholders. The highest percentage of family ownership reaches 89%, and the lowest percentage is 0%, which means there is no family ownership in the share structure. The highest family ownership is found at Maspion Bank, which is controlled by the Salim Group. Some banks that are not owned by family companies are owned by the central and regional governments such as BNI, BRI, Bank Mandiri, BPD Jabar Banten, BTN, BRI Agro, Bank Danamon, and Bumiputera Bank.

Table 3. First model regression result.		Table 4. Second model regression result.	
Variable	**Coefficient**	**Variable**	**Coefficient**
LERNER	0.587709***	LERNER	0.599825***
OS	0.114972***	OS	0.252226*
SIZE	0.051294***	OS*MLS	0.218342*
NIE	−0.470466***	SIZE	0.043062***
		NIE	−0.457388***

Results from FEM panel data estimations from Table 3 to explain the impacts of family ownership and competition on financial stability. Table 4 explain the additional impact come from MLS to family ownership. The dependent variable is Z-score, which is an accounting-based bank-level indicator of financial soundness, calculated using the same method by Bharath and Shumway (2008). LERNER is a bank-level indicator of bank competition calculated as the difference between price and marginal cost as a percentage of price using the stochastic frontier analysis approach. OS is family ownership of the bank calculated from share percentage out of total share. MLS is second largest shareholder calculated from share percentage out of total share. SIZE is the natural logarithm of total assets in millions of IDR. NIE is non-interest income gathered from annual, service and check fees.

*** Indicate significance at the 1% levels, respectively. Robust standard errors are in parentheses.

** Indicate significance at the 5% levels, respectively. Robust standard errors are in parentheses.

* Indicate significance at the 10% levels, respectively. Robust standard errors are in parentheses.

Seen in the results of Table 3, the coefficients for the competition (LERNER), there is a positive value of 0.58 basis points. For the Lerner Index, the higher value means less competition. This means that if banks' competition is lower, then the more stable they are in Indonesia. These results are in line with the competition–fragility theory and with the results of research from Beck and colleagues (2013). Not only that, we also found that having the family as the highest shareholder has a positive relation with banks' stability in Indonesia. We can see that the family coefficient is 0.11, meaning that family ownership can reduce potential risks that will harm bank stability, such as agency problems.

Regarding the existence of the second largest shareholder (MLS), we found that it will strengthen the impact of family ownership on bank stability. Table 4 shows that MLSs' existence increases the value of family ownership. The MLSs will further reduce the potential risk of agency problems proven by a positive result in Table 4. This result is different from the first assumption, which is that MLSs will cause an agency problem.

This result also indicates that MLSs will use their power to act as a "watch dog" for the first largest shareholder; MLSs will monitor and correct the decision of the family as the largest shareholder. Besides that, MLSs can also help the family to make better decisions, which results in increasing bank stability.

5 CONCLUSION

Based on the results of the research and analysis, the conclusions that can be taken are as follows:

1. The relationship between stability and competition in banks in Indonesia from 2007 to 2017 has a positive relation to bank stability. The results support the competition–fragility theory.
2. The relationship between stability and family ownership of banks in Indonesia from 2007 to 2017 indicates that family ownership has a positive relation with bank stability.

3. The result also indicates that the existence of MLSs will further enhance the influence of the family on the family ownership–bank stability relation.

REFERENCES

Afriani, C., Utama, S., and Amarullah, F. 2016. *Corporate Governance and Ownership Structure: Indonesia Evidence. Corporate Governance (Bingley)*, 17: 165–191.

Attig, N., Guedhami, O., and Mishra, D. 2008. Multiple large shareholders, control contests, and implied cost of equity. *Journal of Corporate Finance.* 14: 721–737.

Beck, T., Demirgüç-Kunt, A., and Levine, R. 2006. *Bank concentration, competition, and crises: First results. Journal of Banking and Finance.* 30: 1581–1603.

Beck, T., Demirgüç-Kunt, A., and Levine, R. 2008. *Finance, Firm Size, and Growth. Journal of Money, Credit and Banking.* 40: 1379–1405.

Beck, T., Demirgüç-Kunt, A., and Maksimovic, V. 2010. *Bank Competition and Access to Finance: International Evidence. Journal of Money, Credit and Banking.* 36: 627–648.

Beck, T., De Jonghe, O., and Schepens, G. 2013. *Bank Competition and Stability: Cross-Country Heterogeneity. Journal of Financial Intermediation.* 22: 218–244.

Beck, T., Demirgüç-Kunt, A., and Levine, R. 2013. Bank concentration, competition and crises: 1581–1603. *Journal of Banking and Finance.* 22: 218–244.

Berger, A., Klapper, L., and Turk-Ariss, R. 2009. *Bank Competition and Financial Stability. Journal of Financial Services Research.* 22: 218–244.

Bharath, S. T. and Shumway, T. 2008. Forecasting default with the Merton Distance to Default Model. *Review of Finance.* 21: 1339–1369.

Bloom, N., and Van Reenen, J. 2006. Measuring and explaining management practices across firms and countries. *The Quarterly Journal of Economics.* 122: 1351–1408.

Carletti, E., and Vives, X. 2008. Regulation and competition policy in the banking sector. *Center for Financial Studies.*

Casson, M. 1999. The economics of the family firm. *Scandinavian Economic.* 47: 10–23.

Claessens, S., and Laeven, L. 2004. What drives bank competition? Some international evidence. *Journal of Money, Credit and Banking.* 36: 563–583.

Claessens S., and Joseph, F. 2002. *Corporate Governance in Asia: A Survey. International Review of Finance.* 3: 2.

Demsetz, H., and Villalonga, B. 2001. Ownership structure and corporate performance. *Journal of Corporate Finance.* 7: 209–233.

Fernández de Guevara, J., Maudos, J., and Perez, F. 2005. *Market Power in European Banking* Sectors. *Journal of Financial Services Research.* 27: 109–137.

Flannery, M. J. and Rangan, K. P. 2008. What caused the bank capital build-up of the 1990s? *Review of Finance.* 12: 391–429.

Fu, X., Lin, Y., and Molyneux, P. 2014. *Bank Competition and Financial Stability in Asia Pacific. Journal of Banking and Finance.* 38: 64–77.

Gomej Mejia, L., Takács Haynes, K., and Takács Haynes, M. 2007. Socioemotional Wealth and Business Risks in Family-Controlled Firms: Evidence from Spanish Olive Oil Mills. *Administrative Science Quarterly.* 52: 106–137.

Jiang, F., Kim, K. A., Nofsinger, J. R., and Zhu, B. 2017. A pecking order of shareholder structure. *Journal of Corporate Finance.* 44: 1–14.

Jiménez, G., Lopez, J. A., and Saurina, J. 2013. *How Does Competition Affect Bank Risk Taking? Journal of Financial Stability.* 9: 185–195.

Laeven, L. and Levine, R. 2008. Complex ownership structures and corporate valuations. *Review of Finance Studies.* 21: 579–604.

Laeven, L. and Levine, R. 2009. Bank governance, regulation and risk taking. *Journal of Financial Economics.* 93: 259–275.

Lee, C.-C. and Hsieh, M.-F. 2014. Bank reforms, foreign ownership, and financial stability. *Journal of International Money and Finance.* 40: 204–224.

Maudos, J., and Fernández de Guevara, J. 2004. Factors explaining the interest margin in the banking sectors of the European Union. *Journal of Banking & Finance.* 28: 2259–2281.

Maudos, J., and Solís, L. 2009. The determinants of net interest income in the Mexican banking system: An integrated model. *Journal of Banking & Finance.* 33: 1920–1931.

Maury, B. and Pajuste, A. 2005. Multiple large shareholders and firm value. *Journal of Banking and Finance*. 29: 1813–1834.

Miller, D., Le Breton-Miller, I., and H. Lester, R. 2007. Are family firms really superior performers? *Journal of Corporate Finance*. 13: 829–858.

Morck, R., and Yeung, B. 2000. Agency problems in large family groups. *Entrepreneurship: Theory and Practice*. 27: 367–382.

Soedarmono, M. T. 2013. Bank competition, crisis and risk taking: Evidence from emerging markets in Asia. *International Financial Markets, Institutions and Money*. 23: 196–221.

Stein, J. 1989. Efficient capital markets, inefficient firms: A model of myopic corporate behavior. *The Quarterly Journal of Economics*. 104: 655–669.

Wang, D. 2006. Founding family ownership and earnings quality. *Journal of Accounting Research*. 44: 619–656.

Wibowo, B. 2016. Stabilitas Bank, Tingkat Persaingan Antar Bank dan Diversifikasi Sumber Pendapatan: Analisis Per Kelompok Bank di Indonesia. *Jurnal Manajemen Teknologi*. 15: 172–195.

Yeyati, E., and Micco, A. 2007. Concentration and foreign penetration in Latin American banking sectors: Impact on competition and risk. *SSRN Electronic Journal*. 31: 1633–1647.

Analysis of the effect return on asset, return on equity, non performing loan and loan to depositratio on capital adequacy ratio in Indonesian banks listed on Indonesia stock exchange 2013-2017

I. Yunita & F. Hilmi
Faculty of Economics and Business, Telkom University, Bandung, Indonesia

ABSTRACT: The purpose of this study is to analyse the effect of Return on Asset, Return on Equity, Non-Performing loan, and Loan to Deposit Ratio on Capital Adequacy Ratio. Samples are determined based on the purposive sampling method, as many as five companies. Secondary data taken in the form of bank financial reports starting from 2013 to 2017. Data analysis techniques in this study using panel data regression. CAR as a dependent variable, ROA, ROE, NPL, and LDR as independent variables. Processing data using EViews. The results provide evidence that ROA, ROE, NPL and LDR have a simultaneous significant effect on CAR on banking companies registered in IDX for the period 2013-2017. NPL has a significant positive effect on CAR. ROA, ROE, and LDR have no significant negative effect on CAR.

1 INTRODUCTION

According to article 1 of Law No. 10 of 1998 concerning amendments to Law No.7 of 1992 concerning Banking, banks are defined as follows "Banks are business entities that collect funds from the public in the form of deposits and distribute them to the public in the form of loans or other forms in order to improve living standards many people."

Companies listed on the IDX are divided into several sectors. Currently there are 10 sectors listed on the Indonesia Stock Exchange (IDX), namely the Agriculture, Mining, Basic Industry, Miscellaneous Industries, Consumer Goods, Property, Infrastructure, Finance, and Trade (idx.co.id). In this study, the object of research to be examined is from the financial sector, namely the bank subsector. In the banking sub-sector there is a population of 46 Indonesian banking companies that are listed as public companies (issuers) on the Indonesia Stock Exchange (IDX).

The importance of the banking sector according to Joliana, (2013), is an important sector that supports many other industrial sectors. If the banking sectors is in a troubled situation, then the industrial sectors supported by banks will certainly be affected and the country's economy will automatically be disrupted

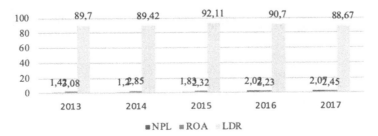

Figure 1.   NPL, LDR, ROA by Indonesian banking statistics (2013-2017).

2 THEORIES AND RESEARCH METHODOLOGY

2.1 Theories

2.1.1 Capital Adequacy Ration (CAR)

Determination of the Capital Adequacy Ratio (CAR), the Central Bank (Bank Indonesia) stipulates the obligation to provide minimum capital that must be owned by each commercial bank, which is stated with a capital adequacy ratio (CAR). In accordance with the standards set by the Bank for International Settlements (BIS), the minimum CAR of each bank is at least 8%. (Darmawi, 2011)

$$CAR = \frac{Capital}{Risk\ Weighted\ Asset} \times 100\% \tag{1}$$

2.1.2 Return On Asset (ROA)

Return On Assets (ROA) is a company's financial ratio related to aspects of earnings or profitability. ROA functions to measure the effectiveness of a company in generating profits by utilizing assets owned. The greater the ROA owned by a company, the more efficient use of assets will increase profits. Large profits will attract investors because the company has a high return. (Wardiah, 2013)

$$ROA = \frac{Net\ Income}{Average\ Total\ Assests} \times 100\% \tag{2}$$

2.1.3 Return On Equity (ROE)

Return On Equity (ROE) is a ratio that shows the ratio between profit (after tax) and capital (core capital) of the bank, this ratio shows the percentage level that can be generated. ROE is an indicator of the ability of banks to manage available capital to obtain net income. ROE can be obtained by calculating the ratio between after-tax earnings and total equity. (Pandia, 2012)

$$ROE = \frac{Net\ Income}{Average\ Total\ Equity} \times 100\% \tag{3}$$

2.1.4 Non-Performing Loan (NPL)

The ratio that can also be used as a factor that influences capital adequacy (CAR) is the asset quality ratio. One calculation of the asset quality ratio used according to SEBI/No.7/10/DPNP dated March 13, 2005, one of which is Non Performing Loans (NPL).

This ratio shows the quality of credit assets which if the collectability is not smooth, doubtful and stalled from the total credit, the bank faces problem loans. According to Bank Indonesia Regulation Number 6/10/PBI/2004 dated 12 April 2004 concerning the Soundness Rating System for Commercial Banks, "the higher the NPL value (above 5%), the bank is not healthy".

$$NPL = \frac{Non - Performing\ Loan}{Total\ Loans} \times 100\% \tag{4}$$

2.1.5 *Loan to Deposit Ratio (LDR)*

LDR is the financial ratio of banks related to liquidity aspects. The LDR shows time deposits, current accounts, savings, etc. that are used to meet customer loan requests. The LDR is also called the credit ratio to total third party funds used to measure third party funds channelled in the form of credit. Credit distribution is the bank's main activity. Therefore, the bank's main source of income comes from this activity. The greater the distribution of funds in the form of credit compared to deposits or deposits from the public in a bank, the greater the risk that must be borne by the bank concerned. A high LDR ratio indicates that a bank lends all its funds (loan-up) or is relatively illiquid (illiquid). Conversely, a low ratio indicates a liquid bank with excess funding capacity that is ready to be lent. (Wardiah, 2013)

$$LDR = \frac{Loan}{Deposit} \times 100\% \tag{5}$$

From the research framework, it can be seen that the independent variables are Return on Assets (ROA), Return on Equity (ROE), Non-Performing Loans (NPL) and Loan to Deposit Ratio (LDR). The dependent variable in this study is the Capital Adequacy Ratio (CAR). This study aims to determine the effect of independent variables on the dependent variable simultaneously and partially.

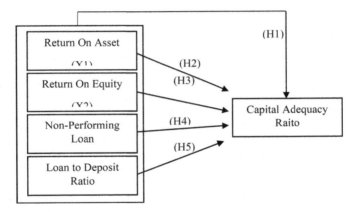

Figure 2.   Research framework by Andini and Yunita (2015).

Therefore, the hypotheses are aim as temporary answers of the research question, the hypotheses are:

H1: ROA, ROE, NPL, LDR simultaneously has positive effect on CAR of banking companies listed in IDX for period 2013 to 2017

H2: ROA has a positive effect on CAR of banking companies listed in IDX for period 2013 to 2017

H3: ROE has a positive effect on CAR of banking companies listed in IDX for period 2013 to 2017

H4: NPL has a negative effect on CAR of banking companies listed in IDX for period 2013 to 2017

H5: LDR has a negative effect on CAR of banking companies listed in IDX for period 2013 to 2017

2.2 Relationship ROA, ROE, NPL and LDR on CAR

2.2.1 Relationship ROA ON CAR
According to Nazaf, (2014) ROA has a positive influence on CAR. The greater the ROA of a bank, the greater the level of profit achieved by the bank and the better the bank's position in terms of asset use. So that CAR which is an indicator of bank health is increasing.

2.2.2 Relationship ROE on CAR
According to Evelina, (2012) ROE has a positive influence on CAR. The higher ROE achieved by the bank shows higher net income after tax, which means that the possibility of accumulated retained earnings increases, so that equity will increase and CAR will also increase.

2.2.3 Relationship NPL on CAR
According to Nazaf, (2014) NPL has a negative influence on CAR where an increase in the value of NPL can decrease the profit to be received by the bank. The higher the NPL, the more available capital in the bank, which causes a decrease in the value of CAR.

2.2.4 Relationship LDR on CAR
According to Nazaf, (2014) that the LDR has a negative effect on CAR where the increase in LDR is due to the growth in the amount of loans given is higher than the growth in the amount of funds collected by the bank so that it will decrease the CAR value of a bank.

2.3 Research methodology

The data was principally collected through Indonesia Stock Exchange. The data was collected from 5 banking companies for period 2013 to 2017. In this research, the data gathered from the site of the Indonesia Stock Exchange (www.idx.com). E-views 10 software was used in the analysis. This study is using quantitative research using statistical method form analysis from the data collection. In this research carried out with descriptive research. The investigation type of this research is causal research. The reason is because the researcher will examine the influence of return on asset, return on equity, non-performing loan, and loan to deposit ratio on capital adequacy ratio. In this research, the author has no interference without intervening the data in the environment of the organization. The unit analysis of this research is banking company listed in Indonesia Stock Exchange for period 2013 to 2017. The time horizon of this research is cross sectional. Panel data regression model is a statistical tool that examine the cross section and time series in the research. This method of analysis expects to provide the right conclusion with this study. The data panel regression analysis used in this research is:

$$Y_{it} = \beta_0 + \beta_1 X_{1it} + \beta_2 X_{2it} + \beta_3 X_{3it} + \beta_4 X_{4it} + U_{it}$$

Where:
Y_{it} = *Capital Adequacy Ratio* (CAR)
β_0 = Constant
$\beta_{(1,2,3,4)}$ = Independent Variable Regression Coefficient
X_{1it} = *Return On Asset* (ROA)
X_{2it} = *Return On Equity* (ROE)
X_{3it} = *Non Performing Loan* (NPL)
X_{4it} = *Loan to Deposit Ratio* (LDR)
U_{it} = *Error Term*

3 RESULT

3.1 Panel data regression

Selection of panel data regression model in this research based on the results of chow test and Hausman test. The result is random effect model has the best fit to this study.

Based on the Table 1 above, the researcher formulated a panel data regression model equation that explained the effect analysis of ROA, ROE, NPL, and LDR on CAR banking companies listed Indonesia Stock Exchange period 2013-2017, namely:

$$CAR = 0,179673 - 0,317123 \text{ ROA} - 0,087474 \text{ ROE} + 2,437546 \text{ NPL} - 0,021530 \text{ LDR}$$

3.1.1 Simultaneous influence test (F-test)

The F-test (simultaneous) is done to test whether the independent variable (X) simultaneously has a significant effect on the dependent variable (Y). If the value of Prob (F - Statistic) ≥ 0,05 (5% significance level) then H0 is accepted which means that all independent variables simultaneously do not significantly influence the dependent variable. But if the value of Prob (F - statistic) <0,05 then H0 is rejected which means all independent variables simultaneously have a significant effect on the dependent variable.

Based on Table 2 the Prob (F - statistic) value of 0.001560 is obtained. These results indicate that the value of Prob (F - Statistic) is smaller than the significance level (0,001560 <0,05) which means that H0 is rejected, H1 is accepted. Based on F test shows that all independent variables (ROA, ROE, NPL, and LDR) simultaneously have a significant effect on the dependent variable (CAR).

3.1.2 Analysis coefficient of determination

Based on Table 2 the adjusted R-squared (adjusted R^2) value is 0,479872. From these results it can be concluded that the contribution of the independent variable (ROA, ROE, NPL, and LDR) to the dependent variable (CAR) is 47,9872% and 52,0128% is determined by other variables not analysed in this study.

Table 1. Panel Data Regression

Variable	Coefficient	Std. Error	t-Statistic	Prob.
C	0.1796731	0.15878572	1.13154453	0.271211
ROA	-0.31717227	0.40298520	-0.78693405	0.440546
ROE	-0.0874741	0.11172201	-0.7829625	0.442821
NPL	2.4375459	0.83705236	2.91205908	0.00861
LDR	-0.0215302	0.18116707	-0.11884174	0.906586

Table 2. Simultaneous influence test (F-test).

Weighted Statistics			
R-squared	0.566560	Mean dependent var	0.029044
Adjusted R-squared	0.479872	S.D. dependent var	0.023497
S.E. of regression	0.016946	Sum squared resid	0.05743
F-statistic	6.535613	Durbin-Watson stat	1.076205
Prob(F-statistic)	0.001560		
Unweighted Statistics			
R- squared	-1.453293	Mean dependent var	0.190900
Sum squared resid	0.044298	Durbin-Watson stat	0.139536

3.1.3 *Partial influence (T-test)*

T-test (partial) is done to determine the significant or insignificant influence of one independent variable (X) partially on the dependent variable (Y). Partial testing in this research uses a significance level of 5% or 0.05. Based on Table 2, the conclusions of partial influence test (t-test) are:

1. Effect of ROA (X1) on CAR (Y)
   From the processed data obtained the value of Prob ROA $0.440546 \geq 0.05$ then H0 is accepted and β (-) = 0.3171227, meaning that ROA has a negative effect that is not significant to the CAR of the banking subsector.
2. Effect of ROE (X2) on CAR (Y)
   From the processed data obtained the Prob (ROA) value of $0.4428 \geq 0.05$ then H0 is accepted and β (-) = 0.087474, meaning that ROE has a insignificant negative effect on CAR.
3. Effect of NPL (X3) on CAR (Y)
   From the processed data obtained the value of Prob NPL $0.0086 \geq 0.05$ then H0 is rejected and β = 2.437546, meaning that the NPL partially has a positive effect significant to the CAR of the banking subsector.
4. Effect of LDR (X4) on CAR (Y)
   From the processed data obtained the value of LDR Prob $0.9066 < 0.05$ then H0 is accept and β (-) = 0.021530, meaning that LDR partially has a negative effect insignificant to the CAR of the banking subsector company.

4 DISCUSSION

4.1 *Effect of ROA on CAR*

Based on the results of testing in this study, Return On Assets (ROA) has a probability value of 0.440546 where this value is bigger than the significance level of 5% ($0.440546 \geq 0.05$) and shows a negative value. This means that partially ROA has a negative effect that is not significant on the Capital Adequacy Ratio (CAR) in the banking subsector companies listed on Indonesia Stock Exchange (IDX).

4.2 *Effect of ROE on CAR*

Based on the results of testing in this study, Return On Equity (ROE) has a probability value of 0.442821 where this value is bigger than the significance level of 5% ($0.442821 \geq 0.05$) and shows a negative value. This means that partially ROE has a negative effect not significant on the Capital Adequacy Ratio (CAR) in the banking subsector of the Indonesia Stock Exchange.

4.3 *Effect of NPL on CAR*

Based on the results of testing in this study, Non-Performing Loans (NPL) has a probability value of 0.008618 where this value is lower than the significance level of 5% ($0.008618<0.05$) and shows a positive value. This means that the NPL partially has a positive effect that is significant on the Capital Adequacy Ratio (CAR) in the banking subsector company registered with Indonesia Stock Exchange (IDX).

4.4 *Effect of LDR on CAR*

Based on the results of testing in this study, it shows that the Loan to Deposit Ratio (LDR) has a probability value of 0.906586 where this value is greater than the significance level of 5% ($0.906586 \geq 0.05$) and shows a negative value. This means that the LDR partially has a negative effect that insignificant on the Capital Adequacy Ratio (CAR) in the banking subsector companies listed on Indonesia Stock Exchange (IDX).

5 CONCLUSION

This research aims to determine the effect of Return On Asset (ROA), Return On Equity (ROE), Non-Performing Loan (NPL), and Loan to Deposit Ratio (LDR) on Capital Adequacy Ratio (CAR) which is proxies by of financial listed in Indonesia Stock Exchange (IDX) banking sub-sector period 2013-2017.

1. Return On Assets (ROA), Return on Equity (ROE), Non-Performing Loans (NPL), and Loan to Deposit Ratio (LDR) simultaneously have a significant effect on the Capital Adequacy Ratio (CAR) of banks registered in Indonesia Stock Exchange (IDX) 2013-2017 research period with the contribution of independent variables (ROA, ROE, NPL, and LDR) to the dependent variable (CAR).
2. Return On Assets (ROA) has no significant negative effect on the Capital Adequacy Ratio (CAR) in banks listed on the Indonesia Stock Exchange (IDX) in the 2013-2017 research period.
3. Return On Equity (ROE) has no significant negative effect on the Capital Adequacy Ratio (CAR) in banks listed on the Indonesia Stock Exchange (IDX) in the 2013-2017 research period.
4. Non-Performing Loans (NPL) has significant positive effect on the Capital Adequacy Ratio (CAR) in banks listed on the Indonesia Stock Exchange (IDX) in the 2013-2017 research period.
5. The loan to deposit ratio (LDR) has a no significant negative effect on the Capital Adequacy Ratio (CAR) in banks listed on the Indonesia Stock Exchange (IDX) in the 2013-2017 research period.

REFERENCES

Barus, A. C. (2011). Analisis Profitabilitas dan Likuiditas terhadap Capital Adequacy Ratio (CAR) pada Institusi Perbankan Terbuka di Bursa Efek Indonesia. *Jurnal Wira Warta Ekonomi Mikroskil*, 1(1), 1–12.

Darmawi, H. (2011). *Manajemen Perbankan*. Padang: Bumi Aksara.

Evelina, E. (2012). Pengaruh Rasio Profitabilitas Terhadap Kesehatan Permodalan Bank Swasta Nasional di Bursa Efek Indonesia. Berkala Ilmiah Mahasiswa Akuntansi, 1(3), 98–104.

Joliana, I. (2013). *Pengaruh Loan to Deposit Ratio, Non Performing Loan, Return On Equity, Interest Margin on Loan, dan Biaya Operasional terhadap Beban Operasioanl terhadap Kecukupan Modal pada Perusahaan Perbankan yang Terdaftar di Bursa Efek Indonesia (periode 2005-2011)*. Medan: Universitas Sumatera Utara.

Pandia, F. (2012). *Manajemen Dana dan Kesehatan Bank (Cetakan Ke-1)*. Jakarta: Rineka Cipta.

Nazaf, F. L. (2014). Pengaruh Kualitas Aset, Likuiditas, dan profitabilitas terhadap Tingkat kecukupan Modal Perbankan (Studi Empiris Pada Perusahaan Perbankan yang Terdaftar di BEI periode 2008-2012. *Jurnal Akuntansi*, 2(2), 1–26.

Wardiah, M. L. (2013). *Dasar-Dasar Perbankan (Cetakan Ke-1)*. Bandung: Pustaka Setia.

Managing Learning Organization in Industry 4.0 – Rachmawati & Hendayani (eds)
© 2020 Taylor & Francis Group, London, ISBN 978-0-367-81920-0

Sharia micro insurance model for small fishermen

T. Kurnia & A. Alhifni
Department of Sharia Economics, University of Djuanda Bogor, West Java, Indonesia

ABSTRACT: This study aimed to determine the micro sharia insurance model that was suitable for fishermen, especially small fishermen. The data analysis method used Miles and Huberman. The result showed that the suitable sharia insurance model for small fishermen was the sharia micro-insurance model through IMFs (ASML). The ASML model involved four parties, i.e. small fishermen, IMFs, Sharia Insurance Institutions, and Government. The ASML model made it easier for fishermen to make premium payments based on the trend of small fishermen's income. The ASML model also regulated the reduction of costs incurred by insurance institutions because it shared with the IMFs and government.

1   INTRODUCTION

West Java is one of the provinces which has large marine resources. This is indicated by a large number of fishermen in West Java which reached 183,000 fishermen (Ministry of Research, Technology and Higher Education, 2016). Pangandaran and Palabuhan Ratu areas are the mainstays for the marine tourism sector and sea catches. Both of these regions are recorded a large contribution to the economy of the people of the region (Nurhayati; 2013). The number of fishermen and their contribution is inversely proportional to the lives of fishermen who are still far from prosperous (Fadilah, 2014:71), (Suwardjo, et. al., 2010:1). According to Satria in Muflikhati (2010: 12), coastal communities, including fishermen, are still the group of people who are considered the poorest among other groups of the community (the poorest of the poor). Some of the problems that cause fishermen poverty include natural, cultural, technical and structural problems (Retnowati, 2011).

Natural disasters are always closely related to the life and safety and health of fishermen, while also causing damage and loss of operational fishing equipment. Small fishermen also experience damage or loss not only when operating at sea, but when they are on land. When high tides often reach the shore that sweeps ships and fishing gear stored along the coast. In addition, theft is the cause of loss of fishing gear for fishermen. In dealing with loss and damage experienced by small fishermen, generally, if the damage is small, it can be overcome by repairing the equipment, but for severe damage, most fishermen must buy back the fishing gear. In order to be able to buy back, small fishermen will borrow money from moneylenders, who will eventually be entangled in a never-ending cycle of debts.

This shows that the level of fishermen's need for institutions that can guarantee repairs for damage or loss experienced by fishermen is very high. Therefore, the government through the Ministry of Maritime Affairs and Fisheries is trying to find solutions for small fishermen. Goverment program is limited and has several provisions, so that not all fishermen can take advantage of this facility, only fishermen who meet the criteria will get the program. Another alternative that can be used by fishermen in dealing with collateral problems is to have insurance independently. However, there are other obstacles faced by fishermen, namely insurance institutions, in general, tend to choose a relatively low-risk segmentation. Meanwhile, the activities carried out by fishermen have a high risk, so that insurance institutions will not provide insurance to fishermen. Therefore, a sharia insurance model must be developed that can connect the needs of small fishermen and the interests of insurance companies.

2 LITERATURE REVIEW

2.1 *Insurance and sharia insurance*

According to the Law of the Republic of Indonesia No. 40 of 2014 concerning Insurance, Insurance is an agreement between two parties, namely an insurance company and a policyholder, which is the basis for receiving premiums by insurance companies as compensation for: a). providing reimbursement to the insured or policyholder due to loss, damage, costs incurred, loss of profits, or legal liability to third parties who may have suffered the insured or policyholder because an unstable event occurred; or b). providing payments based on the death of the insured or payment based on the life of the insured with benefits the amount of which has been determined and/or based on the results of fund management.

According to article 3 of Law No. 2 of 1992 concerning the Law on Insurance Business, the form of the insurance business is divided into two types, namely insurance business and insurance business supporting business. The elements related to insurance are as follows:

a. Insurer and Insured as the parties.
b. Premium, which is a certain amount of funds that must be paid by the policyholder to the guarantor.
c. Certain events, namely events that do not necessarily occur.
d. Compensation, the purpose of holding insurance is intended to provide compensation, but the compensation is known in Insurance Loss only.

2.2 *Fisherman*

Basically, the life of a fisherman is not much different from other people's lives. Fishermen are part of the Indonesian people who live in coastal areas and their lives depend on marine products. Generally, fishermen are divided into two parts, namely traditional fishermen and modern fishermen. Traditional fishermen are fishermen who have low income and limited economic capacity, so fishing activities are carried out traditionally because of limited capital (Alpharesy, et. al., 2012:2), (Sudarso, 2010:1).

While modern fishermen are fishermen who do fishing activities in a modern way. The modern method in question is to use large vessels, adequate safety or security systems and modern equipment such as trawlers, which are a kind of large and long net for fishing, in which the users wear buoys as safety protectors when using them. This type of modern fishermen is less than traditional fishermen whose lives are below the poverty line because most fishing communities in Indonesia are traditional fishermen, including fishermen in Pangandaran and Pelabuhan Ratu, West Java. (Fahmi, 2011: 125). Meanwhile, Widodo (2014), fishermen are divided into four groups, such as subsistence fishermen, native/indigenous/aboriginal fishermen, recreational/sport fishermen and commercial fishermen.

2.3 *Economic condition and fishermen welfare level*

Poverty can be known from income that is not proportional to expenditure. Fisherman expenditure is divided into two parts, namely food and non-food expenditure. Food expenditure is an expenditure that covers daily needs (staples), such as expenses to buy rice, vegetables, fruits and so forth. While non-food expenditure is expenditure for clothing, housing and household facilities including residence, clothing, footwear, headgear, taxes and so on (Firdaus, et. al., 2013:54Conversely, the lower the level of welfare of a person then shows himself in a poor position and can not enjoy the happiness concerned with satisfaction with the fulfillment of life's needs optimally. Based on the indicators of the level of welfare, the fishermen have not been said to be prosperous because they do not yet have satisfaction and good quality of life, including fishermen in Pangandaran and Pelabuhan Ratu, West Java (Fadilah, et. al. 2014:72).

3 METHODOLOGY

3.1 *Participants*

The study was conducted on fishermen in the Palabuhan Ratu and Pangandaran areas as many as 10 fishermen in each area. The method used for gathering data is in-depht interview, in depht interview are conducted directly to sources (fishermens and insurance institutions) in Focus Group Discussion (FGD). The fishermen chosen to be the resource persons in the FGD were fishermen who had at least 20 years of fishing experience. Selected resource persons were resource persons with at least 5 years of experience in each field. The FGD was conducted to get a comprehensive picture of the two sides, both from the fishermen and from the financial institutions involved concerning the model to be prepared. FGD divided into three stages. First stage, is the sources is explaining about all their experiences, including income, risk and hard time that their faced. Secoundly is discussion session, this part researcher asking question that related with research to the sources. In-depth interview is done by asking an open question that enable sources to answer a broader question. The last session is conclusion from discussion, this part of session is to adumbrate all the session and have agreement between researcher an souces that all the data is true and can be answered also agreement for publishing. Data obtained from the FGD results of experience, opinion and knowledge related with operations at sea for fishermen and operations in each institusion for institution sources.

3.2 *Measurements*

This study was a qualitative approach. The study was conducted using the Miles and Huberman method. Miles and Huberman is an interactive data analysis while in the field and is carried out continuously so that accurate data is obtained. Some steps taken in analyzing the data include (Sugiyono, 2016:337): Data Reduction, data display and verification.

3.3 *Data analysis*

Data analysis in these research uses analysis data tehniques on sites development by Miles and Hubermen. Collecting data were raised in a matrix it is presented exclusively a descriptive data around events or specific experience that codes data before and afterward (Miles and Hubermen, 2007). Data analysis in this research is implimented at the time of data gathering since April 2019 until August 2019. By the time of the interview, have done an analysis of the answers that are being interviewed. When answer presented by a source less satifactory, researcher continue to ask question, thus obtaining credible data. Data analysis is structured whithin the site affirmed that the colomn in a matrix of time compiled for a term, in phase forms, which can be seen a certain gene when occurs.

4 RESULT AND DISCUSSION

4.1 *Eligibility of small fishermen in following insurance*

Therefore, the eligibility of fishermen in taking insurance becomes a criterion for small fishermen to determine whether small fishermen are suitable or not to meet the criteria in the following insurance. In determining the eligibility of fishermen in taking insurance, the criteria that can be used as a reference for the feasibility of fishermen can be seen from several points in the insurance policy, these criteria are: guaranteed risk, type of insurance required by fishermen, premiums, and claims. Based on Miles and Huberman's analysis using the status dynamics matrix can be seen in Table 1.

Table 1. Status dynamics matrix: Sharia insurance based on fishermen perspective.

Difficulty faced	Problems faced (as seen by researchers)	How to overcome	How to solve
Guaranteed risk: the view of small fishermen towards the risks faced while at sea	Lack of understanding of risk because these activities are used to being carried out, and perceptions about any accidents can happen to anyone.	Providing an understanding of the impact of risks both on the fishermen household economy/ income and their associated impacts	Conducting socialization carried out by the government and related parties continuously (P), approaching fishermen who affect other fishermen (fishermen leaders) (S)
Premium: premium payment and payment due	Small fishermen income which fluctuates and the amount of premium if paid in one time, moreover the premium is loss unlike savings	Setting premium payments for fisherman insurance and providing subsidies for fisherman insurance with certain conditions.	Government policy in providing subsidies related to fisherman insurance (P). Involving IMFS that can manage fisherman premium management (S)
Claim: disbursement of assistance from insurance companies to fishermen	Disbursement of complicated claims that are not understood and favored by small fishermen, limited knowledge	Providing an explanation and there is assistance in the handling of claims	Involving IMFS in handling claims which also play a role as a companion (S).
Insurance products; personal accident insurance and boat insurance	Payment of double premiums and difficulties in claiming each premium	Providing special insurance products for fishermen	Special products for small fishermen (I). Government policy on special insurance for small fishermen (P)

* P = Procedure * S = Structure * I = Innovation

Insurance products needed by small fishermen include work accident insurance and insurance for equipment work tools. Work accidents are a high potential for fishermen, especially those caused by natural factors such as wind and sea waves. Similarly, loss and damage to the work tools of fishermen are also largely caused by natural factors. Some fishermen have difficulty if they have to take both types of insurance products, therefore, insurance companies can create special insurance products for small fishermen so that they can be an option for small fishermen.

4.2 Insurance company

Sharia insurance companies play an important role related to sharia insurance for fishermen, therefore, the analysis continues on the response of sharia insurance companies to the fishermen's needs for sharia insurance. Status dynamics matrix from the perspective of insurance companies can be seen in Table 2.

The selection of sharia insurance institutions as insurance institutions for fishermen is due to several factors, among others, sharia insurance institutions based on the *tabaru* principle (help each other). Currently, sharia insurance companies are still very limited in issuing insurance products for fishermen, several large-scale insurance companies issue insurance, but only for the framework of large boats or ships. For this reason, this can be resolved by government intervention in encouraging the provision of insurance products for small fishermen.

4.3 Sharia insurance model for small fishermen

The sharia insurance model for small fishermen is needed because there are no insurance products specifically intended for small fishermen. This model involves a synergy between institutions and

Table 2. Status dynamics matrix: Sharia insurance for small fishermen based on sharia insurance company perspective.

Difficulty faced	Problems aced (as seen by researchers)	How to overcome	How to solve
Guaranteed risk: the view of small fishermen towards the risks faced while at sea	The high level of risk faced by small fishermen, boats without protective equipment. Habits of fishermen who go to sea without meeting work safety	Providing the training and safety equipment for small fishermen.	The government provides periodic safety training for small fishermen (P). The government provides work safety equipment (P)
Premium: premium payment and payment due	Fluctuating fishermen's income and fishermen's ability to pay	Managing volatile fishing income	Involving IMFS as a unit link manager and or sharia insurance (S) link
Claim: disbursement of assistance from insurance companies to fishermen	Fisherman characteristics	Improving the characteristics of small fishermen	The government or related institutions provide training or assistance (P).
Insurance products; personal accident insurance and boat insurance	Low demand for insurance products from fishing groups	Increasing fishermen's need for insurance	Government policy towards sharia insurance for fishermen (S).

is built on difficulties both in terms of small fishermen and insurance companies. This model is called ASML which is Micro Sharia Insurance for small fishermen. Micro sharia insurance is suitable insurance for small fishermen because the existence of micro-insurance in Indonesia is aimed at low-income rural and urban communities, such as micro traders, factory workers, farmers, fishermen, and others. This is because the majority of the community has limitations in gaining access to insurance products (Micro Insurance Development Team, 2013). The conceptual difference between microinsurance and other forms of insurance, namely microinsurance has fewer assets and lower premiums (Njuguna, 2013:132). Micro-insurance schemes can also be an important component of a more comprehensive social protection system, including micro-insurance schemes can play a role in empowering and participating members (Rahim et. al., 2013:3). The ASML model can be seen in Figure 1 as follows:

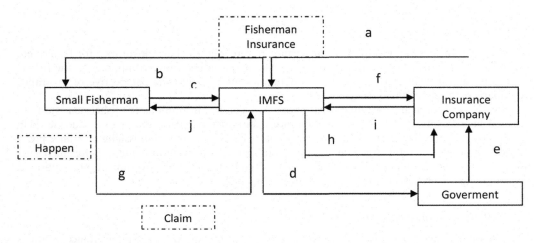

Figure 1. ASML model.

Description of figure:

a. The insurance company proposes a collaboration with IMFS with an MOU that is agreed by both parties as a unit of sharia insurance unit and offered fishermen insurance products.
b. IMFS offers and explains fisherman insurance products to small fishermen.
c. Small fishermen submit applications to become members of sharia insurance policies for fishermen insurance products, and agree to the IMFS provisions as sharia insurance unit links
d. IMFS submits a request for subsidy assistance to the government for small fishermen who will submit a fisherman insurance policy.
e. The government proposes the name of the fisherman and includes the amount of subsidy for fisherman insurance to the specified sharia insurance company.
f. IMFS pays fishermen to insurance companies
g. If an accident occurs, small fishermen can submit claims to IMFS, IMFS assists fishermen in fulfilling the claim requirements.
h. If the requirements are met then IMFS continues the claim of small fishermen to the sharia insurance company
i. The insurance company analyzes the submission of a claim if it has been approved then the payment of the claim is given to IMFS
j. IMFS provides claim disbursement funds to small fishermen.

Fisherman insurance product is a product specially prepared for small fishermen, to overcome the problem of fulfilling the requirements for sharia insurance companies, this product is a product proposed by the government. The government together with sharia insurance companies formulate insurance products for fishermen. This product is a product that contains collateral for personal accidents and boat loss along with equipment. In sharia microinsurance products for fishermen, subsidies are provided by the government to fishermen who meet predetermined criteria. Meanwhile, the ASML model is a model formed for fishermen by optimizing the synergy between sharia microfinance institutions, sharia insurance, and the government.

5 CONCLUSION

Small fishermen are professional groups that have a high level of risk in carrying out their work. Risks faced include the risk of personal accidents and the risk of loss and loss of fishing gear. Therefore, small fishermen need a guarantee for the risks inherent in the work. The ASML model is a model developed by synergizing between IMFS, insurance companies, and the government. The ASML model minimizes the difficulties faced by small fishermen related to the administration required and fluctuations in the income of small fishermen. IMFS functions as a manager of fishermen's income and assists with the administration of small fishermen both in submitting sharia insurance policies. The government is an incentive for small fishermen to raise awareness of personal safety and minimize fishermen's losses through policies and subsidies. Meanwhile, sharia insurance is an institution that provides sharia insurance products for small fishermen.

REFERENCES

Alpharesy, M. A., Anna, Z. & Yustiati, A. 2012. Analisis Pendapatan dan Pola Pengeluaran Rumah Tangga Nelayan Buruh di Wilayah Pesisir Kampak Kabupaten Bangka Barat. *Jurnal Perikanan dan Kealautan* 3(1): 11–16.
Catatan Teknis atas Laporan Tengah Periode Kegiatan Pemantauan dan Evaluasi Asuransi Mikro di Indonesia. *Mengurangi Kerentanan Ekonomi Melalui Asuransi Mikro: Potensi dan Tantangan di Indonesia.*

Direktorat Jenderal Perikanan Tangkap. 2014. *Statistik Perikanan Tangkap di Laut Menurut Wilayah Pengelolaan Perikanan Republik Indonesia (WPP RI)*. Jakarta: Direktorat Jenderal Perikanan Tangkap.

Firdaus, M. & Witomo, C. M. 2014. Analisis Tingkat Kesejahteraan dan Ketimpangan Pendapatan Rumah Tangga Nelayan Pelagis Besar di Sendang Biru, Kabupaten Malang, Jawa Timur. *Jurnal Sosial Ekonomi Kelautan dan Perikanan* 9(2): 155–168.

Kementerian Kelautan dan Perikanan Republik Indonesia. 2016. *Peraturan Menteri Kelautan dan Perikanan Republik Indonesia Nomor 16/PERMEN-KP/2016 Tentang Kartu Nelayan.*

Kementerian Kelautan dan Perikanan Republik Indonesia.Kementrian Kelautan dan Perikanan (KKP). 2016. *Peraturan Menteri dan Perikanan Nomor 18 Tahun 2016.* Kementrian Kelautan dan Perikanan (KKP).

Miles, Mattew B dan Huberman Amichael. 2007. *Analisis Data Kualitatif Buku Sumber Metode-Metode Baru*. Terjemahan Tjejep Rohendi Rohesi. Jakarta: Universitas Indonesia.

Muflikhati, I., Hartoyo, Sumarwan, U., Fahrudin, A. & Puspitawati, H. 2010. Kondisi Sosial Ekonomi dan Tingkat Kesejahteraan Keuarga: Kasus di Wilayah Pesisir Jawa Barat. *Jurnal Ilmu. Kel. & Kons.* 3(1) Januari: 1–7.

Njuguna, A. G. 2013. Risk Management Practices: A Survey of Micro-Insurance Service Providers in Kenya. *International Journal of Financial Research* 4(1): 132–150.

Retnowati, E. 2011. Nelayan Indonesia Dalam Pusaran Kemiskinan Struktural (Perspektif Sosial, Ekonomi dan Hukum). *Jurnal Perspektif* 16(3): 149–159.

Sugiyono. 2015. *Metode Penelitian Kuantitatif Kualitatif dan R&D*. Bandung: Alfabeta.

Widodo. 2014. Karakteristik Sosial Budaya dan Ekonomi Nelayan Kecil di Wilayah Pesisir Desa Puger Wetan Kecamatan Jember 10(1).

Tim Asuransi Mikro Bank Dunia. 2016. *Panduan Pelatihan Dasar-Dasar Asuransi Mikro Indonesia: Perlindungan yang diperlukan untuk Masyarakat.* Tim Asuransi Mikro Bank Dunia.

Tim Pengembangan Asuransi Mikro. 2013. *Grand Design: Pengembangan Asuransi Mikro Indonesia.* Otoritas Jasa Keuangan.

The role of social capital as mediator financial literacy and financial inclusion in productive age in DKI Jakarta

A.S. Dewi
Telkom University, Bandung, Indonesia

A. Ilmalhaq
Faculty of Economics and Business, Telkom University, Bandung, Indonesia

ABSTRACT: DKI Jakarta is a province that has the highest level of financial literacy and financial inclusion in Indonesia, has not proven that the people already have sufficient knowledge about the financial products and the high use of financial service institutions. The existence of social capital is expected to be a mediator in increasing literacy and financial inclusion. This study aims to examine the role of social capital as a mediator between financial literacy and financial inclusion in productive age in DKI Jakarta. The research method uses a survey method with a quantitative approach. Data collection techniques by distributing questionnaires to samples taken using non-probability sampling techniques, with a total sample of 400 productive age respondents in DKI Jakarta. This research adopts and uses the Sobel test for mediation analysis. The results show that social capital is a significant mediator in the relationship between financial literacy and financial inclusion of productive age in DKI Jakarta. Social capital mediates partially (partial mediation) i.e. independent can to influence directly dependent variable without involving mediating variables.

1 INTRODUCTION

DKI Jakarta is a province that has the highest level of financial literacy and financial inclusion. However, the high level of financial literacy is not directly proportional to financial behavior. The example is the high level of consumption in productive age in DKI Jakarta evidenced by the highest level of credit card ownership in DKI Jakarta is 22,3%. This is reinforced by the credit card market in Indonesia, which is quite large, with 17 million credit cards in circulation. After further investigation the owner is only 7.5 million, meaning that one person can have between 2-3 credit cards (Tribunjateng.com, 2016).

When someone knows finance, that person will aware of the benefits and use of services from the financial industry (Bongomin, et al., 2016). Therefore, financial literacy has a relationship with financial inclusion. The high level of financial inclusion in DKI Jakarta is still relatively low because of the uneven financial inclusion sector. The lowest sector of financial inclusion in DKI Jakarta is the capital market which is only 1,2% (OJK, 2016). Data from (KSEI, 2016) states that DKI Jakarta is the province with the highest number of investors in 2015 and 2016, amounting to 165,373 SID with 59,285 SID active investors. If this data compared with the total population of DKI Jakarta, it is only around 2,38% of the total population of DKI Jakarta.

Various efforts to support increased financial literacy and financial inclusion have been undertaken by the government through the OJK and Association of Financial Services Institutions (Otoritas Jasa Keuangan, 2017). Programs that have been planned and realized have not a significant impact, because if based on level of financial literacy and financial inclusion in Indonesia (67,82%) is still low when compared with 5 ASEAN countries (Radio Republik

Indonesia, 2017). This must be a serious concern for OJK to be able to cooperate with other relevant parties, as with social capital.

Putnam in (Bongomin, et al., 2016) explained that social capital can positively influence education and contribute to economic development. In Jakarta, there is social capital that supports and involved in increasing financial literacy and financial inclusion, it is Kocek. The Financial Smart Community (Kocek) is a community that places for learning, and sharing financial knowledge, tips, and information to the public (Cerdas Keuangan, 2013).

2 LITERATURE

Financial Literacy

Financial literacy is knowledge, skills, and beliefs that influence attitudes and behaviors to improve the quality of decision making and financial management to achieve prosperity (Otoritas Jasa Keuangan, 2017).

Financial Inclusion

According to draft OJK number/POJK.07/2016 that financial inclusion is the availability of access for the public to utilize financial products or services in financial service by needs and abilities of community to realize prosperity (Peraturan Otoritas Jasa Keuangan (POJK), 2016).

Social Capital

Social capital is which states the relationship between individuals, informal groups or communities with government and relations between countries that have the nature of mutual trust, mutual benefit, by social needs and lead to higher relations which gives an increase in productivity without economic improvement is social capital (Pramono, 2012). Based on the theoretical basis and framework of thought above, the research hypothesis is:

H_1: *Financial Literacy has a significant influence on Social Capital in productive age.*
Putnam in (Bongomin, et al., 2016) explained that social capital can positively influence education and contribute to economic development.

H_2: *Social Capital has a significant influence on Financial Inclusion in productive age.*
Putnam in (Bongomin, et al., 2016) explained that social capital can facilitate knowledge and skills through interactions in social networks that have an important role in increasing financial inclusion.

H_3: *Social Capital mediates the relationship between Financial Literacy and Financial Inclusion in productive age.*
According to (Bongomin, et al., 2016) social capital is a significant mediator in the relationship between financial literacy and financial inclusion. Through full mediation of social capital can be a relationship boosts between financial literacy and financial inclusion.

3 METHODOLOGY

3.1 *Participants*

The sample of this study refers to the population of productive age in DKI Jakarta, with range from 15-55 years according to BPS DKI Jakarta data in 2015 amounting to 5.584.841 people. The sampling technique used *Non-Probability Sampling*. Determination of the number of samples is calculated using Slovin formula. The results with the Slovin formula is 399,98 ~ 400. It can be concluded that the minimum sample size in this study is 400 productive age respondents in DKI Jakarta.

3.2 *Measurements*

The research method uses a survey method with a quantitative approach. The total respondents taken for the validity test were 30 respondents with a significance level of 5% (R Tabel = 0,361).

The results show that all questionnaire question item are valid (32 items). The reliability test is used Cronbach's Alpha. The results show that the positive Cronbach's Alpha value is 0.60 which is equal to 0.949. This can be interpreted that all the question items in this study are reliable and feasible to use.

3.3 *Data analysis*

In this study, the technic analysis is used *descriptive analysis, Pearson correlation analysis*, and mediation analysis using *Sobel Test*. For *Descriptive analysis* use the following formula:

$$Persentase = \frac{NilaiKumulatifItem}{NilaiFrekuensi} \times 100\% \tag{1}$$

Pearson correlation analysis use the following formula:

$$r = \frac{\sum XY}{\sqrt{(X^2)(\sum Y^2)}} \tag{2}$$

The significance of the estimated indirect effect of the independent variable (Financial Literacy) on the dependent variable (Financial Inclusion) according to Sobel Test (Baron and Kenny, 1986) is:

$$Sab = \sqrt{b^2 Sa^2 + a^2 Sb^2 + Sa^2 Sb^2} \tag{3}$$

Where a = independent variable; S_a = *standard error* a; b = dependent variable; S_b = *standard error* b.

4 RESULTS AND DISCUSSION

Correlation and regression analyses

According to Sarwono (2012) correlation is an analytical technique included in one of the measurement techniques of associations or relationships *(Measure of Association)*. The result indicated in Table 1:

Table 1 .   Correlation and regression analysis result.

Correlations

		Literasi_Keusangan	Modal_Sosial	Inklusi_Keuangan
Literasi_Keuangan	Pearson Correlation	1	.400[**]	.458[**]
	Sig. (2-tailed)		.000	.000
	N	400	400	400
Modal_Sosial	Pearson Correlation	.400[**]	1	.450[**]
	Sig. (2-tailed)	.000		.000
	N	400	400	400
Inklusi_Keuangan	Pearson Correlation	.458[**]	.450[**]	1
	Sig. (2-tailed)	.000	.000	
	N	400	400	400

** Correlation is significant at the 0.01 level (2-tailed).

The result revealed that financial literacy is indicated a significant and positive relationship with social capital, with strength of correlation tends to be weak (r= 0,400, Sig < 0,05). Furthermore, results from the table also indicated that there is a significant and positive relationship between financial literacy and financial inclusion among productive age, with strength of correlation tends to be weak (r= 0,458, sig = 0,05). The result between social capital with financial inclusion is indicated significant and positive relationship (r=0,450, sig = 0,05), with moderate correlation strength tend to low. Based on the results of *Pearson's correlation analysis* can be concluded that there is a correlation between each variable in this study. So this study is eligible to measure the mediator of by the research of Baron & Kenny (1986) is an independent variable assumed to cause a mediator if two variables are correlated.

4.1 *Normality test*

The normality test aims to test whether, in the regression model, the dependent variable and independent variable are normally distributed (Sujarweni, 2015).. The results of normality test are:

The results are indicated in Table II. The results revealed that data normally distributed. Because score of Sig (Asymp.Sig) is 0,054 > 0,05.

4.2 *Multicollinearity test*

Multicollinearity test aims to test whether the regression model found a correlation between the independent variable or not (Sujarweni, 2015). The results of multicollinearity test are:

The results are indicated in Table III. The results revealed that financial literacy and Social Capital (M) did not occur multicollinearity (Tolerance= 0,840> 0,1 and VIF =1,190<10).

4.3 *Mediation analysis*

A test to establish mediation by social capital in financial literacy and financial inclusion is used Sobel Test. The result indicated in Table IV:

Results from the table revealed that financial literacy is indicated a significant effect on social capital (Coefficient = 0,4368, p<0,05). Based on the results of the analysis, H1 is accepted that financial literacy has a significant effect on social capital. The coefficient of financial literacy on financial inclusion in model 2 = 0,4856, with p< 0,05. While the financial literacy coefficient on financial inclusion in model 3 is 0,3505 with p < 0,05. This indicated that there is a significant effect between financial literacy and financial inclusion in model 2 and model 3. It can be concluded that the significant effect of financial literacy on financial inclusion in model 3 is lower than model 2 (Baron & Kenny, 1986). Therefore, the coefficient of social capital in financial inclusion was 0,3095 with p< 0,05. This indicated that there is a significant effect between social capital and financial inclusion. So, H2 is accepted that social capital has a significant effect on financial inclusion. Based on the results of the mediation effects analysis, it can be concluded that this study has fulfilled the requirements for

Table 2 . Normality test result.

One-sample Kolmogorov-Smirnov Test		
		Unstandardized Residual
N		400
Normal Parameters[a,b]	Mean	.0000000
	Std.Deviation	5.53465414
Most Extreme Differences	Absolute	.067
	Positive	.047
	Negative	-.067
Kolmogorov-Smirnov Z		1.346
Asymp.sig.(2-tailed)		.054

a. Test distribution is Normal.
b. Calculated from data.

Table 3. Multicollinearity test result coefficients[a].

Model		Unstandardized Coefficients		Standardized Coefficients			Collinearity Statistics	
		B	Std. Error	Beta	t	Sig.	Tolerance	VIF
1	(Constant)	8.385	2.099		3.994	.000		
	X	.350	.049	.330	7.180	.000	.840	1.190
	M	.309	.045	.318	6.924	.000	.840	1.190

    a. Dependent Variable: Y

Table 4. Mediation analysis result.

	Modal Sosial (M)			Inklusi Keuangan					
	Model 1			Model 2 (Total Effect Model)			Model 3		
Predictor	Coeff	SE	p	Coeff	SE	P	Coeff	SE	p
Constant	30,0126	1,8108	0,0000	17,6728	1,7072	0,0000	8,3849	2,0991	0,0001
Literasi Keuangan	**0,4368**	0,0502	0,0000	**0,4856**	0,0473	0,0000	**0,3505**	0,0488	0,0000
Modal Sosial							**0,3095**	0,0447	0,0000

Notes: n=400; 95%

Table 5. Sobel's Z test result.

Normal Theory tests for Indirect Effect

Effect	SE	Z	p	Effect Size
0,1352	0,0301	5,4179	0,0000	0,1273

establishing mediation, in the first equation independent variable, then in the second equation is independent variable has an effect on dependent variable, and in third equation the mediator variable has the effect on dependent variable.

Based on table V, the indirect effect of financial literacy on financial inclusion through social capital is 0,1352, with a Z value is 5,4179 and p-value < 0,05, effect size value is 0,1273. For more details can be seen in the following picture

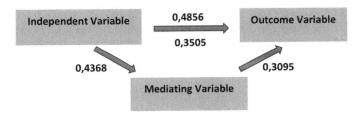

Figure 1. Sobel's Z test result.

Based on the Sobel test in table IV shows that the indirect effect between financial literacy and financial inclusion through social capital is 0,1352. While the direct effect between financial literacy and financial inclusion is 0,3505. The value of social capital can be seen from the value of effect size (Effect size = 0,1273). If the effect size is in the range 002 – 0,15, it is included in the weak category. Sobel Z value is 5,4179, which indicated that the indirect effect of financial literacy on financial inclusion through social capital will be significant if > from a critical point the effectiveness of mediation effects for significant level 5% (Z= 5,4179 > 1,96). So, the result is the independent variable can influence directly dependent variable without involving mediating variables at a productive age in DKI Jakarta (Partial mediation).

5 CONCLUSIONS AND RECOMMENDATIONS

5.1 Conclusions

This study found that financial literacy has a significant effect on social capital, social capital has a significant effect on financial inclusion, and social capital can mediate partially the relationship between financial literacy and financial inclusion at productive age in DKI Jakarta.

5.2 Recommendations

Regulators are expected to be able to utilize social capital like the community to improve financial literacy and financial inclusion in DKI Jakarta and Indonesia. The community can be a driver for increasing financial literacy and financial inclusion in society. For the researcher, further research can also examine other mediating variables that are more significant in helping to mediate the relationship between financial literacy and financial inclusion.

REFERENCES

Baron, R. M., & Kenny, D. A. 1986. The moderator-mediator variable distinction in social psychological research: Conceptual, strategic, and statistical considerations. *ResearchGate*: 1173–1182.

Bongomin, G. O., Ntayi, J. M., Munene, J. C., Nabeta, I. N. 2016. Social Capital: Mediator of Financial Literacy and Financial Inclusion in rural Uganda. *Emerald Insight*: 304.

Cerdas Keuangan. (2013, Agustus 31). *KoCek Events*. Diambil kembali dari cerdaskeuangan.com: https://cerdaskeuangan.com/kegiatan/school-of-innovations-smart-financial-planning-for-youth

KSEI. 2016. *Annual Report KSEI 2016*. Jakarta: KSEI.

OJK. 2016. *Survei Nasional Literasi dan Inklusi Keuangan 2016*. Diambil kembali dari Otoritas Jasa Keuangan: www.ojk.go.id

Otoritas Jasa Keuangan. 2017. *Strategi Nasional Literasi Keuangan Indonesia (Revisit 2017)*. Jakarta: Otoritas Jasa Keuangan.

Peraturan Otoritas Jasa Keuangan(POJK). 2016. *Peraturan Otoritas Jasa Keuangan (POJK) Republik Indonesia*. Jakarta: Undang-Undang Dasar RI.

Radio Republik Indonesia. (2017, Maret 31). *OJK: Tingkat Inklusi Keuangan Indonesia Masih Rendah*. Diambil kembali dari rri.co.id: http://rri.co.id/post/berita/377553/ekonomi/ojk_tingkat_inklusi_keuangan_indonesia_masih_rendah.html

Sarwono, J., & Martadiredja, T. 2008. *Riset Bisnis Untuk Pengambilan Keputusan*. Yogyakarta: C.V Andi Offset.

Sujarweni, V. W. 2015. *Statistik Untuk Bisnis dan Ekonomi*. Yogyakarta: Pustaka Baru Press.

Tribunjateng.com. (2016, Maret 19). *Sebanyak 17 Juta Kartu Kredit di Indonesia Hanya Dimiliki Oleh 7,5 Juta Orang*. Diambil kembali dari jateng.tribunnews.com: http://jateng.tribunnews.com/2016/03/19/sebanyak-17-juta-kartu-kredit-di-indonesia-hanya-dimiliki-oleh-75-juta-orang

Self-assessment systems, tax auditing, and tax evasion

E. Kartiko, L. Nurlaela, H.S. Hanifah, M. Romdhon & S. Yantika
Accounting Department, Faculty of Economy, University of Garut, Garut Regency, Indonesia

ABSTRACT: This research comprised a quantitative study with a descriptive and verification approach that aimed to determine the effect of: (1) self-assessment systems on tax evasion and (2) tax audit on tax evasion. The population of this study was 34,577 non-employed taxpayers registered at the Garut tax office. The sample used in this study consisted of 100 respondents. The respondents were taxpayers who do business with or represent small-micro-medium enterprises and the techniques used for obtaining data included convenience sampling with a questionnaire. The instrument trials were analyzed using validity and reliability testing. The data analysis technique used in the study was multiple linear regression analysis. The results showed that: (1) self-assessment systems and tax audits affect tax evasion, (2) self-assessment systems haves a negative effect on tax evasion, and (3) tax auditing has a positive effect on tax evasion.

1 INTRODUCTION

The tax ratio of the Indonesian gross domestic product (GDP) still lags behind that of other countries. It is only 10.78% of the agreed standard tax ratio of 15%. One of the factors that has led to this suboptimal situation is tax evasion. The taxpayer's active role can be seen from how obedient the taxpayer is in paying his tax. Tax authorities can increase services to taxpayers in order to encourage compliance with paying taxes, which will ultimately increase tax revenues. In line with the study of Guerra and Harrington (2018), cross-national experimental research showed that populations with high levels of moral tax demonstrated higher avoidance rates than those with low moral tax levels. Thus, the study found that self-reported moral tax could not predict actual tax avoidance. The following is the submission data of annual non-employee taxpayer SPT in KPP Pratama Garut as a research object, as seen in Table 1.

According to Table 1, the SPT compliance ratio is not optimal. The taxpayers listed annually have increased, but there is a difference between the most flat taxpayer and the realization of the SPT. Government measures have been put in place to increase taxpayer compliance by implementing an official assessment system, which is the system by which tax officers take account of taxpayers' payable taxes. However, the voting system was considered less effective in its implementation, so the government began to implement the self-assessment system, which is a tax-withholding system that authorizes taxpayers to self-calculate their tax. Tax evasion occurs often, which is an illegal action to reduce the income tax. Government tax audit is an effort to prevent this (Dell'anno & Davidescu, 2019). The importance and consequences of tax evasion change over the top taxes in the business cycle until they require different policy actions.

2 LITERATURE REVIEW

2.1 *Self-assessment system*

The self-assessment system is a tax system that trusts taxpayers to calculate their tax. The implementation of taxation activities currently uses an electronic administration system so as to facilitate filing, calculating, and paying taxes.

Table 1. Non-employee taxpayer compliance ratio 2014–2018.

Uraian	2014	2015	2016	2017	2018
WP Terdaftar	21.114	24.335	26.671	30.725	34.577
WP Terdaftar Wajib SPT	14.001	10.625	11.084	10.786	11.689
Realisasi SPT	1.334	1.876	2.034	2.732	7.661
Rasio Kepatuhan	10%	18%	18%	25%	66%

*Source: KPP Pratama Garut (2018) (data were reprocessed in 2019)

2.2 Tax audit

Rahayu Kurnia (2013) states that supervision is important in implementing the self-assessment system, which adheres to the law with its activities of collecting and processing data and of carrying out legislation in order to test compliance with tax obligations. Higher levels of audits have a major deterrent effect on the avoidance of individual income tax and vice versa (Eleftheriou, 2018).

2.3 Tax evasion

Tax evasion occurs when an active taxpayer illegally reduces, eliminates, or manipulates tax debts and avoids paying taxes as they should be paid according to statutory regulations (Hikmawati & Sri, 2017).

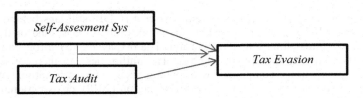

Figure 1. Research model.

3 RESEARCH METHODS

3.1 Research methods

Researchers used a statistical data analysis tool, which is a descriptive method of verification. The independent variables were the self-assessment system (X1) and tax auditing (X2) while the dependent variable was tax evasion (Y). The population of this research was taxpayers registered in the Office of the Tax Service KPP Pratama Garut, as many as 34,577 taxpayers. In determining the number of samples researchers used the Slovin formula with a fault-tolerant limit of 10%. The samples consisted of 100 taxpayers.

3.2 Data sources

The data sources used in this study were primary data obtained directly at the Office of the Tax Service KPP Pratama Garut by using questionnaires and secondary data obtained from various sources such as articles, books, the Internet, databases, websites, or related documents.

3.3 Normality test

Table 2. Test result normality, one-sample Kolmogorov-Smirnov test unstandardized residual.

		Unstandardized Residual
N		100
Most Extreme Differences	Absolute	0.052
	Positive	0.049
	Negative	−0.052
Test Statistic		0.052
Assymp. Sig. (2-tailed)		0.200[c,d]
a. Test distribution is normal		
b. Calculated from data		

* Source: SPSS Output version 23 (processed in 2019)

The results of the test obtained a Kolmogorov-Smirnov Z value of 0.200 with a significance value (ASYMP. Sig 2-Failed) of 0.200. That is, the significance is more than 0.05%, so the residual value is normal. In addition, using the Kolmogorov-Smirnov test, the normality of data was obtained using normal chart plots. Figure 2 shows the normal chart plot.

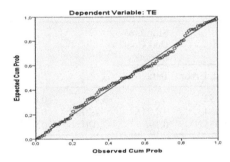

Figure 2. Test of normality.

3.4 Heteroscedasticity test

Table 3. Heteroscedasticity test results.

Model		Unstandardized Coefficients		Standardized Coefficients	t	Sig.
		B	Std. Error	Beta		
1	(Constant)	−0.015	2,423		−0.006	0.995
	SAS	−0.005	0.084	−0.006	−0.056	0.956
	Tax Audit	0.035	0.034	0.102	1.005	0.317

* Source: SPSS Output version 23 (processed in 2019)

According to Table 3, we can conclude that the entire variable has a significance value of > 0.05, meaning heteroscedasticity does not occur.

3.5 Multi-collinearity test

Table 4. Multi-collinearity test results.

Model	Unstandardized Coefficients		Standardized Coefficients	T	Sig.	Collinearity Statistics	
	B	Std. Error	Beta			Tolerance	VIF
1 (Constant)	15,096	3,852		3,919	0.000		
SAS	−0.296	0.134	−0.215	−2,214	0.029	0.992	1.008
Tax Audit	0.111	0.055	0.196	2,020	0.046	0.992	1.008

* Source: SPSS Output version 23 (processed in 2019)

According to Table 4, the tolerance value of all independent variables (the self-assessment system and tax auditing) is bigger than 0.10 and the VIF value of all independent variables is smaller than 10, meaning there is n o correlation between the independent variables. We can conclude that there is no multi-collinearity.

3.6 Autocorrelation test result

Table 5. Autocorrelation test results.

Model	Unstandardized Coefficients		Standardized Coefficients	T	Sig.	Collinearity Statistics	
	B	Std. Error	Beta			Tolerance	VIF
11 (Constant)	15,096	3,852		3,919	0.000		
SAS	−0.296	0.134	−0.215	−2,214	0.029	0.992	1.008
Tax Audit	0.111	0.055	0.196	2,020	0.046	0.992	1.008

* Source: SPSS Output version 23 (processed in 2019)

Table 5 shows a Durbin-Watson (DW) test value of 2.158. Then the value is compared to DL and Du. The DW values can be seen from the DW table with a = 5%. Then DL = 1.63 and the value du = 1.72 with n = 100. Thus, after being taken into account compared to the DW table, the DW value of 2.158 is between DL and 4-du – i.e., 1.715 < 2.158 < 2.285. We can thus conclude that there is no autocorrelation on the regression model in this study.

4 DISCUSSION

4.1 Hypothesis testing and multiple linear regression analyses

4.1.1 Simultaneous test F (F-Test)
The F-test was used to test the effect of the self-assessment system and tax auditing of simultaneous and partial tax evasion. Test criteria at significant rates = 0.05 (α = 5%) If F counts > f table, then H0 is rejected; otherwise if the F counts the < F table, then H0 is received. F-test results can be seen in Table 6.

From test results F can be obtained the value of F count of 4.916 with significance 0.009. In this study using the significance of 0.05. F table obtained by the calculation of f table = F (k; n-k) = (2.98). Thus we can conclude that the calculated value of > F table with

Table 6. Multiple regression test results (F-test).

Model		Sum of Squares	Dr	Mean Square	F	Sig.
1	Regression	49,030	2	24,515	4.916	0.009[b]
	Residual	483,720	97	4,987		
	Total	532,750	99			

* Dependent variable: TE, B. Predictors: (Constant), tax check, self-assessment system

a significantly smaller (0.009 < 0.05). This suggests that self-assessment systems and tax auditing jointly affect tax evasion.

4.1.2 Partial test (T-Test)
The T-statistic test shows how far the influence of one independent variable describes the variation of the dependent variable when the value probability < 0.05 or T count > T table and vice versa. T-test results on this research can be seen in Table 7.

Table 7. Multiple linear regression test result (t-Test).

Model	Unstandardized Coefficients		Standardized Coefficients		
	B	Std. Error	Beta	t	Sig.
1 (Constant)	15,096	3,852		3,919	0.000
Self-assessment system	−0.296	0.134	−0.215	−2,214	0.029
Tax auditing	0.111	0.055	0.196	2,020	0.046
a. Dependent Variable: TE					

* Source: SPSS Output version 23 (processed in 2019)

According to Table 7, the multiple linear equations used are column B, the first line indicating constants (a) and the next row indicating the independent variable coefficient. The regression models used in Table 7 were as follows: $Y = 15.096 + (−0.296) + 0.111 + E$.

A constant value with a regression coefficient in Table 7 can be described as follows:

A. Constants of 15.096 indicate that if the self-assessment system and tax auditing are assumed to not undergo a change (constant), then the value of tax evasion is 15.096.

B. The self-assessment system variable coefficient of −0.296 with a significance rate of 0.029 is smaller than 0.05. This means H1 is accepted so that it can be said the self-assessment system negatively affects tax evasion because the level of significance gained is < 0.05 (0.029 < 0.05) and the calculated T value of > 0.1985 (−0.296 < 0.1985). A negative relationship indicates that the implementation of a less effective self-assessment system will increase tax evasion.

C. The variable coefficient of tax auditing was 0.111 with a significance rate of 0.046, which is smaller than 0.05. This suggests that H2 is accepted; tax auditing has a positive effect on tax evasion because the level of significance obtained is < 0.05 (0.046 < 0.05), and the calculated T value is > 0.1985 (0.111 < 0.1985). Intensive implementation of tax auditing will decrease tax evasion.

4.1.3 Coefficient analysis determinations (R2)
According to Table 8, the value of R Square is 0.092, meaning the self-assessment system and tax audit variables have an effect on tax evasion of 9.2%, while the remaining 90.8% is influenced by other factors that were not researched in this study.

Table 8. Coefficient of determination.

Model	R	R Square	Adjusted R Square	Std. Error of the Estimate
1	0.303[a]	0.092	0.073	2,233

a. Predictors: (Constant), Pemeriksaan Pajak, self-assessment system
b. Dependent Variable: TE

* Source: SPSS Output version 23 (processed in 2019)

5 CONCLUSION

1) The self-assessment system negatively affects tax evasion at KPP Pratama Garut. A negative relationship indicates that the implementation of a poor self-assessment system will increase tax evasion. Conversely, the more effective a self-assessment system, the more it will lower tax evasion.
2) Tax auditing has a positive effect on tax evasion at KPP Pratama Garut; if the examination is done intensively, then tax evasion will decrease.
3) The self-assessment system and tax auditing affect tax evasion at KPP Pratama Garut.

ACKNOWLEDGMENTS

1. The Seventh International Seminar and Conference on Learning Organization
2. Garut University, Faculty of Economy
3. Office Tax KPP Pratama Garut
4. Employee of tax office KPP Pratama Garut

REFERENCES

Dell'anno, R., & Davidescu, A. A. (2019). Estimating the shadow economy and tax evasion in Romania: A comparison of different estimation approaches. *Economic Analysis and Policy.* 63(C). 130–149. doi: 10.1016/j.eap.2019.05.002

Eleftheriou, K. (2018). Spillovers of tax audits in Greek tourist destinations. *Annals of Tourism Research.* 76(c). 334–337. doi: 10.1016/j.annals.2018.10.002

Guerra, A., & Harrington, B. (2018). Attitude–behavior consistency in tax compliance: A cross-national comparison. *Journal of Economic Behavior & Organization.* 156(C). 184–205. doi: 10.1016/j.jebo.2018.10.013

Mardiasmo. 2013. *Perpajakan Edisi Revisi.* Yogyakarta: Andi Offset.

Rahayu Kurnia, Siti. 2013. *Perpajakan Indonesia: Konsep dan Aspek Formal.* Yogyakarta: Graha Ilmu.

Rahim, Hikmawati. (2017). The influence of the tax system and tax audits on tax evasion by corporate taxpayer. *Study & Accounting Research* 14(2). 62–69.

The effect of corporate social responsibility disclosure on earnings per share in listed mining in Indonesia

D. Hanni & A. Krisnawati
Faculty of Economics and Business, Telkom University, Indonesia

ABSTRACT: In recent years, the role of business in emerging countries has turned into a socially responsible approach. One such nation is Indonesia, which has laws regulating corporate social responsibility (CSR) activities for operating companies. Companies have responded to these changes and rules by implementing CSR activities. Corporate social responsibility activities have a relationship with company performance because costs are involved in carrying out these activities. A company's performance can be measured by financial aspects, such as earnings per share (EPS). This study aimed to determine the relationship of corporate social responsibility discloser (CSRD) to EPS with control variables consisting of leverage and firm size. This research used quantitative methods with panel data analysis techniques. The research sample comprised twenty-nine companies in the mining sector listed on the Indonesia Stock Exchange (IDX) during the period 2013 to 2017. The results of this study indicated that CSRD has no significant effect on EPS and the control variables simultaneously with CSRD also have no significant effect on EPS. Even so, it is still recommended that companies continue implementing and increasing CSR activities.

1   INTRODUCTION

Globalization has spurred the growth and excellence of company competencies in maintaining business. The role of business in emerging countries has changed from the classic approach emphasizing profit maximization to a socially responsible approach. Business is not only responsible to shareholders but must also be responsible to workers, the community, the public, and the government. In Indonesia, corporate social responsibility (CSR) has been regulated in Law Number 40 of 2007 concerning limited liability companies (article 74 paragraph 1), which states that "companies that carry out their business activities in the fields and/or relating to natural resources are required to carry out Social and Environmental Responsibility." According to the Pernyataan Standar Akuntasi Keuangan (PSAK), CSR activities for disclosure of social responsibility activities or nonfinancial reporting activities is still mandatory for companies operating in Indonesia. According to Ta and Bui (2018), several nongovernmental organizations and countries have introduced standards and regulations to guide companies to develop and display CSR information, one of which is the Global Reporting Initiative (GRI).

A company's performance can be measured through financial aspects and can be seen from several ratios, such as return on assets (ROA), return on equity (ROE), earnings per share (EPS), net profit margin (NPM), and Tobin's Q (Jitaree, 2015). Ta and Bui (2018) state that the practice and disclosure of CSR will lead to better financial performance. The disclosure of CSR has a positive impact on a company's financial performance because social responsibility is considered a corporate protector against negative information that damages a company's reputation, so as to protect financial results (Adeneye, 2015). Hery (2015) states that EPS is a ratio to measure the success of company management in providing benefits for investors who have common stock in a company. The source of funding has a relationship with a company's financial performance. The source of funding can be seen from leverage analysis, which can be measured through a debt-to-equity ratio (DER) (Jitaree, 2015). In addition, financial performance or financial results are also influenced by company size, which can be calculated through several attributes, one of which is a company's total assets (Asnawi and Wijaya, 2005).

Ahmed, Zakaree, and Kolawole (2016) state that CSR disclosure has a significant effect on financial performance as seen from EPS. The disclosure of CSR will have a positive impact on the return of company assets and company capital where the financial performance can be seen through financial ratios, such as ROA and ROE (Matuszak and Różańska, 2017). However, according to Ngoc (2018), CSR disclosure does not produce economic benefits for companies in the short term in the industry based on measurement through ROA, so CSRD has no significant effect on financial performance.

This study was based on Ahmed and colleagues' (2016) research on the effect of CSR disclosure on EPS in manufacturing companies registered in Negeria using the EPS dependent variable with leverage control variables and firm size. The difference in this study was to replace the manufacturing sector with the mining sector, which is listed on the Indonesia Stock Exchange (IDX) for the period 2013 to 2014.

2 LITERATURE REVIEW

2.1 Corporate Social Responsibility Disclosure (CSRD)

The formation of corporate social responsibility is based on two interrelated theories, namely legitimacy and stakeholders. Legitimacy explains that companies must operate within the limits and norms set by the community and meet the expectations of the community (Jitaree, 2015). Based on the theory of legitimacy, the determinants of CSRD are factors related to social pressure (Coffie, Aboagye, and Musah, 2018). Interaction between internal and external stakeholders with the company makes the concept of social responsibility and disclosure develop and merge into a single unit (Ahmed et al., 2016). Through CSRD activities, companies have the opportunity to provide explanations on various matters that support the welfare of shareholders, employees, suppliers, consumers, regulators, and surrounding communities that are directly related to the daily activities of the company (Anggraeni and Chaerul, 2018). One of the guidelines for CSR disclosure is to use the Global Reporting Initiative (GRI) standard. The GRI is an organization that provides guidelines for companies in reporting business sustainability (sustainability report). Guidelines for CSR reporting use GRI Fourth Generation (GRI G.4).

2.2 Earnings Per Share (EPS)

One technique used to assess company performance is financial ratio analysis. The ratio used is the profitability ratio, one of which is EPS. Husnan and Pudjiastuti (2015) state that investors often focus on EPS when conducting analyses.

The EPS formula is:

$$Earnings\ Per\ Share\ (EPS) = \frac{Profit\ before\ Tax - Preferred\ Stock\ Dividends}{Number\ of\ Ordinary\ Shares} \quad (1)$$

2.3 Leverage

According to Periansya (2015), leverage is a ratio used to measure the extent to which a company's assets can be financed with debt or financed by outside parties. The debt-to-equity Ratio (DER) is a ratio used to measure the level of leverage against the total shareholder equity owned by a company.

The DER formula is:

$$Debt\text{-}to\text{-}Equity\ Ratio\ (DER) = \frac{Total\ Debt}{Capital} \times 100\% \qquad (2)$$

2.4 *Company size*

According to Jitaree (2015), company size is one of the most important attributes that can affect company activities. According to Prasetyantoko (2008), measurement of the size of a company can be illustrated through the size of the company's total assets.

The company size formula is:

$$Size = Ln\ (Total\ Assets) \qquad (3)$$

2.5 *Hypothesis*

Based on various factors and on previous research that has been presented before, the hypothesis of this study was as follows:

H1: CSRD has a significant effect on EPS in the mining sector.

H2: CSRD with leverage and company size has a significant effect on EPS in the mining sector.

3 METHODOLOGY

This study used a quantitative method by measuring the effect of CSRD on EPS with control variables of leverage and firm size. Sampling was conducted using a nonprobability sampling method with a purposive sampling technique. The sample of this study comprised twenty-nine companies in the mining sector listed on the Indonesia Stock Exchange (IDX) during the period 2013 to 2017. Data collection used secondary data from annual reports and audited and published financial reports. Data analysis techniques in this study used panel data, which are a combination of time-series and cross-sectional data.

4 RESULT AND DISCUSSION

4.1 *Estimating model selection*

The initial step of the calculation in this study was to determine the most appropriate model between the common, fixed, and random effect models in estimating panel data. First, it used a chow test that compares common and fixed effects. The results showed that the value of the prob. Chi-square cross-section was less than 0.05, so the fixed effect model was chosen. Then the Hausman test was performed because the results obtained previously were fixed effects. The Hausman test compares fixed and random effects. The results showed that the better model was the random effects model. Therefore, further tests need to be done – namely the Lagrange test, which shows the most appropriate results using the random effects model. Finally, this study used a random effect model in estimating panel data.

The panel data regression results using the random effect model as follows in the mining sector can be seen in Table 1.

Table 1. Result of measurement.

Dependent variable: EPS
Method: Panel EGLS (Cross-section random effect)
Period included: 5
Cross-sections included: 29
Total panel (balanced) observation: 145
Swam and Arora estimator of component variances

Variable	Coefficient	Std. Error	t-Statistic	Prob.
CSRD	934.5149	681.3018	1.371661	0.1723
CSIZE	9.465354	25.38661	0.372848	0.7098
LEVERAGE	6.265686	5.248054	1.193907	0.2345
C	−423.5508	729.9968	−0.580209	0.5627
Weighted statistics				
R-squared	0.026767	Mean dependent var.		28.26935
Adjusted R-squared	0.006059	S.D. dependent var.		255.4992
S.E. of regression	254.7239	sum squared res.		9148682.
F-statistic	1.292627	Durbin-Watson stat.		1.661202
Prob. (F-statistics)	0.279381			

From the foregoing calculation, an equation is formed as follows:

EPS = −423.5508 + 934.5149 CSRD + 9.465354 CSIZE + 6.265686 LEVERAGE (DER).

The intercept or constant coefficient of −423.5508 shows that if the value of CSRD, CSIZE, or company size and leverage is constant (0), then the EPS value is −423.5508. Negative constant values are not a problem because what should be considered is the slope value. In addition, if there is an increase in CSRD of 1%, there will be an increase in EPS growth of 934.5149 units, for every 1% increase in CSIZE, the EPS will increase by 9.465354 units, and if there is a 1% increase in leverage, then the EPS will have an increase of 6.265686 units assuming other variables are considered fixed.

4.2 Hypothesis test

The next step is to analyze the inner model by looking at the R-square value used to explain the effect of exogenous latent variables on endogenous latent variables. The R-squares values obtained are 0.75, 0.50, and 0.20, and we can conclude that the model is strong, moderate, and weak. This value will present the amount of variance of the construct explained by the model.

H1: CSRD has a significant effect on EPS in the mining sector. In this study the results obtained for the mining sector were 1.9766 (with a significance level of 0.05). Based on Table 4, the t-count is smaller than the t-table, so statistically it can be said that CSRD has no significant effect on EPS. The results of this study are in accordance with the results of Mansaray (2017), which showed that CSRD had no significant effect on financial performance in several industries, such as the transportation industry, sales and manufacturing, and health and pharmacy. In addition, according to previous research by Kamatra and Kartikaningdyah (2015), CSRD only affects ROA and NPM, but does not significantly influence ROE and EPS in the mining sector or the basic and chemical industry sectors. However, it is different from the results of Ahmed and colleagues' research (2016), which states that CSRD has a significant effect on EPS.

This unaffected result can be caused by the low average value of CSRD in the mining sector, which is 26.77%, where this value is relatively small because the ideal value of CSRD is 100%. Only two out of twenty-nine companies were able to achieve a CSRD index of more than 50% – namely Timah (Persero) Tbk. (TINS) and Bukit Asam Coal Mine (Persero) Tbk. (PTBA). In addition, the low average value of social responsibility disclosures can also be caused because CSR disclosure is information that will provide added value to companies that disclose it according to financial accounting standards or PSAK.

H2: CSRD with leverage and company size has a significant effect on EPS in the mining sector. In this study F-table results obtained for the mining sector were 2.67 (with a significant level of 0.05). Based on Table 4, the calculated value of the F-count is 1.292627, where the value is smaller than the F-table. This shows that statistically CSRD with leverage and company size together do not have a significant effect on EPS. The results of this study are consistent with the results of previous research by Ahmed and colleagues (2016), which shows that CSRD with leverage and company size together does not significantly influence EPS. However, in contrast to the results of Kamatra and Kartikaningdyah's (2015) research, CSRD with leverage and company size together has a significant effect on EPS.

In the mining sector it can be said that the average value of CSRD is very low, only 26.77%. Then the average value of leverage (DER) is 1.22, which means that the structure of the debt value of each company is greater than the capital it has. Several mining companies have debts but do not have the capital or their capital value is negative. This will certainly affect the company's capital structure, which will result in revenue results and returns on profits obtained.

5 CONCLUSION

Based on research, analysis, and discussion, we can conclude that corporate social responsibility disclosure (CSRD) has no significant effect on earnings per share (EPS) in the mining sector, as well as that CSRD with the presence of leverage control variables and company size together also had no significant effect on EPS in the mining sector during the period 2013 to 2017. As for suggestions for further research, scholars can add a longer period of time and a larger number of samples, so the results obtained are more actual. This can also be accomplished by adding variables (independent and control) and indicators and/or adjusting the reference indicators that can be used in calculating CSR disclosures. In the aspect of prediction, one way to optimize a company's financial performance is earnings per share or EPS, so a good EPS disclosure is needed in order to increase the value of CSRD. In addition to increasing CSRD, company assets need to be managed properly because they can affect the company's capital structure and have an impact on the value of the company's assets and leverage itself.

REFERENCES

Adeneye, Y. B. 2015. Corporate social responsibility and company performance. *Journal of Business Studies Quarterly* 7(1), 151–166.

Ahmed, M. N., Zakaree, S., & Kolawole, O. O. 2016. Corporate social responsibility disclosure and financial performance of listed manufacturing firms in Nigeria. *Research Journal of Finance and Accounting* 7(4), 47–58.

Anggraeni, D. Y., & Chaerul, D. D. 2018. Pengujian Terhadap Kualitias Pengungkapan CSR di Indonesia. *Junal Ekonomi dan Keuangan* 2(1), 22–41.

Asnawi, S. K., & Wijaya, C. 2005. *Riset Keuangan: PengujianPengujian Empiris*. Jakarta: Gramedia Pustaka Utama.

Coffie, W., Aboagye O. F., & Musah, A. 2018. Corporate social responsibility disclosures (CSRD), corporate governance and the degree of multinational activities: Evidence from a developing economy. *Emerald Insight* 8(1), 106–123.

Hery. 2015. *Analisis Laporan Keuangan: Pendekatan Rasio Keuangan*. Yogyakarta: Center for Academic Publishing Service.

Husnan, S., & Pudjiastuti, E. 2015. Dasar-Dasar Manajemen Keuangan. Yogyakarta: UPP STIM YKPN.

Jensen, M. C., & Meckling, W. H. 1976. Theory of the firm: Managerial behavior, agency costs and ownership structure. *Journal of Financial Economics* 3: 305–360.

Jitaree, W. 2015. Corporate social responsibility disclosure and financial performance: Evidence from Thailand. *Research Online University of Wollongong*.

Kamatra, N., & Kartikaningdyah, E. 2015. Effect of corporate social responsibility on financial performance. *International Journal of Economics and Financial Issues* 5(special issue), 157–164.

Mansaray, Amidu P & Liu Yuanyuan & Sesay Brim. 2017. The impact of corporate social responsibility disclosure on financial performance of firms in Africa - *International Journal of Economics and Financial Issues* 7(5), 137–146.

Matuszak, L., & Różańska, E. 2017. An examination of the relationship between CSR disclosure and financial performance: The case of Polish banks. *Accounting and Management Information Systems* 16(4), 522–533.

Ngoc, N. B. 2018. The effect of corporate social responsibility disclosure on financial performance: Evidence from credit institutions in Vietnam. *Canadian Center of Science and Education* 14(4), 109–122.

Otoritas Jasa Keuangan. 2016. Undang-Undang No. 40 Tahun 2007 Tentang Perseroan Terbatas.

Periansya. 2015. Analisis Laporan Keuangan. Palembang. *Politeknik Negeri Sriwijaya*.

Prasetyantoko, A. 2008. *Corporate Governance: Pendekatan Institusional*. Jakarta: PT. Gramedia Pustaka Utama.

Ta, H. T. T., & Bui, N. T. 2018. Effect of corporate social responsibility disclosure on financial performance. *Asian Journal of Finance & Accounting* 10(1), 40–58.

Managing Learning Organization in Industry 4.0 – Rachmawati & Hendayani (eds)
© 2020 Taylor & Francis Group, London, ISBN 978-0-367-81920-0

The influence of financial attitudes, financial literacy, and parental income on personal financial management (A case study of students of Bandung)

A.S. Dewi & U.D. Salwani
Faculty of Economics and Business, Telkom University, Bandung, Indonesia

ABSTRACT: This study aimed to determine the effect and relationship between financial attitudes, financial literacy, and parental income on students' financial management in Bandung. The research employed quantitative methods by using the Cronbach's alpha to test reliability, and Pearson's correlation to test validity. Data collection was carried out by distributing questionnaires to 400 students in Bandung. The sampling technique used in this study was a nonprobability sampling technique. In line with previous studies, the result showed that financial attitudes impact personal financial management, financial literacy impacts personal financial management, and parental income doesn't impact personal financial management.

1 INTRODUCTION

Financial attitude is needed to increase financial literacy and inclusion, and college students' financial attitudes indicate their level of financial literacy. Otoritas Jasa Keuangan predicted that financial literacy in early 2018 reached 31% among Indonesians, an increase of 29.7% from 2017.

The low financial literacy of students will affect their use of the income their parents provide. Most students currently rely more on their parents' income in daily life because many students still do not have their own income. Parental income is the result of parents who have an income obtained from work. Parental income is one of the factors in student financial management.

Financial attitudes, financial literacy, and parental income influence student financial management. Students are consumptive individuals in controlling their finances and therefore a good financial literacy is needed so financial attitudes can help in making both long-term and short-term financial decisions. Parents' income also becomes a factor for students to manage their finances because students tend to make the most of their parents' income for their daily needs. Poor student financial management can affect not only student finances but also academic, mental, and physical health, and even students' ability to find work after graduation (Bodvarsson and Walker, 2004).

2 LITERATURE REVIEW

One of the pioneering studies in research on behavioral finance was conducted by Herdijonoand Damanik (2016). Previous research indicated that financial attitudes affect personalfinancial management, financial knowledge has no effect on personal financial management, and income does not affect personal financial management.

In another study, Mien and Taho (2015) investigated the relationship among financial attitudes, financial knowledge, locus of control, and personal financial management. This study indicated that financial attitude and financial knowledge were significantly positively related to financial management, and locus of control had negative effect on financial management.

3 METHODOLOGY

3.1 *Participants*

We examined three main research questions: whether financial attitude, financial literacy, and parental income have a significant influence on personal financial management. This study used three independent variables – X1 (financial attitude), X2 (financial literacy), and X3 (parental income) – and used Y for the dependent variable (personal financial management). This study also examined whether students' personal financial management in Bandung can be categorized as good. The sampling technique used in this study was a nonprobability sampling technique. Four hundred respondents were used as samples in this study with the limitation of Bandung students age fifteen to twenty-four years, with the determination of the sample using the Slovin method.

$$n = \frac{N}{1 + Ne^2} \tag{1}$$

$$n = \frac{240.943}{1 + 240.943(5\%)^2}$$

$$n = \frac{240.943}{603,3575}$$

$$n = 399,33 \sim 400$$

3.2 *Measurements*

In this study, we used the Pearson test for validity and Cronbach's alpha for reliability in order to test a sample of thirty respondents, and $\alpha = 5\%$. The validity test yielded a significant value above 0.361 and Cronbach's alpha was above 0.60. To see the effects simultaneously, this study used the F-test method and the partial effect method with the T-test. To see the relationship between the independent and dependent variables, we used a multiple linear regression test:

$$Y = a + b_1 X_1 + b_2 X_2 + b_3 X_3 \tag{2}$$

3.3 *Data analysis*

After a multiple linear regression test with the results of constants of personal financial management variables (3.264), financial attitudes (0.110), financial literacy (0.595), and income (0.177), the following equation was produced:

$$Y = 3.264 + 0.110X_1 + 0.595X_2 + 0.177X_3 \tag{3}$$

The X variables of financial attitudes, financial literacy, and parental income had a direct relationship with the Y variable, personal financial management. This caused an increase in financial attitudes, financial literacy, and parental income; it also affected personal financial management by values of 0.110, 0.595, and 0.177 for each variable X.

Hypothesis Testing

1. Silmutan Test F
An F-test was needed to determine whether financial attitudes, financial literacy, and parental income have a simultaneous influence on personal financial management, with the following hypothesis:

Table 1. Anova.

Model		Sum of Squares	df	Mean Square	F	Sig.
1	Regression	6,581.106	3	2,193.702	141.093	0.000[b]
	Residual	6,156.966	396	15.548		
	Total	12,738.073	399			

a. Dependent Variable: PMK
b. Predictors: (Constant), PD, LK, SK

H_1: Financial attitudes, financial literacy, and parental income have a significant, simultaneous influence on personal financial management.

From the results of the F-test in this study the sigma value of 0.000 is known. Thus we can conclude that the value of sigma in this study is smaller than 0.005. Then the independent variable has a significant effect on the dependent variable simultaneously so that H_1 is accepted.

2. Partial T-test
A T-test was needed to determine whether financial attitudes, financial literacy, and parental income have a partial effect on personal financial management, with the following hypotheses:

Table 2. Coefficients[a].

Model		Unstandardized Coefficients		Standardized Coefficients			Correlations			Collinearity Statistics	
		B	Std. Error	Beta	t	Sig.	Zero-order	Partial	Part	Tolerance	VIF
1	(Constant)	3.264	1.559		2.093	0.037					
	SK	0.110	0.045	0.103	2.437	0.015	0.467	0.122	0.085	0.679	1.473
	LK	0.595	0.039	0.652	15.396	0.000	0.713	0.612	0.538	0.680	1.471
	PD	0.177	0.127	0.049	1.394	0.164	0.072	0.070	0.049	0.989	1.011

a. Dependent Variable: PMK

Financial Attitude

H_2: Financial attitudes have a significant partial influence on personal financial management.

In this study, the significant value for financial attitudes is 0.015 with a probability value of 0.05. From the results of the T-test, we can conclude that H_2 is accepted for financial attitudes. There is a significant partial influence between financial attitudes on personal financial management.

Financial Literacy

H_3: Financial literacy has a significant partial attitude on personal financial management.

In this study, the significant value for financial attitudes is 0.000 with a probability value of 0.05. From the results of the T-test, we can conclude that H_3 is accepted for financial literacy. Financial literacy has a significant partial influence on the behavior of personal financial management.

Parental Income

H_4: Parental income has a significant partial influence on personal financial management.

In this study, the significant value for parental income is 0.164 with a probability value of 0.05. From the results of the T-test, we can conclude that H_4 is not accepted for parental income. Parental income has no significant partial influence on personal financial management.

Determination Coefficient Test

The coefficient of determination is used to measure how much financial attitudes, financial literacy, and parental income influence personal financial management.

From these results, we know that the coefficient of determination is 51.7%, where the value indicates that the overall variable X affects Y by 51.7%, while 48.3% is influenced from outside.

Table 3. Model summary[b].

					Change Statistics				
Model	R	R-Square	Adjusted R-Square	Std. Error of the Estimate	R-Square Change	F Change	df1	df2	Sig. F Change
1	0.719[a]	0.517	0.513	3.943	0.517	141.093	3	396	0.000

a. Predictors: (Constant), PD, LK, SK
b. Dependent Variable: PMK

4 RESULTS AND DISCUSSION

Attitudes toward Personal Financial Management

From the results of regression calculations, we can see that financial attitudes have a direct relationship with personal financial management of 0.110. Partially, financial attitudes have an influence also on personal financial management of 0.015. This is in line with Herdijono and Damanik (2016), who state that financial attitudes have a significant influence on personal financial management.

Financial Literacy and Personal Financial Management

The results of the financial literacy regression test show a direct effect on personal financial management of 0.595. Partially, financial literacy has a significant influence on personal financial management.

Parental Income and Personal Financial Management

Based on the regression results, we can see that parental income has a direct effect on personal financial management, but partially based on the t-test, parental income does not have a significant effect on personal financial management. The results of this study are in line with Herdijono and Damanik (2016), who states that parental income does not have a significant effect on personal financial management.

Financial Attitudes, Financial Literacy, and Parental Income and Personal Financial Management

Based on the results of the F-test, we can see that financial attitudes, financial literacy, and parental income have a simultaneous influence on personal financial management. This result is also known from the determination coefficient of 51.7%, from which we can conclude that financial attitude, financial literacy, and parental income has an effect on personal financial management of 51.7%, and the remaining 48.3% is influenced by other factors.

5 CONCLUSIONS AND RECOMMENDATIONS

Based on this research, we can conclude that:

a. Financial attitudes, financial literacy, and parental income have a significant simultaneous influence on the personal financial management of Bandung students.
b. Financial attitudes have a significant influence on personal finance management in Bandung students.
c. Financial literacy has a significant influence on personal finance management in Bandung students.

d. Parental income does not have a significant influence on the personal financial management of Bandung students. This is because parents' income does not affect the size of the distribution of student allowances so that it will cause similar student financial management.

REFERENCES

Arif, A. Y. 2019. Pengertian Pendapatan Adalah: Jenis dan Perbedaan Dengan Penghasilan. Tersedia: https://rocketmanajemen.com/definisi-pendapatan/ [February 18, 2019].

Badan Pusat Statistik Kota Bandung. 2017. Kota Bandung Dalam Angka 2017. Tersedia: https://bandungkota.bps.go.id/publication/2017/08/11/7cf46753e6cb9992a7e401b6/kota-bandung-dalam-angka-2017

Bodvarsson, Örn & Walker, Rosemary. (2004). Do parental cash transfers weaken performance in college?. *Economics of Education Review*. 23. 483–495.

Bongomin, G.O.C., Ntayi, J.M., Munene, J.C., Nabeta, I.N. 2016. Social capital: Mediator of financial literacy and financial inclusion in rural Uganda. *Review of International Business and Strategy* 26(2), 291–312.

Cude, B.J., Lawrence, F.C., Lyons, A.C., Metzger, K., LeJeune, E., Marks, L., Machtmes, K. 2011. *College students and financial literacy: What they know and what we need to learn*. Eastern Family Economics and Resource Management Association.

Dew, J., & Xiao, J. J. 2011. The Financial Management Behavior Scale: Development and validation. *Journal of Financial Counseling and Planning* 22(1), 43–59.

Furnham. 1984. Many sides of the coin: Psychology of money usage. *Personality and Individual Differences* 5(5), 501–509.

Herdijono, I., & Damanik, L. A. 2016. *Pengaruh* financial attitude, financial knowledge, parental income *terhadap* financial management behavior. *Jurnal Manajemen dan Teori Terapan* 9(3) (December), 226–241.

Hung, A., Parker, A., & Yoong, J. 2009. Defining and measuring financial literacy. RAND Roybal Center for Financial Decision Making, WR-708, September 2009.

Ida, & Dwinta, C. 2010. *Pengaruh* locus of control, financial knowledge, income *terhadap* financial management behavior. *Jurnal Bisnis dan Akuntansi* 12(3) (December 2010), 131–144.

Kurniawan, A,. 2019. Pengertian Pendapatan Menurut Para Ahli Beserta Jenisnya. Tersedia: www.gurupendidikan.co.id/pengertian-pendapatan-menurut-para-ahli-beserta-jenisnya. [February 18, 2019].

Mien, & Thao. 2015, July. Factors affecting personal financial management behaviors: Evidence from Vietnam. Proceedings of the Second Asia-Pacific Conference on Global Business, Economics, Finance and Social Sciences. Paper ID: VL532.

Otoritas Jasa Keuangan. 2013. Literasi Keuangan. Tersedia: www.ojk.go.id/id/kanal/edukasi-dan-perlindungan-konsumen/Pages/Literasi-Keuangan.aspx. [February 18, 2019].

Rajna, A., Sharifah, W. P., Al Junid, S., & Moshiri, H. 2011. Financial management attitude and practice among medical practitioners in public and private medical service in Malaysia. *International Journal of Business and Management* 6(8), 105–113.

Sandy, F. 2017. (2017, April 11). OJK Dorong Masyarakat Punya Sikap Keuangan. Sindonews. Tersedia: https://ekbis.sindonews.com/read/1196294/178/ojk-dorong-masyarakat-punya-sikap-keuangan-1491913240 [February 18, 2019].

Widodo. 2017. *Metodelogi penelitian populer dan praktis*. Jakarta: Rajagrafindo Persada.

Managing Learning Organization in Industry 4.0 – Rachmawati & Hendayani (eds)
© 2020 Taylor & Francis Group, London, ISBN 978-0-367-81920-0

Effect of solvency, operating cash flow, board of directors size and audit quality on financial distress

M.R. Nazar & F.S. Balqis
Telkom University, Bandung, Indonesia

ABSTRACT: This study aims to determine the effect of Solvency, Operating Cash Flow, Board of Directors Size and Audit Quality on the prediction of financial distress in food and beverage subsector companies listed on the Indonesia Stock Exchange in the 2014-2018 period. The dependent variable of this study is financial distress. The independent variables in this study are Solvency, Operating Cash Flow, Board of Directors Size and Audit Quality. The population used in this study were all food and beverage subsector companies listed on the Indonesia Stock Exchange for 5 years. This study used a purposive sampling sample selection technique and obtained 16 companies with 80 samples. The data analysis technique in this research is quantitative analysis using descriptive statistics and logistic regression analysis methods. Based on research that has been done, Solvency, Operating Cash Flow, Board of Directors Size and Audit Quality affect financial distress.

1 INTRODUCTION

The company has a goal to get profits and increase profits every year. Therefore every company must have the hope to be able to operate in a very long time. But this desire may be difficult to achieve for one reason or another. One of the reasons why a company stops operating is the company's inability to pay off its obligations. The food and beverage sector in 2017 accounted for the largest Gross Domestic Product in the manufacturing sector. Although the biggest contributor, in 2017 in this sector several companies experienced Financial Distress. One sign of a company experiencing financial distress is when the company has negative net income and negative earnings per share for years. According to Widhiari and Merkusiwati (2015) through earning per share, the entity's profit gained in the period concerned and can implicitly explain how the company's performance in the past and the prospects of the company concerned.

2 LITERATURE REVIEW AND RESEARCH HYPOTHESIS

2.1 *Agency theory*

Agency theory was first stated by Jensen and Meckling (1976). According to Jensen and Meckling (1976) in Warsono et al. (2009: 10) company managers are "agents" while shareholders are "principals". Shareholders or principals give decision-making responsibilities to managers as agents of shareholders.

2.2 *Solvency ratio*

Solvency ratio or also called leverage ratio is a ratio that is used to measure the extent to which a company's assets are financed with debt (Hery, 2016: 162). According to Fahmi (2011, 116), the solvency ratio is a ratio that shows how the company can manage its debt to obtain profits and is also able to repay its debt and this ratio measures the company's ability to meet its obligations in the long run. In this study, the indicator is

$$\frac{total\ liabilities}{total\ shareholder's\ equity} \tag{1}$$

2.3 Cash flow statement

In general, cash flow information helps to assess a company's ability to meet its obligations, pay dividends, increase capacity, and obtain funding for Subramanyam & Wild, (2010: 92). Cash flow statements are used by management to evaluate operational activities that have taken place, and plan future investment and financing activities, while for creditors and investors, cash flow statements are used to assess the level of liquidity and the potential of the company in generating profits (Hery, 2016: 88). The cash flow statement distinguishes the sources and uses of cash flows into three classification activities, namely operating activities, investing activities and funding activities. In this study, the indicator is

$$\frac{Operating\ Cash\ Flow}{Total\ Liabilities} \tag{2}$$

2.4 Board of directors

According to Article 1 of Law No. 40 of 2007, the board of directors is the organ of the company in power and is responsible for managing the company for the interests of the company in accordance with the aims and objectives of the company, and represents the company inside or outside the court in accordance with the provisions of the articles of association. The board of directors is a company organ that has the authority to take care of the company so that the company can achieve its goals. In this study, the indicator is

$$\sum board\ of\ directors\ member \tag{3}$$

2.5 Audit quality

Audit quality according to Yadiati & Mubarok (2017: 113) is the accuracy of the information reported by the auditor by the audit standards used by the auditor including accounting violation information in the client's company financial statements. Yadia & Mubarok (2017: 112) defines audit quality as a function of the auditor's ability to detect material misstatements or technical abilities and report errors (auditor independence). Meanwhile, according to (Francis, 2004) audit quality is the opposite of audit failure (audit failure). Audit quality according to (Yadiati & Mubarok, 2017) is the possibility that an auditor does not issue a reasonable opinion without exception in financial statements that contain material errors. In this study, Companies that use KAP Big Four are given the value of 1 and companies that use KAP other than Big Four are given the value of 0.

2.6 Financial distress

According to Widhiari and Merkusiwati (2015), financial distress is a condition experienced by companies before bankruptcy or liquidation caused by the process of declining financial position of the company. According to (Dewi, Khairunnisa, & Mahardika, 2017) explained that bankruptcy is a situation where the company is unable to fulfill its obligations because the company does not have or lack of funds to run and manage the company so that it does not get the targeted profit. According to Altman and Hotchkiss (2006: 13), the most fundamental reason a company experiences financial distress is because of the company's managerial inability. The company went bankrupt for several reasons, but the most frequent thing was due to the inability of management to manage the company. The deteriorating business that is being run by the company can also make the company experience financial distress. In this study, Companies that have negative earnings per share are given a value of 1 and companies that have positive earnings per share are given a value of 0.

2.7 *Research hypothesis*

Based on the theories that have been described and the framework of thought that has been described, the hypotheses of this study are as follows:

H_1: Solvency (SOLV), Operating Cash Flow (CFO), Board of Directors Size (DIR), Audit Quality (AUDIT) simultaneously influence the prediction of Financial Distress (FD) in food and beverage subsector companies listed on the Indonesia Stock Exchange in the 2014-2018 period.

H_2: Solvency (SOLV) partially has a positive effect on the prediction of Financial Distress (FD) in food and beverage subsector companies listed on the Indonesia Stock Exchange in the 2014-2018 period.

H_3: Operating Cash Flow (CFO) partially has a negative effect on the prediction of Financial Distress (FD) in the food and beverage sub-sector companies listed on the Indonesia Stock Exchange in the 2014-2018 period.

H_4: The size of the Board of Directors (DIR) partially negatively affects the prediction of Financial Distress (FD) in the food and beverage subsector companies listed on the Indonesia Stock Exchange in the 2014-2018 period.

H_5: Audit Quality (AUDIT) partially has a negative effect on the prediction of Financial Distress (FD) in food and beverage subsector companies listed on the Indonesia Stock Exchange in the 2014-2018 period.

3 RESEARCH METHODS

The population in this study are food and beverage subsector companies listed on the Indonesia Stock Exchange in the 2014-2018 period. There are 80 samples in this study. In this study, data analysis is an activity after the data has been collected. In data analysis, the activities carried out are grouping data based on variables, tabulating data, presenting data from variables that have been studied, doing calculations to answer the problem formulation and doing calculations to test hypotheses. In this study, data analysis techniques used are descriptive statistics, panel data regression analysis, and statistical hypotheses.

4 RESEARCH RESULT

4.1 *Descriptive statistical analysis*

In this study, there are outlier data. Outliners are cases or data that have unique characteristics that look very different from other samples and appear in the form of extreme values in either a single variable or a combination (Ghozali, 2018: 40). To detect outliers can be done by determining the boundary value that will be categorized as outlier data, namely by converting data values into standardized scores or called z-scores. After issuing outlier data, the sample in this study was 65 samples. Here are the results from descriptive statistics:

Table 1. Descriptive statistics test results.

	N	Min	Max	Mean	St. Deviasi
SOLV	65	0.16354	3.02864	1.1076775	0.51313807
CFO		-0.19797	1.09643	0.2368500	0.30004920
DIR		2	10	5.35	2.175
AUDIT		0	1	0.46	0.502
FD		0	1	0.14	0.348
Valid N (listwise)					

Source: SPSS 23 output, data processed by the author (2019)

4.2 *Logistic regression analysis*

4.2.1 *Test the overall model*
After testing the Overall Model, there is a difference between the values of -2LogL Block Number = 0 and -2LogL Block Number = 1, which is the difference of 11,315 where this indicates that the model is fitted with the data and it is proven that the variable Solvency, Operating Cash Flow, Board of Directors Size and Quality Auditing can significantly improve the model.

4.2.2 *The feasibility of the regression model*
After testing The Feasibility of the Regression Model, in this study the Hosmer and Lemeshow Test has a Chi-square value of 2.139 and a significance level of 0.952. Because the significance value is 0.952> 0.05, H0 is accepted and hypothesis testing is acceptable.

4.2.3 *Coefficient of determination*
After testing Coefficient of Determination, in this study, the value of NagelKerke R Square is greater than the value of Cox & Snell R Square. When the value of Nagelkerke R Square is greater than the value of Cox & Snell R Square then this shows that the ability of independent variables namely solvency (SOLVABILITY), operating cash flow (CFO), board size (DIRECTORS) and audit quality (AUDIT) affect the dependent variable or financial distress (FD) of 54.6% and the remaining 45.5% is explained by other variables not examined in this study.

4.2.4 *Simultaneous testing results*
After testing the simultaneous model, in this study, the Chi-square value is 23,331 with a degree of freedom 4 and a significance level of 0,000. It can be concluded that the sig value of 0.000 is smaller than 0.05, then H01 is rejected and Ha1 is accepted, which indicates that the variable solvency (SOLVABILITY), operating cash flow (CFO), board size (DIRECTORS) and audit quality (AUDIT) have a significant effect. on financial distress.

4.2.5 *Partial testing results*
In addition to the simultaneous test, this research also conducted a partial test. The result of partial test output in this study is as follows:

$$Ln\frac{FD}{1-FD} = 2.367 + 0.739\ SOLV - 9.543\ CFO - 1.054\ DIR + 2.164\ AUDIT \qquad (4)$$

Or downgraded to

$$FD = \frac{1}{1 + 2.367 - (0.739\ SOLV - 9.543\ CFO - 1.054\ DIR + 2.164\ AUDIT)} \qquad (5)$$

269

5 CONCLUSION

5.1 *Effect of solvency on financial distress*

In the logistic regression test that has been done, the solvency variable as measured by the debt to equity ratio has a regression coefficient of 0.739. The regression coefficient of 0.739 indicates a direct relationship between the solvency variable and the financial distress variable as the dependent variable. The solvency variable has a significance value of 0.522. The significance value is greater than 0.05. This indicates that the Ha1 hypothesis was rejected and H01 was accepted. So it can be concluded that the solvency variable partially does not affect financial distress. This is not following the hypothesis that has been made where solvency has a positive effect on financial distress. This is because even though the company's debt is large, if the company's ability to carry out its obligations is good, then the company is certainly not affected by financial distress. The results of the study that the solvency ratio calculated by the debt to equity ratio does not affect financial distress supports this research (Sopian & Rahayu, 2017) & (Sucipto & Muazaroh, 2016).

5.2 *Effect of operating cash flow on financial distress*

In the logistic regression test that has been done, the operating cash flow variable measured by the ratio of operating cash flows has a regression coefficient of -1.054. The regression coefficient of -1.054 shows the opposite relationship between the operating cash flow variable and the financial distress variable as the dependent variable. Operating cash flow variable has a significance value of 0.010. The significance value is smaller than 0.05. This indicates that the Ha1 hypothesis was accepted and H01 was rejected. So it can be concluded that the solvency variable partially has a negative effect on financial distress. This is consistent with the hypothesis that has been made that operating cash flow has a negative effect on financial distress. This is because operating cash flows that have a positive value indicate the receipt of sales can cover all routine operating expenses. When the company can cover all expenses, it indicates that the company's financial situation is no problem or avoiding financial distress. The results of the study that operating cash flow has a negative effect on financial distress support the research conducted by (Halim, 2017).

5.3 *Effect of board of directors on financial distress*

In the logistic regression test that has been done, the variable size of the board of directors as measured by the number of board of directors has a regression coefficient value of -9.543. The regression coefficient value of -9.543 shows the opposite direction relationship between the board size variable and the financial distress variable as the dependent variable. The variable size of the board of directors has a significance value of 0.029. The significance value is smaller than 0.05. This indicates that the Ha1 hypothesis was accepted and H01 was rejected. So it can be concluded that the variable size of the board of directors partially has a negative effect on financial distress. This is consistent with the hypothesis that has been made that the size of the board of directors has a negative effect on financial distress. This is because when the number of boards of directors is large, it will produce various views on the decision-making policies related to the company. The results of the study that the size of the board of directors negatively affect financial distress support research that has been conducted by Hanafi & Breliasiti (2016).

5.4 *Effect of audit quality on financial distress*

In the logistic regression test that has been done, the audit quality variable is measured by seeing whether the company chooses the big four auditors to audit it or chooses a non-big four auditor, has a regression coefficient of 2,164. The regression coefficient of 2,164 indicates a direct relationship between audit quality variables with financial distress variables as the dependent variable. The audit quality variable has a significance value of 0.070. The significance value is greater than 0.05. This indicates that the Ha1 hypothesis was rejected and H01 was accepted. So it can be concluded that the audit quality variable partially does not affect financial distress. This is not following the hypothesis which states that audit quality negatively affects financial distress. This is because when a company chooses a big four KAP as a public accountant who audits their company, auditors in the big four KAP cannot predict the possibility of the company being affected by financial distress. The results of the study that audit quality does not affect financial distress support the research conducted by Brahmana (2007).

REFERENCES

Altman, E. I. & Hotchkiss, E. 2006. *Corporate Financial Distress and Bankruptcy*. New Jersey: John Wiley & Sons, Inc.

Brahmana, R. K. 2007. Indentifying FInancial Distress Condition in Indonesia Manufacture Industry. *Birmingham Business School, University of Birmingham, United Kingdom.*

Dewi, R. J., Khairunnisa & Mahardika, D. P. 2017. Analisis Pengaruh Likuiditas, Leverage, dan Operating Capacity Terhadap Finan.

Fahmi, I. 2011. *Analisis Laporan Keuangan*. Bandung: CV Alfabeta.

Fitri, Z. H. 2018) *PENGARUH RASIO LIKUIDITAS, RASIO LEVERAGE, RASIO PROFITABILITAS DAN SALES GROWTH TERHADAP FINANCIAL DISTRESS*. Bandung: Skripsi Telkom University.

Ghozali, I. 2018. *Aplikasi Analisis Multivariate Dengan Program IBM SPSS 25*. Semarang: Badan Penerbit Universitas Diponegoro.

Halim, M. 2017. Penggunaan Laba dan Arus Kas untuk Memprediksi Kondisi Financial Distress (Studi Empiris pada Perusahaan Manufaktur yang Terdaftar di BEI tahun 2013-2014. *Jurnal Ilmiah Akuntansi Indonesia.*

Hanafi, J. & Breliasiti, R. 2016. Peran Mekanisme Good Corporate Governance dalam Mencegah Perusahaan Mengalami Financial Distress. *Jurnal Oline Insan Akuntan* 1(1): 195–220.

Hery. 2016. *Analisis Laporan Keuangan*. Jakarta: PT Grasindo.

Sopian, D. & Rahayu, W. P. 2017. Pengaruh Rasio Keuangan dan Ukuran Perusahaan Terhadap Financial Distress (Studi Empiris pada Perusahaan Food and Beverage di Bursa Efek Indonesia). *Competitive Jurnal Akuntansi dan Keuangan.*

Subramanyam, K. R. & Wild, J. J. 2010. *Analisis Laporan Keuangan*. Jakarta: Salemba Empat.

Sucipto, A. W. & Muazaroh. 2016. Kinerja rasio keuangan untuk memprediksi kondisi financial distress pada perusahaan jasa di Bursa Efek Indonesia periode 2009-2014. *Journal of Business and Banking* 6(1): 81–98.

Warsono, S., Amalia, F. & Rahajeng, D. K. 2009. *CORPORATE GOVERNANCE CONCEPT AND MODEL*. Yogyakarta: Center for Good Corporate Governace.

Widhiari, N. L. & Merkusiwati, N. K. 2015. PENGARUH RASIO LIKUIDITAS, LEVERAGE, OPERATING CAPACITY DAN SALES GROWTH TERHADAP FINANCIAL DISTRESS. *E-Jurnal Akuntansi Universitas Udayana* 11(2): 456–469.

Yadiati, W. & Mubarok, A. 2017. *Kualitas Pelaporan Keuangan: Kajian Teoritis dan Empiris*. Jakarta: Kencana.

Street vendor management in Indonesia and Thailand

R. Pasciana, P. Pundenswari & G. Sadrina
Faculty of Social and Political Sciences, Garut University, Garut, Indonesia

ABSTRACT: Street vendors are an urban global phenomenon that exists almost all over the world. In members of the Association of Southeast Asian Nations (ASEAN), including Indonesia and Thailand, the informal economic sector is quite dominant. Street vendors are considered a source of problems, such as disruption of pedestrians, traffic jams, and making the city look dirty. This study aimed to compare the management of street vendors in Indonesia and Thailand. The research method was qualitative. Data collection techniques were carried out through a literature review derived from published journals related to this research. The results of this study indicated that community culture and government policies present differences related to the management of street vendors in Indonesia and Thailand. Indonesia can adopt several things from Thailand, such as management and strategies to encourage street vendors to be directed by the government for the common good.

1 INTRODUCTION

The informal sector dominates the economy in Southeast Asia. The average employment generated by the informal economy for countries in Southeast Asia is 75.2% (Team, P. M., 2019). Although the informal sector is a source of people's livelihoods, its existence is considered problematic. The problem is the sector's interference in cities' neatness, beauty, and convenience. Indonesia and Thailand are both Southeast Asian countries that have a big number of street vendors and have the same problems as a consequence. For most urban planners and policy makers, the existence of informal-sector actors, especially street vendors and urban slums, is a disturbance to the beauty and regularity of cities (Wydaningrum, 2009). Street vendors represent a complex issue that Indonesia and Thailand must face.

Therefore this research question was, "How does the management of street vendors in Indonesia compare with the management of street vendors in Thailand?" Specifically, this research focused on big cities in both countries.

2 LITERATURE REVIEW

2.1 *Concept of management*

Management comes from the word *manage*, according to the large dictionary of the contemporary Indonesian language written by Peter Salim and Yenny Salim (2012: 695), and it "means to lead, control, regulate, and strive to be better, more advanced and so forth and responsible for certain jobs." Furthermore according to Hasibuan (2012: 1), "management is the science and art of regulating the process of utilizing human resources and other resources effectively and efficiently to achieve a certain goal." "Managers carry out management functions, namely planning, coordinating, and controlling. And also with planning, managers carry out the functions of organizing management, leadership, and control" (Fayol, 2010: 179). Based on these descriptions, we can understand that management is an activity that regulates or manages within an organization with a series of processes that consist of management functions and involve several available resources to achieve the desired results or objectives.

2.2 *Concept of street vendors*

Street vendors is a term used to refer to merchants who use carts. The term is often used for street traders in general because street traders use five legs. The five legs are the two legs of the trader plus the three "legs" of the cart (which are actually three wheels or two wheels and one foot) (Ali and Alam, 2016: 185). In general, traders are suppliers of goods and services in an urban area. McGee defines street vendors as "the people who offer goods or services for sale from public places, primarily streets and pavements" (cited in Ali and Alam, 2016: 186). Thai food is an important part of Thailand's tourism industry. Campaigns such as "Amazing Thai food" or the "Thai street food festival 2014" highlight authentic local Thai dishes so as to build a better reputation for Thailand (Thailand, 2016). A lot of Thai food is sold by street vendors, especially in Bangkok, Thailand's capital city. But on July 30, 2013, the street vendors were informed of the Bangkok Metropolitan Administration's plan to clear the streets for pedestrians. The clearance took place in August 2014 and affected about 300 street vendors on the street of Tha Chang (Boonjubun, 2017). The situation is no different in Indonesia; Indonesia has the same problem with street vendors. Street vendors are one of the attractions of tourism because shopping is a popular activity among tourists, and one of the most common encounters Western tourists have while traveling in developing countries is street vendors who sell foods, accessories, fashion, etc. (Dallen and Timothy, 1997).

3 METHODOLOGY

This study used a qualitative method that aims to describe and analyze the street vendor phenomenon.

3.1 *Participants*

Participants in this study were street vendors in Indonesia and Thailand.

3.2 *Data analysis*

A bibliometric analysis was carried out through analysis derived from a compilation of journals that cover street vendors in Indonesia and Thailand.

4 RESULTS AND DISCUSSION

4.1 *Management of street vendors in Indonesia*

The Indonesian government has implemented policies on street vendors such as Presidential Regulation No. 125 of 2012 and Minister of Home Affairs Regulation No. 41 of 2012. Local regulations have also been passed, but the problems street vendors face are very complex, especially in big cities like Jakarta and Bandung, which are the economic centers of Indonesia and have many street vendors. The local governments continuously carry out management and restructuring efforts, and the two cities have different policies.

Tanah Abang is the largest textile market in Southeast Asia and is one of Indonesia's icons. The government has tried to make arrangements for street vendors there; in fact, street vendors still sell on the side of the road and on the sidewalk, and their number is increasing. One of the efforts the DKI provincial government has made to organize the Tanah Abang area involved relocating traders to Sky Bridge. As CNN Indonesia (2019) has reported, Sky Bridge is a facility built and provided by the DKI Jakarta provincial government in order to accommodate pedestrian traffic from Tanah Abang Station to the market in the Jatibaru area. The bridge was inaugurated on January 2019 and is a relocation site for street vendors located on Jatibaru Street, Tanah Abang. Sky Bridge has provided as many as 446 stalls for traders to sell. To maintain security, cleanliness, and lighting traders using the Sky Bridge area are required to pay a service charge of Rp. 500,000 per month (CNN Indonesia, 2018). After the street vendors were relocated to Sky Bridge, the sidewalks in Jakarta began to be well arranged, but after a while, street vendors were still selling around Tanah Abang and under Sky Bridge under the excuse of not getting a place in Sky Bridge and the location was quiet with sellers. The government immediately followed up on the problem by relocating street vendors to a temporary relocation site, the sidewalk. The governor of DKI Jakarta has begun to allow street vendors to sell on the sidewalk with a number of conditions, and, of course, this raises both pros and cons.

Not only Jakarta but also the city of Bandung experienced these problems related to street vendors, but in contrast to Jakarta, the arrangement of street vendors in Bandung was closer to reality. The process of structuring street vendors is carried out by the Bandung city government with a preventive approach and consists of several stages starting with data collection, structuring with a zoning system, guidance, and supervision. The arrangement of street vendors in the Cicadas area of Bandung takes place in an orderly manner. One thing that deserves appreciation is that the traders and the officials cooperate with each other in the structuring (Nursyabani, 2019). The street vendors even dismantle their stalls themselves and are very enthusiastic. The demolition had been long awaited because the vendors wanted to be organized and the street vendors in Cicadas were aware that their trading activities on the sidewalk were causing problems and violating pedestrian rights. Therefore they agreed to organize so that pedestrian rights would be respected and street vendors could sell safely and calmly. The Cicadas area is expected to become a new shopping destination and an attraction in Bandung.

4.2 *Management of street vendors in Thailand*

Comparable to numerous other Asian nations, Thailand incorporates a long history of street vendors. In Bangkok, the biggest city in Thailand, street vendors have given nearby Thai residents cheap and helpful access to a wide range of products and an opportunity to make a living. In recent years, street vendors have been considered as a way to sustain enterprise, as well as adding to the tourist attractions of Bangkok by bringing dynamic quality and essentialness to the city (Secretariat, 2014).

In Bangkok, street vendors must register with the Bangkok Metropolitan Administration in order to carry out their activities legally. Registered street vendors are issued a license card that must be renewed every year at a cost of 100 baht. In 2013, more than 20,000 street vendors were registered in Bangkok. Registered street vendors must pay a monthly fee to the Bangkok Metropolitan Administration for the cleaning and maintenance of the streets they occupy (Secretariat, 2014).

Street vendors cannot operate on Mondays, because it has been set as the cleaning day. The Public Cleansing and Public Parks Section, the City Law Enforcement Section, and the vendors work together to clean sidewalks and public spaces on the second and third Mondays of the month. If vendors cannot help in cleaning, they must provide cleaning equipment such as brooms, soap, or baking soda. The Bangkok Metropolitan Administration specifies the trading hours for street vendors. The trading hours vary across different areas. In several areas, street vendors are allowed only after rush hours. For example, street vendors on Ratchadamri Road and in the Tha Phrachan area must vacate their respective sidewalks from 5:00 pm to 7:00 pm each day in order to make sure that the sidewalks are free for use by pedestrians. Ratchadamri Road is in the central business district and Tha Phrachan is a district popular with tourists and university students (Kusakabe, 2014).

One manifestation of the informal economic sector in the capital of Thailand is its numerous streets stuffed with street vendors. Soi Rangnam is one of them – a street 700 meters long with two shopping malls, a night market, dozens of restaurants, twelve bars, five massage parlors, a park, a university, and a high school. The sidewalks of the street buzz with street vendors at all hours of the day, and the place is locally called a dinner area, attracting large crowds after dark. The location is similar to other street markets; however, it is larger and more well-known than most (Batreau and Bonnet, 2015). Soi Rangnam has three types of street vendors: (1) Fixed, registered vendors do not have permanent stalls but always set up their shops at the same place. In Soi Rangnam, fixed vendors sell food or drinks, except for one who sells beauty products. (2) Mobile, unregistered vendors sell from a pushcart or a motorcycle and move from spot to spot. Some sell only in Soi Rangnam and alternate between a few locations in the street, following the customers or avoiding the municipal policemen. Others cover a larger area and follow a circuit from vending zone to vending zone, stopping on the way only to serve a customer who hails them. (3) Market vendors operate in a market area in front of a local mall, the majority of whom sell clothes or accessories (Batreau and Bonnet, 2015).

After a tug of war with the street vendors' association, the Thai government finally issued regulations for street vendors.

a) Street vendors must operate in the designated area and at the specified time. Each zone has its own schedule and slots are limited so that only a number of vendors are allowed to operate in certain areas.
b) The government has decided on an open zone for business since 2005 in Soi Rangnam.
c) Street vendors can settle legally in the suburbs only when sidewalks are not crowded.
d) Vendors must pay once a year a "cleaning fee" of BHT 100 (USD 3.3), and street vendors selling food must be inspected by the district health department.
e) Status as an official vendor does not guarantee vendors protection against eviction and does not provide compensation if the local government wishes to reclaim the land.
f) Street vendors pay monthly fees proportional to the size of their business and how long they use the location each day. This is significantly more expensive than cleaning costs.
g) Street vendors must ensure that their cart environment is clean and orderly. (Merdeka, 2019)

Comparison of street vendor management in Indonesia and Thailand can be seen in the following table.

Table 1. Comparison of street vendor management in Indonesia and Thailand.

	Indonesia	Thailand
Regulation	Policies and regulations are quite good regarding the arrangement of street vendors, but regarding licensing and location they cannot be said to be effective because of the vagueness of these aspects. For oversight mechanisms, there is still third-party interference. (CNN Indonesia, 2018)	Policies and regulations are clear and detailed regarding the arrangement of street vendors in terms of structuring mechanisms, location of licensing and supervision. (Secretariat, 2014)
Implementation	In Bandung, the policy has been implemented well and goes according to plan; there is cooperation with both parties. But for Jakarta, it has not been well realized, plus there is a lack of supporting facilities and infrastructure. Policies are limited to	The policy that has been implemented can be said to run well; in addition to applying some requirements for street vendors, the government also provides orderly infrastructure and facilities that support street vendors to sell well. (Lefevre, 2018)

(Continued)

275

Table 1.   (*Continued*)

	Indonesia	Thailand
	the direction of the leader, so that it seems to bear the responsibility and there is no harmony and full awareness to develop the policy. (Pamungkas, 2016)	
Community Cultures	Consists of several aspects including: 1. Behavior a. Traders In Jakarta, the behavior of traders is not the same. Precisely in Tanah Abang, there is a lack of discipline, in terms of both time and place. Lack of understanding and awareness of the policies that apply means some rules are violated. In Bandung, the behavior of traders in Cicadas can be said to be good because they know their position to sell. (Nursyabani, 2019) b. Buyers The behavior of buyers in both cities has the same characteristics where buyers will shop at street vendors because prices are easily accessible and obtained. (Wydaningrum, 2009) 2. Habits a. Traders In Jakarta, the habits of traders still need to be changed and improved, because these habits make the conditions on the street disorderly and many rules are broken. For example, traders often cover parts of the road or sidewalk to sell so pedestrians feel disturbed, and traders sometimes do not maintain the cleanliness of the environment in which they sell. They throw away garbage, for example, coffee and hot water, which is dumped on the sidewalk. This can damage the sidewalk and ruin environmental aesthetics. In Bandung, traders' habits can be said to be good; for example, in the Cicadas area, they have an awareness of pedestrians' rights and this makes street vendors easier to organize. Street vendors also maintain the location where they sell. (Nursyabani, 2019) b. Buyers Habits of buyers in the two cities have in common that they will continue to shop from street vendors if they have an easier mind-set and low prices and there is a bargaining system. As long as street vendors operate, the habit of throwing	Consists of several aspects including: 1. Behavior a. Traders Traders in Thailand are familiar with the policies set by the government, one of which is discipline in time and place so that the local vendors in Thailand sell obediently and in an orderly fashion. Considering that street vendors in Thailand are also one of the tourist attractions, merchants serve tourists well. (Kusakabe, 2014) b. Buyers Similar to traders, buyers recognize their position as those who buy both with local and with foreign buyers. They equally obey the applicable rules. (Wydaningrum, 2009) 2. Habits a. Traders In Thailand, traders have good habits so that they have unity with the policies set by the government. Traders' habits include maintaining environmental aesthetics and the hygienic nature of the food sold. (Kusakabe, 2014; Seneviratne, 2018) b. Buyers Buyers also have a good habit – that is, buyers help in maintaining the cleanliness of the surrounding environment, so that the environment of selling remains clean and neatly arranged. (Batreau and Bonnet, 2015; Merdeka, 2019)

(*Continued*)

Table 1. (*Continued*)

Indonesia	Thailand
garbage around is still difficult to change. Buyers throw away packaging around the sidewalks and roadsides, making the road look dirty. In Jakarta, there is still a lack of understanding and awareness of the regulations that are set so that many street vendors are disorderly and violate regulations. Such as ignoring the cleanliness of the place that has been provided by the government and the specified trading time. (Wydaningrum, 2009)	

5 CONCLUSIONS AND RECOMMENDATIONS

Good management is management that can achieve its objectives. Management strategies that work in Thailand may not work in Indonesia because street vendors have their respective characteristics and rules and regulations must be accepted by all traders. But Indonesia can adopt some tactics from Thailand such as management strategies that encourage street vendors to be directed by the government for the common good. That means the government must ensure that street vendors can sell without causing problems or disrupting traffic and pedestrians. It takes harmony and one thought to make policy and then implement it. Street vendors are expected to be a tourist attraction with local wisdom that can be an attraction for each city and can be one source of revenue for the government.

Recommendations for further research are: (1) Conduct research not only through literature studies but also through observation and interviews. (2) In addition, research can also be carried out by analyzing further the comparison of Indonesian and Thai street vendors in terms of economics or spatial planning.

REFERENCES

Ali, F. & Alam, A. S. 2016. *Studi Kebijakan Pemerintah*. Bandung: PT Refika Aditama.

Batreau, Q., & Bonnet, F. 2015. *Managed informality: Regulating street vendors in Bangkok*. From Wiley Online Library: 4–5. www.onlinelibrary.wiley.com/doi/full/10.1111/cico.12150

Boonjubun, C. 2017. Conflicts over streets: The eviction of Bangkok street vendors. *Elsevier* 70: 22–31.

Budi, T. 2018, January 22. PKL Berharap Pemerintah Kucurkan Kredit Tanpa Agunan. Retrieved October 7, 2019, from Okezone Economy: http://economy.okezone.com/read/2018/01/22/320/1848603/pkl-berharap-pemerintah-kucurkan-kredit-tanpa-agunan.

CNN Indonesia. 2018, November 13. Tak Diusir, PKL Tanah Abang Sebut Era Anis Lebih Nyaman. Retrieved October 2, 2019, from CNN Indonesia: www.cnnindonesia.com/nasional/20181113175309-20-346252/tak-diusir-pkl-tanah-abang-sebut-era-anies-lebih-nyaman.

CNN Indonesia. 2019, August 23. Suara PKL Tanah Abang Soal Kelanjutan Nasib Usai Putusan MA. Retrieved September 30, 2019, from CNN Indonesia: www.cnnindonesia.com/nasional/20190821082556-20-423262/suara-pkl-tanah-abang-soal-kelanjutan-nasib-usai-putusan-ma

Dallen J. & Timothy, G. W. 1997. Tourists Selling to Indonesian Street Vendors. *Pergamon: Annals of Tourism Research* 24(2): 322–340.

Fayol, H. 2010. *Manajemen Public Relation*. Jakarta: PT Indeks.

Hasibuan. 2012. *Manajemen Sumber Daya manusia*. Jakarta: PT Bumi Aksara.

Kusakabe, K. 2014. Street Vending Policies and Practices: A Case Study of Bangkok. Genewa: International Labour Office, 1–13.

Lefevre, A. S. 2018, September 18. Bangkok's Street Vendors Decry Evictions As Authorities Clean Up. Retrieved September 30, 2019, from Reuters: www.reuters.com/article/us-thailand-streetvendors/bangkoks-street-vendors-decry-evictions-as-authorities-clean-up-idUSKCN1LX2NH.

Merdeka. 2019, August 28. Penataan PKL, Berkaca Dari Pengalaman Bangkok Hingga New York. Retrieved October 4, 2019, from Merdeka.com: www.merdeka.com/dunia/penataan-pkl-berkaca-dari-pengalaman-bangkok-hingga-new-york.html.

Nursyabani, F. 2019, August 13. PKL Cicadas Bongkar Sendiri Lapaknya Saat Penataan. Retrieved October 2, 2019, from Ayo Bandung: www.ayobandung.com/read/2019/08/13/60450/pkl-cicadas-bong kar-sendiri-lapaknya-saat-penataan.

Pamungkas, B. 2016. Pedagang Kaki Lima Dan Pengembangan Kota: Analisa Kebijakan Pengelolaan Pasar Malam PKL Kota Jakarta Dan Kuala Lumpur. *Prosiding Seminar Nasional INDOCOM-PAC*: 1–2.

Salim, P., & Salim, Y. 2012. *Kamus Bahasa Indonesia*. Jakarta: Modern English Press.

Secretariat, L. C. 2014. Hawker policy in Thailand. *Fact Sheet* 1.

Seneviratne, K. 2018, February 15. Sustainable Livelihoods Behind Street Vending In Thailand. Retrieved September 30, 2019, from In-Depth News: www.indepthnews.net/index.php/the-world/asia-pacific/1677-sustainable-livelihoods-behind-street-vending-in-thailand.

Sukarna. 2011. *Dasar-dasar Manajemen*. Bandung: Mandar Maju.

Team, P. M. 2019, March 1. *State of employment in informal sectors of Southeast Asia*. Retrieved October 7, 2019, from People Matters Global: www.peoplemattersglobal.com/article/global-perspective/state-of-employment-in-informal-sectors-of-southeast-asia-20968

Thailand, T. A. 2016, November 3. Retrieved from Amazing Thailand: www.tourismthailand.org/home.

Wydaningrum, N. 2009. Kota dan Pedagang Kaki Lima. *Jurnal Analisis Sosial* 14: 6–8.

The role of youth in managing village-owned enterprises (BUMDES) through digital marketing

B. Albab & A.I. Munandar
Faculty of Strategic and Global Study, University of Indonesia, Jakarta, Indonesia

ABSTRACT: Youth play a role in carrying out national development, and entrepreneurship is also an inherent role for youth, based on statistical data released by the Central Agency on Statistics (Badan Pusat Statistik) (BPS) in 2018. Per the BPS, 87.44% of youth own cell phones, 34.01% of youth are computer users, and 73.37% of youths are Internet users. Based on data obtained from the Ministry of Villages, Disadvantaged Regions and Transmigration states that village-owned enterprises (BUMDES) as of November 2018 reached 41,000 units spread over 74,957 villages, and the government is supporting digital marketing through the BUMDES Go Digital program. The purpose of this study was to learn the extent of the role of youth in managing BUMDES through digital marketing. This research used a qualitative method that consisted of an analysis of the role of rural youth entrepreneurship in the management of BUMDES through digital marketing by taking a sample in Garut Regency of Indonesia. Steps taken in general were to ask questions according to research, collect data from related parties, analyze data inductively from specific or general discussions, and interpret the meaning of the data obtained. The results of this study indicated that youth have begun to be involved in the management of BUMDES, but their competence in management is still lacking, the use of technology and marketing using digital marketing systems is still not optimal, and the central government provides no specific assistance in managing BUMDES using the digital marketing system. This study recommended that the government provide training to young people in order to improve their competence in managing BUMDES, and training and guidance in the use of technology in running digital marketing systems.

1 INTRODUCTION

Youth have a responsibility to play an active role in national development, as stated in article 16 of Law No. 40 of 2009 on youth – namely, "Youth play an active role as a moral force, social control, and agents of change in every aspect of national development." Article 17 paragraph 3 letter b states that the active role of youth as agents of change is manifested in developing economic resources.

Youth are very close to technological developments; based on statistics from Indonesian youth in 2018, 87.44% of youth had cell phones and 93.02% of youth had used cell phones during the previous three months. In addition, around 34.01% of youth used computers and 73.27% of youth used the Internet during the previous three months. These data make clear that Indonesian youth are very close to technology, so they can support their beneficiaries in carrying out their work and entrepreneurship in particular (BPS, 2018).

Village-owned enterprises (BUMDES) are business entities managed by a village in order to advance the village's economy and develop the potential of local natural resources and communities. The number of BUMDES, based on data from the Ministry of Villages, Disadvantaged Regions and Transmigration, had reached 41,000 units and spread in 74,957 villages in Indonesia at the end of November 2018 (Zuraya, 2018).

In this case the government continues to make efforts to develop the local economy by adjusting to the advancement of existing technology, including marketing support through digital marketing systems. One of the government programs is BUMDES Go Digital, which is held in collaboration with the Ministry of Villages, Disadvantaged Regions and Transmigration with e-commerce in Indonesia (Kominfo, 2018).

Based on the potential of young people, who are close to the use of technology, and the government's support for BUMDES management through digital marketing, researchers examined the extent to which the role of youth in digital marketing systems meets its existing potential. This research was conducted in Garut Regency, West Java Province, Indonesia.

2 LITERATURE REVIEW

2.1 *Entrepreneurship*

Entrepreneurship is the ability to create business activities, and it requires innovation. Entrepreneurship seeks to achieve certain principles and targets. The targets of entrepreneurship are as follows: (1) the younger generation, generally schoolchildren, school dropouts, and prospective entrepreneurs; (2) economic actors consisting of small entrepreneurs and cooperatives; (3) government agencies that carry out community organization and group activities (Basrowi, 2011).

2.2 *Youth*

Law No. 40 of 2009 concerning youth states that youth are Indonesian citizens who enter an important period of growth and development from age sixteen to thirty. Under the youth law, youth development in Indonesia has its own orientation, among others, to create young people who are intelligent, creative, innovative, independent, democratic, responsible, and competitive, and who have leadership, entrepreneurship, pioneering, and nationality based on Pancasila and the Constitution Unitary State of the Republic of Indonesia in 1945.

2.3 *BUMDES*

Village-owned enterprises (BUMDES) are intended to encourage or accommodate all activities that increase community income, both those that develop according to local customs and culture and economic activities that are managed by the community through programs or projects of the central or regional governments. As a village business, the formation of BUMDES is expected to maximize the potential of rural communities in terms of the economy, natural resources, or human resources. Conceptually, BUMDES empowerment is not much different from the concepts of community empowerment that are well known today. Village-owned enterprises (BUMDES) are a government program based on empowerment and decentralization. With the BUMDES program, the government has the spirit to redevelop trust within the community in order to work together to create an economically independent village community (Purnamasari, 2015).

2.4 *Digital marketing*

Digital marketing is defined as a marketing activity that uses Internet-based media. The Internet is a powerful tool for business. Social media allow businesses to reach consumers and build more personal relationships. Social media are divided into two groups according to the nature of connections and interactions: (1) profile-based, namely social media based on profiles that focus on individual members; (2) content-based, namely social media that focus on content, discussion, and comments on the content displayed (Wardhana, 2015).

3 METHODOLOGY

3.1 *Participants*

The researcher determined the resource persons with a purposive technique, meaning selecting speakers who already know and are involved in the management of BUMDES in Garut, West Java, including the two headmen in Garut Regency, youth in the village who are involved in the BUMDES management process, and, finally, members of the Garut regional government.

3.2 *Data analysis*

Data analysis consisted of three main activities – namely, data reduction, data presentation, and drawing conclusions. Before the researchers conducted data reduction, the data obtained were transcribed verbatim. To test its validity, this study used a triangulation method to analyze data, which is to compare data across different times and places.

4 RESULTS AND DISCUSSIONS

This research was conducted in two villages in Garut Regency – Sindangsari and Dungusiku. The number of BUMDES in Garut Regency alone was 354 out of the 421 villages in Garut, and, according to the Office of Community and Village Empowerment (DPMD), which became the responsible agency, BUMDES ensure active management. The products and services offered by BUMDES in Garut itself are diverse, including daily necessities, party equipment, and the management of sports facilities, stalls, products for agriculture, and others.

BUMDES are the responsibility of the headman in each village; the management is carried out by the local village community. In Garut the management of BUMDES has involved local youth, as stated by the local government and also confirmed in the two villages researched. The involvement of youth is indeed a hope for villages to manage BUMDES because of their productive age and their ability to innovate.

The problem found in the field was that the involvement of youth is only temporary and not continuing; they leave BUMDES management because it offers no salary in the form of money. Besides that, village youth are not interested in the type of BUMDES management available in the village; for example, Desa Dungusiku youth feel no longer interested in managing because the BUMDES is managed only as a stall.

The management of BUMDES in Garut Regency is still not optimal. In this study, BUMDES tended to be unstable, meaning they do not survive on the products that were initially managed. This obstacle is due to the absence of maximum benefits.

Product sales of the BUMDES in Garut Regency are mostly still generated through the usual marketing methods, which means that they are sold directly in the market and not through a digital marketing system. Only a few villages market through the digital system. The villages studied still did not use the Internet to market their products and only sold as they normally would, or they used the Internet only for communication between sellers and buyers, not for marketing.

The young people under study preferred to sell BUMDES products in the usual way, the reason being that it is difficult to use digital marketing methods and they lack technological literacy and training in digital marketing, but they believe that marketing using the Internet can be more effective.

Based on the statement of the secretary of the DPMD, the central government has directed every meeting so that BUMDES can be marketed through the Internet, but the term BUMDES Go Digital itself has not yet been socialized. The socialization regarding the management of BUMDES through digital marketing has not yet reached the bottom layer; the two headmen who were asked for information stated that there has been no socialization about digital marketing, let alone the BUMDES Go Digital program.

The closeness of youth to technology, especially in the use of the Internet and Handpone, does not give effect to their desire to innovate in the management of product marketing in Garut. Their desire to join the BUMDES and their hopes of mastering digital marketing already existed, but they received no assistance, so the choice to market normally was still believed to be the way to sell their products.

It is not only socialization that youth expect but also assistance so that they can innovate in marketing and also be effective in running BUMDES. The regional government of Garut confirmed that innovation using the Internet will be increased in 2020.

5 CONCLUSIONS AND RECOMMENDATIONS

The results of this study indicated that youth have begun to be involved in the management of BUMDES, but their competence in management is still lacking, the use of technology and marketing using digital marketing systems is still not optimal, and the central government offers no specific assistance in managing BUMDES using the digital marketing system. We recommend the government provides training to young people in order to improve their competence in managing BUMDES, and training and guidance in the use of technology in running digital marketing systems

REFERENCES

Badan Pusat Statistik. 2018. Statistik Pemuda Indonesia 2018. www.bps.go.id/publication/2018/12/21/ 572f941511d090083dd742d6/statistik-pemuda-indonesia-2018.html. Accessed on April 1, 2019.

Basrowi. 2011. *Kewirausahaan Untuk Perguruan Tinggi*. Bogor: Ghalia Indonesia.

Dede, P., Rahmi, and Shandy, A. 2017. Pemanfaatan digital marketing: Bagi usaha mikro, kecil, dan menengah (UMKM) di Kelurahan Malaka Sari, Duren Sawit. *Jurnal pemberdayaan Masyarakat Madani* (JPMM) 1(1).

Kominfo. 2018. Kemdes PDTT Gandeng e-Commerce Dukung BUMDES Go Digital. www.kominfo. go.id/content/detail/13089/kemendes-pdtt-gandeng-e-commerce-dukung-bumdes-go-digital/0/berita. Accessed on April 1, 2019.

Law No. 40 of 2009 on Youth.

Purnamasari, N. 2015. Badan Usaha Milik Desa (Dalam Alur Regulasi). file:///C:/Users/Admin/Down loads/Documents/22.-BUMDES-dalam-Alur-Regulasi.pdf. Accessed on April 1, 2019.

Wardhana, A. 2015. Strategi digital marketing dan Implikasinya pada Keunggulan Bersaing UKM di Indonesia. *International Journal of Research in Marketing - Prosiding Seminar Nasional*.

Zuraya, N. 2018. Kemdes: Jumlah BUMDES Mencapai 41 Ribu Unit. https://republika.co.id/berita/eko nomi/korporasi/18/11/26/pissvc383-kemendes-jumlah-bumdes-mencapai-41-ribu-unit. Accessed on April 1, 2019.

The effect of university support on entrepreneurial intention: Case study of a Widyatama graduated student

N.P.N.P. Wijaya & G. Apryani
Widyatama University, Bandung City, Indonesia

ABSTRACT: The profile of graduates is important for universities. This can be a selling point for tertiary institutions. Currently the average college has a target to resolve the issue of the waiting period students often undergo before getting a job. The main problem can be overcome if many graduates become entrepreneurs. This study aimed to examine the effect of university support on entrepreneurial interest. By looking at it from the perspective of graduates, a tertiary institution becomes aware of its priorities in shaping entrepreneurial interest in its students. The sample in this study comprised 100 respondents who are alumni of the Widyatama University School of Business and Management. The results of this study indicated a significant influence of university support on entrepreneurial interest.

1   INTRODUCTION

The role of universities in reducing the unemployment rate is crucial. This is because universities have a great responsibility in producing quality students. The data reported by Badan Pusat Statistik show a significant change in the number of unemployed people in Indonesia.

Based on Figure 1, we can see that the number of unemployed in Indonesia has decreased. This is certainly a success of the synergy between the government, the private sector, and, of course, universities. Unemployment is a joint economic problem that these sectors must solve together. For this reason, it is very important that synergy is implemented. Figure 2 shows data on the employment status of workers in Indonesia.

Based on these data (BPS, 2019), the largest number of workforce jobs in Indonesia are positions filled by laborers/employees. In second place comes jobs as entrepreneurs (own business). From these data we can see that entrepreneurship is now the choice of the workforce. As we also know, with someone becoming an entrepreneur he can open up employment opportunities for others. This is also the main solution in reducing unemployment rates.

In their role as labor providers, universities of course have a big responsibility. According to Priyandono (cited in Hartinah, 2016), the average waiting time for a graduate with a bachelor's degree (S1) to get a job is zero to nine months. If within more than nine months the graduate has not gotten a job, it means there is a problem caused by various factors. This is also a thought for Widyatama University. Widyatama University currently has a target of seeing its graduates obtain work or begin their own businesses within three months. Of course this is a task not only for students who will graduate but also for the entire academic community at Widyatama University. To realize this goal, of course, students must have the support of the entire academic community. The choice to become an entrepreneur is one way to realize this goal. For this reason, the writer takes entrepreneurship as the research topic for this paper.

Figure 1. Unemployment rate 2019.

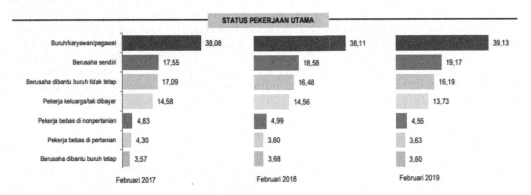

Figure 2. Main job status.

2 LITERATURE REVIEW

Entrepreneurial intention involves a variety of factors that influence it. According to Thomson (cited in Liñán, Rodríguez-Cohard, & Rueda-Cantuche, 2011), entrepreneurial intention could be defined as the self-acknowledged conviction by a person that they intend to set up a new business venture and consciously plan to do so at some point in the future. In his research Thomson also explained the factors that influence entrepreneurial intention, the main thing a person has as a knowledge entrepreneur. Mustafa, Hernandez, Mahon, and Chee (2016) research university support and proactive personality factors that influence entrepreneurial intention. Based on these two studies, we can see that a person's basis for a growing interest in entrepreneurship is knowledge about entrepreneurship. Knowledge about entrepreneurship can be obtained through higher education. The primary means by which academic institutions can raise students' entrepreneurial awareness is through educational programs. Besides educational support, academic institutions can further support the entrepreneurial intention of students through creating an enabling environment for entrepreneurship (Mustafa et al., 2016).

Smith and Beasley (2011), in their research on graduate students, have also stated that universities that have been business based have succeeded in growing students' initial interest in entrepreneurship. Based on research conducted by Turker and Selcuk (2009), a person's education can encourage entrepreneurship. In addition, the chosen field of science was also an encouragement in being an entrepreneur. Therefore, academic institutions might play critical roles in the encouragement of young people to choose an entrepreneurial career. However, they are sometimes accused of being too academic and encouraging entrepreneurship insufficiently (Gibb, 1993, 1996, cited in Turker & Selcuk, 2009).

Based on this literature survey, we can draw a hypothesis:
Ho University support has no influence on entrepreneurial intention.
Ha University support influences entrepreneurial intention.

3 METHOD

This examination utilized a basic direct regression strategy on the grounds that there is a connection between two factors where, in this exploration, the autonomous variables are university support and entrepreneurial intention. The populace in this exploration was the graduate students of the School of Business and Management at Widyatama University, which has five resources. The sample for this research comprised about 100 respondents.

4 RESULT

The questionnaire in this study was distributed to 100 respondents who are graduates of the Widyatama School of Business and Management. This study examined the effect of the university support variable, which is an independent variable, on entrepreneurial interest, which is the dependent variable. Before conducting a regression test to see the effect of the two variables, the researcher conducted validity and reliability tests in order to see the feasibility of the tested variables. Based on the results of the validity and reliability tests, all tested indicators were declared valid and reliable, so the research could proceed to the next test. Table 1 shows the regression results from the variable of the effect of university support on entrepreneurial interest.

Based on the results of the processing as contained in Table 1, we can see that the value of sig. for the influence of university support (X) on entrepreneurial intention (Y) has an effect of 16.8%. For that reason Ho was rejected and Ha was accepted.

Table 1. Regression result for the effect of university support on entrepreneurial intention.

Model Summary[b]

Model	R	R-Square	Adjusted R-Square	Std. Error of the Estimate	Durbin-Watson
1	0.420[a]	0.177	0.168	0.58471	1.840

a. Predictors: (Constant), University Support
b. Dependent Variable: Entrepreneur intention

ANOVA[a]

Model		Sum of Squares	df	Mean Square	F	Sig.
1	Regression	7.183	1	7.183	21.011	0.000[b]
	Residual	33.504	98	0.342		
	Total	40.688	99			

a. Dependent Variable: Entrepreneur intention
b. Predictors: (Constant), University Support

5 DISCUSSION

The results of this study show an influence in the relationship between these two variables. However, the value of the influence that occurs from these two variables is not large, only 16.8%. This shows that in fostering students' interest in entrepreneurship, university support is needed along with other factors. This is in line with the results of previous studies (Wijaya & Ramadhan, 2018) showing that in fostering entrepreneurial interest for a student, the influencing factors can be internal or external. Likewise in this study where the respondents were university graduates. Context, education, and moral support from universities can foster students' entrepreneurial interest. Even though the respondents in this study are alumni, they were fully aware of the importance of the role of the university in shaping entrepreneurial interest.

6 CONCLUSION

The role of the university in creating entrepreneurs is very significant; in order to study it, we need sufficient understanding of what factors influence interest in entrepreneurship. Based on research done with Widyatama graduate students, the results showed that university support affects entrepreneurial intention by 16.8%. We can see that the magnitude of the influence is not too large, which may be due to other factors that can be used as references for further research.

REFERENCES

Hartinah, N. Y. S. 2016. *Analisis Uji Ketahanan Hidup Data Waktu Tunggu Sarjana Dengan Metode Kaplan.*

Indonesia, B. P. S. *STATISTIK.* 2019.

Liñán, F., Rodríguez-Cohard, J. C., & Rueda-Cantuche, J. M. 2011. Factors affecting entrepreneurial intention levels: A role for education. *International Entrepreneurship and Management Journal* 7(2): 195–218. doi.org/10.1007/s11365-010-0154-z

Mustafa, M. J., Hernandez, E., Mahon, C., & Chee, L. K. 2016. Entrepreneurial intentions of university students in an emerging economy: The influence of university support and entrepreneurial intention. *Journal of Entrepreneurship in Emerging Economies.* 8(2),162–179. doi.org/10.1108/JEEE-10-2015-0058

Smith, K., & Beasley, M. 2011. Graduate entrepreneurs: Intentions, barriers and solutions. *Education and Training* 53(8): 722–740. doi.org/10.1108/00400911111185044

Turker, D., & Selcuk, S. S. 2009. Which factors affect entrepreneurial intention of university students? *Journal of European Industrial Training.* 3(22). 142–159. doi.org/10.1108/03090590910939049

Wijaya, N. P. N. P., & Ramadhan, N. 2018. Analisis Faktor Internal Dan Eksternal Yang Berpengaruh. *Entrepreneurship*: 510–517. Prosiding SNKIB VIII - Universitas Tarumanegara. Jakarta.

Author Index